Near-Infrared Spectroscopy

Edited by H. W. Siesler, Y. Ozaki, S. Kawata, H. M. Heise

Related Titles from WILEY-VCH

H. Günzler, A. Williams (Eds.)

Handbook of Analytical Techniques

2001. XVI. 1182 pages
Hardcover. ISBN 3-527-30165-8

J.W. Niemantsverdriet

Spectroscopy in Catalysis

2000. XII. 312 pages
Hardcover. ISBN 3-527-30200-X

B. Schrader

Raman/Infrared Atlas of Organic Compounds

1989. XIV. 1118 pages
Hardcover. ISBN 3-527-26969-X

H. Günzler, H.M. Heise, H.-U. Gremlich

IR Spectroscopy

2002. 360 pages
Softcover. ISBN 3-527-28896-1

Near-Infrared Spectroscopy

Principles, Instruments, Applications

Edited by
H. W. Siesler, Y. Ozaki, S. Kawata, H. M. Heise

Editors

Prof. Dr. H. W. Siesler
Institut für Physikalische Chemie
Universität-GH Essen
Schützenbahn 70
45117 Essen, Germany

Prof. Dr. Y. Ozaki
Kwansei Gakuin University
Department of Chemistry
School of Science
Uegahara
Nishinomiya 662-8501, Japan

Prof. S. Kawata
Department of Applied Physics
Osaka University
Suita
Osaka 565-0871, Japan

Dr. H. M. Heise
Institut für Spektrochemie und Angewandte Spektroskopie
Universität Dortmund
Bunsen-Kirchhoff-Straße 11
44139 Dortmund, Germany

1st Edition 2002
1st Reprint 2004
2nd Reprint 2005
3nd Reprint 2006

Library of Congress Card No.: applied for

British Library Cataloguing-in-Publication Data:
A catalogue record for this book is available from the British Library.

Die Deutsche Bibliothek – CIP-Cataloguing-in-Publication Data
A catalogue record for this publication is available from Die Deutsche Bibliothek

printed in the Federal Republic of Germany
printed on acid-free paper

Typesetting K+V Fotosatz GmbH, D-64743 Beerfelden
Printing and Bookbinding buch bücher dd ag, D-96158 Birkach

ISBN 3-527-30149-6

Contents

Foreword

The increasing demand for product quality improvement and production rationalization in the chemical, petrochemical, polymer, pharmaceutical, cosmetic, food and agricultural industries has led to the gradual substitution of time-consuming conservative analytical techniques (GC, HPLC, NMR, MS) and nonspecific control procedures (temperature, pressure, pH, dosing weight) by more specific and environmentally compatible analytical tools. In this respect, of the different methods of vibrational (mid-infrared, near-infrared and Raman) spectroscopy primarily the near-infrared technique has emerged over the last years – in combination with light-fiber optics, new in- and on-line probe accessories and chemometric evaluation procedures – as an extremely powerful tool for industrial quality control and process monitoring. With this development the wavelength gap between the visible and the mid-infrared region which has over a long period been lying idle is eventually also filled with life and exploited according to its real potential. The period of idling can be mainly contributed to two facts:

- on the one hand conservative spectroscopists did not accept the frequently broad and diffuse overtone and combination absorption bands of the near-infrared region as a complementary and useful counterpart to the fundamental vibrational absorption bands of the Raman and mid-infrared technique
- most of the early users of near-infrared spectroscopy were – with few exceptions – working in the field of agricultural science taking advantage of this new nondestructive analytical tool and having low or no interest to further exploit the spectroscopy behind the data in scientific depth.

This situation has changed progressively since about the mid-eighties and today near-infrared spectroscopy is a recognized technique also practiced by conservative spectroscopists.

Apart from chapters on the theory and instrumentation of near-infrared spectroscopy and a short introduction to chemometrics, the present book offers a diversified view in such a way, that experts in different fields highlight the potential of near-infrared spectroscopy for applications in their respective research and working environment. Although, frequently a large multiplicity of characters contributing to a multi-authored book is not necessarily regarded a positive prerequisite for its success, we hope that the readers will extract useful information which will assist in a faster and better understanding of the molecular origin of the near-infrared spectroscopic data and their correlation with the chemical or physical properties of the investigated samples. Furthermore, it is hoped that this book contributes to a more critical evaluation of near-infrared spectroscopic data thereby extending its implementation not only to quality control and production environments but also to research laboratories.

Essen, September 2001

List of Contributors

L. Bokobza
Laboratoire de Physico-chimie
Structurale et Macromoléculaire
ESPCI, 10 rue Vauquelin
75231 Paris, France

Y. Furukawa
Department of Chemistry
Waseda University
Shinjuku-ku
169-8555 Tokyo, Japan

H. M. Heise
Institut für Spektrochemie
und Angewandte Spektroskopie
Universität Dortmund
Bunsen-Kirchhoff-Straße 11
44139 Dortmund, Germany

S. Kawano
Ministry of Agriculture,
Forestry and Fisheries
2-1-2 Kannendai
305-8642 Tsukuba, Japan

S. Kawata
Department of Applied Physics
Osaka University
Suita
Osaka 565-0871, Japan

Y. Ozaki
Kwansei Gakuin University
Department of Chemistry
School of Science
Uegahara
Nishinomiya 662-8501, Japan

H. W. Siesler
Institut für Physikalische Chemie
Universität-GH Essen
Schützenbahn 70
45117 Essen, Germany

Roland Winzen
Chemometrische Beratung,
Analytische Dienstleistungen,
Softwareentwicklung
Daimlerstraße 70
D-41462 Neuss, Germany

1
Introduction

H. W. Siesler

1.1
General Remarks

Although this book is dedicated to near-infrared (NIR) spectroscopy, its theory, instrumentation and applications can only be understood properly in the context of an overview of all the available vibrational spectroscopic techniques. With this knowledge, the potential of NIR spectroscopy can be much better evaluated and appreciated in terms of its implementation as a quality- and process-control tool in an industrial environment.

At this point, critics as well as enthusiastic users of this technique are addressed. Near-infrared spectroscopy – especially in combination with chemometric evaluation routines – is neither a black-box instrumentation nor should it be used for any analytical problem in an over-enthusiastic fashion without prior detailed feasibility studies. Most disappointments with, and opposition against, this technique can be traced back to a negative experience based on a superficial, quick and dirty application trial which inevitably led to disastrous results. Furthermore, it should always be kept in mind that, in the majority of cases, NIR spectroscopy requires reference techniques to build up calibration routines and to guarantee the proper maintenance of an established calibration with reference to outlier detection and trouble-shooting. Finally, it should be emphasised that near-infrared spectroscopy is not only a routine tool but also has a tremendous research potential, which can provide unique information not accessible by any other technique. Several chapters of this book will certainly prove this statement.

Historically, the discovery of near-infrared energy is ascribed to Herschel in 1800 [1]. The 200th anniversary of this scientific milestone was celebrated in a recent meeting of the Royal Society of Chemistry in London [2]. As far as the development of instrumentation and its breakthrough for industrial applications in the second half of the 20th century were concerned, NIR proceeded in technology jumps (Fig. 1.1). In this respect, credit has largely to be given to researchers in the field of agricultural science, foremost Norris [3, 4], who recognised the potential of this technique from the very early stages of its development. At the same time, with few exceptions [5–7], comparatively low priority has been given to NIR spectroscopy in the chemical industry. This situation is also reflected in the fact

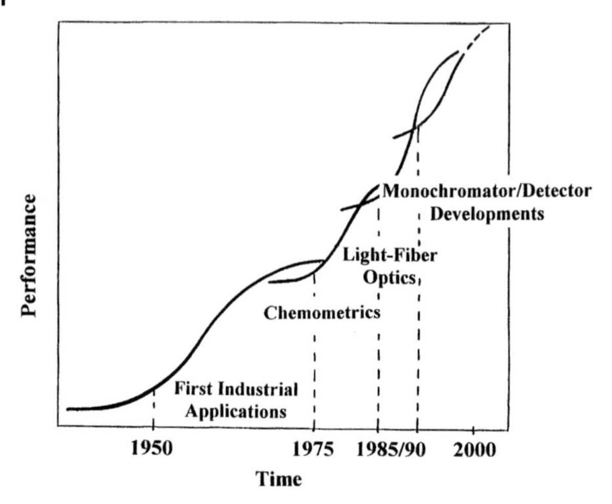

Fig. 1.1 The development of near-infrared spectroscopy in technology jumps

that, for a long time, the near-infrared spectral range was only offered as a low- or high-wavenumber add-on to ultraviolet-visible (UV/Vis) or mid-infrared (MIR) laboratory spectrometers. This situation has changed dramatically since about 1980 when stand-alone NIR instrumentation became widely available. Nevertheless, it needed another decade of intense conviction before near-infrared spectroscopy became a generally accepted technique. Since the early nineties, the availability of efficient chemometric evaluation routines, light-fibre optics coupled with specific probes for a multitude of purposes and the fast progress in miniaturisation – based on new monochromator/detection designs – has launched NIR spectroscopy into a new era for industrial quality and process control.

The following short overview on mid-infrared, near-infrared and Raman spectroscopy is intended to provide a minimum basis for putting the available vibrational spectroscopies in perspective according to their practical applications. Over the last few years, these techniques have developed into indispensable tools for academic research and industrial quality control in a wide field of applications ranging from chemistry to agriculture and from life sciences to environmental analysis. In what follows, the physical principles and some instrumental aspects of MIR, NIR and Raman spectroscopy will be discussed and the advantages and disadvantages of the individual techniques will be outlined with reference to their implementation as industrial routine equipment. For more detailed information the interested reader is referred to the pertinent literature [8–15].

While mid-infrared spectroscopy has been a well-established research tool and structure-elucidation technique in the industrial as well as the academic environment for many decades, for a long time Raman spectroscopy was restricted primarily to academic research. It has gained broader industrial recognition only in combination with the Fourier-Transform (FT) technique and NIR-laser excitation in the late eighties. Just recently, dispersive Raman spectroscopy with charge-coupled device (CCD) detection is undergoing a further renaissance with reference to process-control applications [9, 11]. On the other hand, NIR spectroscopy has only occasionally been used since the early fifties for chemical analysis, although more frequently in the field of agriculture. However, a real breakthrough as a quality- and process-control tool has occurred only within the last decade, since the introduction of efficient chemometric evaluation techniques and the development of light-fibre coupled probes. In actual fact, despite the lack of comparable specific spectral information, NIR spectroscopy is quickly overtaking Raman and primarily mid-infrared spectroscopy as a process-monitoring technique [11]. The main reason for this is the much easier sample presentation and the possibility of separating the sample measuring position and the spectrometer by distances of several hundred meters by the use of light fibres. Although similar arguments hold for Raman spectroscopy, its restriction to only small sample volumes, comparatively low sensitivity, interference by fluorescence and safety arguments are still limiting its industrial application on a really broad scale.

1.2 Basic Principles of Vibrational Spectroscopy

Although the three spectroscopic techniques are very different in several aspects, their basic physical origin is the same: absorption bands in the MIR, NIR and Raman spectra of chemical compounds can be observed as a consequence of molecular vibrations. Considering the simple diatomic oscillator of Fig. 1.2, the vibrational frequency ν (based on a harmonic oscillator approximation) can be correlated with molecular parameters by:

$$\nu = \frac{1}{2\pi}\sqrt{\frac{f}{\mu}} \tag{1.1}$$

where the force constant f reflects the strength of the bond between m and M and the reduced mass μ is given by:

$$\mu = \frac{mM}{m+M} \tag{1.2}$$

Thus, the vibrational frequencies are very sensitive to the structure of the investigated compound and this is the basis for the wide-spread application of MIR (and less frequently) Raman spectroscopy for structure elucidation.

Furthermore, due to the different excitation conditions of MIR, NIR and Raman spectroscopy, the relationships between the absorption intensities and the addressed functionalities of the molecules under examination vary significantly. These specific features lead to extremely different responses of the same molecular vibration to the applied technique. Fig. 1.2 summarises the potential energy (V) transitions, excitation conditions and the most important instrumental as well as qualitative and quantitative aspects of the three spectroscopies. The principal

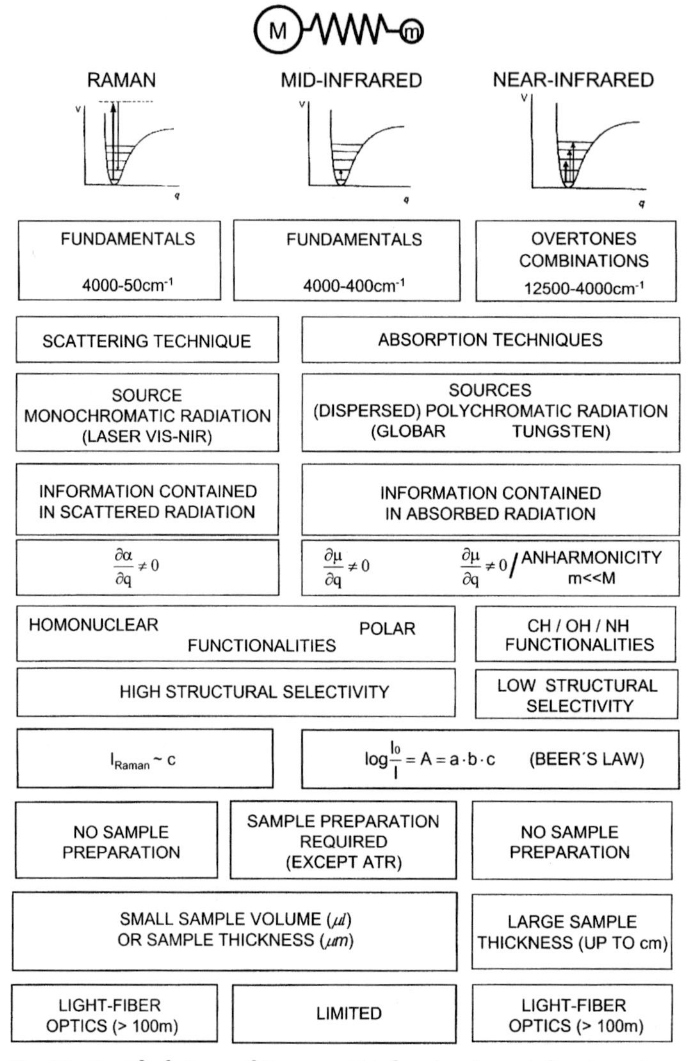

Fig. 1.2 Specific features of Raman, mid-infrared and near-infrared spectroscopy (see text)

difference between the three techniques is that Raman spectroscopy is a scattering technique, whereas mid- and near-infrared spectroscopy are absorption techniques. Thus, whereas scanning MIR and NIR spectrometers operate with a polychromatic source from which the sample absorbs specific frequencies corresponding to its molecular vibrational transitions (mostly fundamental vibrations for the MIR and overtone or combination vibrations for the NIR), in Raman spectroscopy, the sample is irradiated with monochromatic laser light whose frequency may vary from the visible to the NIR region. This radiation excites the molecule to a virtual energy state that is far above the vibrational levels of this anharmonic oscillator for a visible laser and in the range of high overtones for NIR laser excitation (see also Fig. 1.4). From this excited energy level, the molecule may return to the ground state by elastic scattering thereby emitting a Rayleigh line that has the same frequency as the excitation line and does not contain information in terms of the molecular vibration (this case is not shown). If it only returns to the first excited vibrational level by inelastic scattering, the emitted Raman line (the so-called Stokes line) has a lower frequency, and the frequency difference to the excitation line corresponds to the vibrational energy of the MIR absorption. Also not shown is the case of the anti-Stokes line, in which the starting level is the first excited vibrational state and the molecule returns to the ground state by inelastic scattering, thereby emitting a Raman line of higher frequency (here, too, the frequency difference to the excitation line corresponds to the vibrational energy of the MIR absorption) but of lower intensity compared with the Stokes line, due to the lower population probability of the excited state (Boltzmann's law). Commonly, the Stokes lines are used for practical Raman spectroscopy and refer to the same vibrational transitions as MIR spectroscopy.

One of the limiting factors for the application of the Raman technique, however, becomes evident by comparing the intensities of the source and the scattered radiation:

$$I_{\text{Raman}} \approx 10^{-4} I_{\text{Rayleigh}} \approx 10^{-8} I_{\text{source}} \tag{1.3}$$

From these figures it can readily be seen that sensitive detection of the Raman line alongside efficient elimination of the Rayleigh line are experimental prerequisites for the successful application of Raman spectroscopy. As shown in Fig. 1.2, Raman and MIR spectroscopy cover approximately the same wavenumber region with the Raman technique extending into the far-infrared (FIR) region (ca. 50 cm^{-1}) for instrumental reasons (primarily because of the MIR detector cut-off). In some cases, this additional frequency range is valuable, since it often contains absorptions of lattice modes of molecular crystals which may be very characteristic of a specific polymorph.

Another important relation for the comparison of visible- versus NIR-Raman spectroscopy, is the dependence of the scattered Raman intensity I_{Raman} on the fourth power of the excitation frequency ν_{exc}:

$$I_{\text{Raman}} \approx \nu_{\text{exc}}^4 \tag{1.4}$$

The consequences of this relationship with reference to the application of visible- or NIR-Raman spectroscopy for an individual problem will be outlined below.

NIR spectroscopy covers the wavenumber range adjacent to the MIR and extends up to the visible region. NIR absorptions are based on overtone and combination vibrations of the investigated molecule and, due to their lower transition probabilities, the intensities usually decrease by a factor of 10–100 for each step from the fundamental to the next overtone [10]. Thus, the intensities of absorption bands successively decrease from the MIR to the visible region, thereby allowing an adjustment of the sample thickness (from millimetres up to centimetres) depending on the rank of the overtone. This is a characteristic difference from MIR and Raman spectra, where the intensities of the fundamentals vary irregularly over the whole frequency range and depend exclusively on the excitation conditions of the individual molecular vibrations. Thus, for a Raman band to occur, a change of the polarisability α (the ease of shifting electrons) has to take place during the variation of the vibrational coordinate q. Alternatively, the occurrence of an MIR band requires a change of the dipole moment μ during the vibration under consideration. These different excitation conditions lead to the complementarity of the Raman and MIR techniques as structural elucidation tools, because Raman spectroscopy predominantly focuses on vibrations of homonuclear functionalities (e.g. C=C, C-C and S-S) whereas the most intense MIR absorptions can be traced back to polar groups (e.g. C-F, Si-O, C=O and C-O). In a simplified form, NIR spectroscopy requires – in addition to the dipole moment change – a large mechanical anharmonicity of the vibrating atoms (see Fig. 1.2) [16]. This becomes evident from the analysis of the NIR spectra of a large variety of compounds, where the overtone and combination bands of CH, OH and NH functionalities dominate the spectrum, whereas the corresponding overtones of the most intense MIR fundamental absorptions are rarely represented. One reason for this phenomenon is certainly the fact that most of the X-H fundamentals absorb at wavenumbers >2000 cm^{-1} so that their first overtones already appear in the NIR frequency range. The polar groups leading to the most intense fundamental absorptions in the MIR on the other hand, usually absorb at wavenumbers <2000 cm^{-1}, so that their first (and sometimes higher) overtones still occur in the MIR region. Due to the intensity loss for each step from the fundamental to the next overtone, the absorption intensities of these functionalities have become negligible by the time they occur in the NIR range.

The superposition of many different overtone and combination bands in the NIR region causes a very low structural selectivity for NIR spectra compared with the Raman and MIR analogues where many fundamentals can usually be observed in isolated positions. Nevertheless, NIR spectra should also be assigned in as much detail as possible with reference to their molecular origin [17]; this allows a more effective application for research purposes and combination with chemometric evaluation procedures. For the assignment of overtones and combination bands in the NIR to their corresponding fundamentals in the MIR it is recommended that the wavenumber notation be used instead of the wide-spread wavelength (nm or μm) scale. It should be mentioned, however that the wavenumber positions of the overtones deviate with increasing multiplicity from the exact multiples of their fundamentals due to the anharmonicity of the vibrations [10, 15–17].

As far as the quantitative evaluation of vibrational spectra is concerned, MIR and NIR spectroscopy follow Beer's law, whereas the Raman intensity I_{Raman} is directly proportional to the concentration of the compound to be determined (Fig. 1.2). To avoid compensation problems, in most cases quantitative Raman spectroscopy is performed with an internal reference band in the vicinity of the analytical absorption being analysed.

An important issue for the implementation of a technique as an industrial routine tool is the sample preparation required for this technique. In this respect, it can be seen from Fig. 1.2 that Raman and NIR spectroscopy have considerable advantages over MIR spectroscopy, which usually requires individual sample preparation steps before data acquisition. Only the technique of attenuated total reflection (ATR) [18] circumvents time-consuming sampling procedures for MIR spectroscopy.

1.3 Instrumentation

Fig. 1.3 summarises the present state of the most frequently used monochromator/detection principles for the different scanning spectroscopies. As mentioned above, in Raman spectroscopy two techniques are currently in use:

- excitation by a visible-laser, combined with monochromatisation of the scattered radiation by a holographic grating and simultaneous detection of the dispersed, narrow frequency ranges by a CCD detector
- NIR-laser excitation and measurement in a FT spectrometer.

Both alternatives involve compromises and the choice of the applied technique depends on the individual problem. In Fig. 1.4, the trends of the main limiting factors – fluorescence and low scattering efficiency – have been outlined with reference to the two excitation mechanisms. If a molecule is irradiated with visible radiation, it may be excited to an energy level of the next-higher electronic state. Return to the ground state or an excited vibrational level of the original electronic state can easily proceed through fluorescence as shown in Fig. 1.4. Thus, for a large proportion of samples, irradiation with visible light causes strong fluorescence due to additives or impurities (or the sample itself) which will superimpose and, in many cases, inundate the Raman spectrum of the sample. As an example, the effect of fluorescence in the Raman spectrum of thioindigo as measured with Ar^+-ion (488 nm) laser excitation relative to the fluorescence-free Raman spectrum obtained by NIR-laser excitation (Nd:YAG, 1064 nm) is demonstrated in Fig. 1.5. A number of time-consuming approaches have been proposed to circumvent the problem [14] including prior purification or "burning out" of the fluorescence. These approaches have, at best, been only partially successful. Additionally, highly

RAMAN	MID-INFRARED	NEAR-INFRARED
NIR-RAMAN (FT) VIS-RAMAN (CCD)	FT-IR	GRATING FT-NIR AOTF DIODE-ARRAY

Fig. 1.3 The current monochromator/detection principles of scanning MIR, NIR and Raman spectrometers

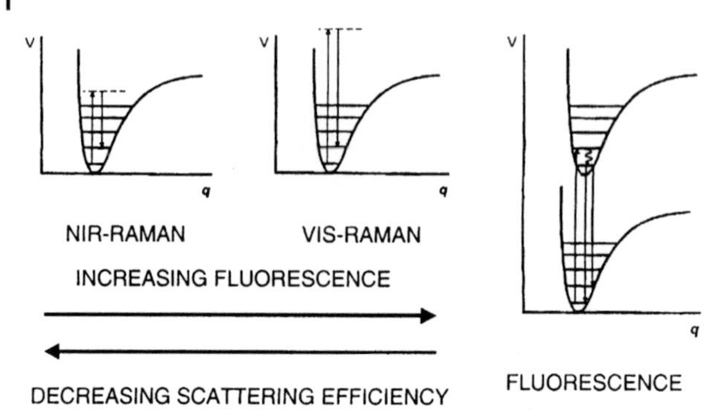

Fig. 1.4 Fluorescence and scattering efficiency in NIR- and visible-Raman spectroscopy [22]

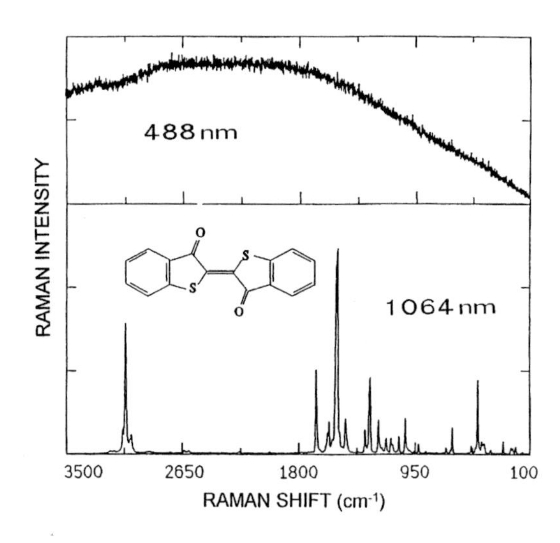

Fig. 1.5 Raman spectra of thioindigo (top: Ar^+-ion laser excitation, bottom: NIR laser excitation) [20]

coloured samples may absorb the Raman photons, thereby preventing them from reaching the detector and leading to thermal degradation of the investigated material. The use of NIR-laser excitation confers a number of advantages on a Raman system. Both fluorescence and self-absorption are very much reduced in the Raman signal, and, due to the lower energy of the excitation radiation, thermal degradation is also less of a problem. However, these advantages are partly neutralised by the disadvantages of using a low-frequency laser as a source. The Raman technique is obviously less sensitive due to the ν^4-dependence of the scattering efficiency (see also Fig. 1.4). Thus, a shift of the excitation line from the visible region (e.g. Ar^+-ion laser, 488 nm/20492 cm^{-1}) to the NIR region (1064 nm/9398 cm^{-1}) reduces the scattering intensity. At 0 cm^{-1}, the sensitivity of a Nd-YAG laser is 23 times lower than that of an Ar laser, by 4000 cm^{-1}, due to the frequency dependence of the sensitivity, this has increased to a factor of 87 [19, 20]. As shown in Fig. 1.3, however, NIR-Raman spectroscopy is performed on FT spectrometers, and the sensitivity loss can be quite efficiently compensated for by the accumulation of multiple scans. As a valuable compromise for suppressing fluorescence and at the same time retaining an acceptable scattering efficiency, excitation with a diode laser at 785 nm (12739 cm^{-1}) is increasingly used [9].

Today, as far as MIR spectroscopy is concerned, it is almost exclusively FT-based instruments that are in routine use (Fig. 1.3). Besides the optical principle of the Michelson interferometer, polarisation interferometers based on a stationary and a moving MIR-transparent (KBr) wedge have been integrated into scanning spectrometers [9]. Contrary to Raman and MIR spectroscopy, scanning NIR spectroscopy offers the largest multiplicity of monochromator principles. Thus, apart from different designs of moving parts, such as grating instruments and FT spectrometers with Michelson and polarisation interferometers (with NIR-transparent quartz wedges) two fast-scanning approaches with no moving parts are available: diode-array systems and acousto-optic tuneable filters [9, 10]. In this context it should be mentioned that miniaturisation has certainly progressed most significantly for NIR spectroscopy. A similar trend is also observable for visible-Raman CCD spectrometers.

1.4 Process-Monitoring

In the last two lines of Fig. 1.2 the most important aspects for the implementation of the individual spectroscopies as process-monitoring tools are addressed. The very small representative sample volume or thickness in Raman and MIR/ATR spectroscopy may certainly lead to problems if special care is not taken to avoid the formation of a stationary layer on the reactor window or on the ATR crystal. In this respect, NIR spectroscopy is the method of choice in view of the comparatively large sample volume/thickness involved in these measurements. The ability to separate the spectrometer from the point of sampling is certainly a great advantage for Raman and NIR spectroscopy. Although light-fibres based on ZrF_4 and AgCl are also available for MIR spectroscopy, it should be mentioned that their cost, attenuation properties and mechanical/chemical stability are still inferior compared with the well-established quartz fibres [10]. Specific probes are available for all three techniques. NIR spectroscopy, especially, offers a wide range of in-line and at-line transmission and diffuse-reflection probes designed for the measurement of liquids and solids. Large differences can also be identified with respect to their ability to measure aqueous solutions. Water is an extremely strong absorber in the MIR and also a strong NIR absorber thereby limiting the available wavenumber regions in both techniques. In contrast, it is a weak Raman scatterer and Raman spectroscopy is therefore recommended for consideration as an analytical tool for aqueous solutions. Care has to be taken, however, with the NIR-Raman FT-technique (1064 nm) because, due to the absorption of the water-overtone vibration, the Raman spectrum may be modified relative to the visible-laser excited Raman spectrum [21].

In conclusion, over the last few years MIR, NIR and Raman spectroscopy have been developed to a point where each technique can be considered a candidate for industrial quality control and process monitoring applications. However, adding up the specific advantages and disadvantages of the individual techniques, NIR spectroscopy is certainly the most flexible alternative.

1.5 References

[1] W. Herschel, *Phil. Trans. Roy. Soc. (London)*, **1800**, *284*.

[2] A. M. C. Davies, *NIR News*, **2000**, *11(2)*, 3.

[3] A. W. Brant, A. W. Otte, K. H. Norris, *Food Tech.* **1951**, *5(9)*, 356.

[4] G. S. Birth, K. H. Norris, *Food Tech.* **1958** *12(11)*, 592.

[5] W. Kaye, *Spectrochim. Acta*, **1954**, *6*, 257.

[6] R. G. J. Miller, H. A. Willis, *J. Appl. Chem.* **1956**, *6*, 385.

[7] R. F. Goddu, D. A. Delker, *Anal. Chem.* **1960**, *32*, 140.

[8] *Infrared and Raman Spectroscopy* (Ed.: B. Schrader), Wiley-VCH, Weinheim, **1995.**

[9] J. Coates, *Appl. Spectrosc. Revs.* **1998**, *33(4)*, 267.

[10] H. W. Siesler, *Makromol. Chem. Macromol. Symp.* **1991**, *52*, 113.

[11] J. J. Workman Jr. *Appl. Spectrosc. Revs.* **1999**, *34(1/2)*, 1–89.

[12] P. R. Griffiths, J. A. de Haseth, *Fourier Transform Infrared Spectrometry*, Wiley-Interscience, New York, **1986.**

[13] H. Günzler, H. M. Heise, *IR-Spektroskopie*, VCH, Weinheim, **1995.**

[14] P. Hendra, C. Jones, G. Warnes, *Fourier Transform Raman Spectroscopy*, Ellis Horwood, Chichester, **1991.**

[15] H. W. Siesler, K. Holland-Moritz, *Infrared and Raman Spectroscopy of Polymers*, Marcel Dekker, New York, **1980.**

[16] W. Groh, *Makromol. Chem.* **1988**, *189*, 2861.

[17] P. Wu, H. W. Siesler, *J. Near Infrared Spectrosc.* **1999**, *7*, 65.

[18] *Internal Reflection Spectroscopy* (Ed.: F. M. Mirabella, Jr.), Marcel Dekker, New York, **1993.**

[19] H. W. Siesler, *Revue de l'Institut Français du Pétrole*, **1993**, *48* (*3*), 223.

[20] A. Hoffmann, PhD-Thesis, University of Essen, **1991.**

[21] W.-D. Hergeth, *Chemie Ingenieur Technik*, **1998**, *70(7)*, 894.

[22] B. Schrader, *Chemie in unserer Zeit*, **1997**, *31(5)*, 229.

2
Origin of Near-Infrared Absorption Bands

L. Bokobza

2.1
Introduction

In the last ten years, near-infrared spectroscopy has become a very popular technique for a wide range of analyses in various industries [1–8]. The usefulness of this technique is mainly attributed to its allowing the rapid and nondestructive analysis of bulk materials. On the other hand, improvements in instrumentation, and especially the development of chemometric software [9–11], have contributed to the tremendous expansion of and to the current state of popularity of this technique.

The statistical treatment extensively used for solving analytical problems has been the major factor in the growing interest in near-infrared spectroscopy. But it is often claimed that such approaches can result in the production of a "black box" technique, in that very little is understood about the assignment of the observed bands or about the rules of the vibrational spectroscopy underlying a NIR spectrum. The purpose of this chapter is to provide the reader with the basic concepts which make the origin of NIR absorption bands understandable.

2.2
Principles of Near-Infrared Spectroscopy

Infrared spectroscopy is used to investigate the vibrational properties of a sample. Molecular vibrations give rise to absorption bands generally located in the mid-infrared range (between 400 and 4000 cm^{-1}) where they are the most intense and the simplest.

Adjacent to the mid-infrared, the NIR region covers the interval between approximately 4000 and 12500 cm^{-1} (2.5–0.8 μm). As will be shown later, this region contains absorption bands corresponding to overtones and combinations of fundamental vibrations.

Infrared radiation absorbed by a molecule causes individual bonds to vibrate in a manner similar to that of a diatomic oscillator. So it is convenient to start from the diatomic molecule as the simplest vibrating system and then extend the concepts developed to polyatomic molecules.

2.2.1 The Diatomic Molecule

2.2.1.1 The Harmonic Oscillator

In the case of an ideal harmonic oscillator, the potential energy V contains a single quadratic term:

$$V = \frac{1}{2}k(r - r_e) = \frac{1}{2}kx^2 \tag{2.1}$$

where k is the force constant of the bond, r is the internuclear distance, r_e is the equilibrium internuclear distance and $x=(r-r_e)$ is the displacement coordinate. The potential energy curve is parabolic in shape and symmetrical about the equilibrium bond length, r_e (Fig. 2.1).

The vibrating mechanical model for a diatomic molecule leads to the classical vibrational frequency, ν:

$$\nu = \frac{1}{2\pi}\sqrt{\frac{k}{\mu}} \tag{2.2}$$

where μ is the reduced molecular mass, such that: $\mu=m_1m_2/(m_1+m_2)$ and m_1 and m_2 are the masses of the two nuclei.

A quantum mechanical treatment shows that the vibrational energy may only have certain discrete values called energy levels. For the harmonic oscillator, these energy levels are given by:

$$E_{\text{vib}} = h\nu\left(\text{v} + \frac{1}{2}\right) \tag{2.3}$$

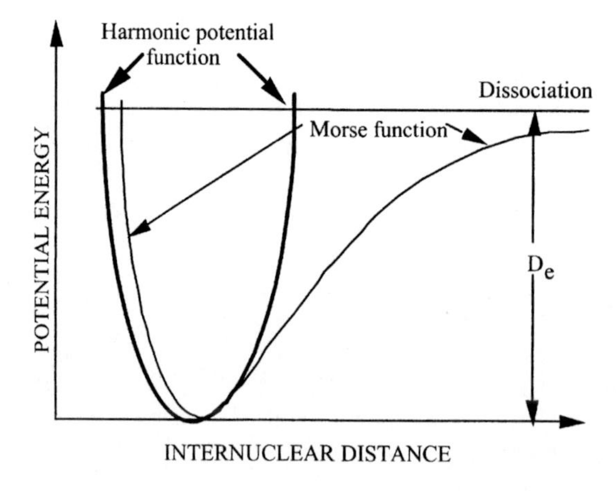

Fig. 2.1 Harmonic and anharmonic potential functions for a diatomic oscillator

where h is Planck's constant, ν is the classical vibrational frequency defined above and v is the vibrational quantum number which can only have integer values 0, 1, 2, 3...

The energy levels G(v), are expressed in wavenumber units (cm^{-1}) by the expression:

$$G(\mathrm{v}) = E_{\mathrm{vib}}/hc = \bar{\mathrm{v}}\left(\mathrm{v} + \frac{1}{2}\right) \tag{2.4}$$

$\bar{\mathrm{v}}$ being the wavenumber of the vibrational transition.

The vibrational energy levels corresponding to different values of v are represented in Fig. 2.2 as horizontal lines. As can be seen, these energy levels are equally spaced. The selection rules that state which transitions are active or allowed can be deduced from examination of the transition moment given by the expression:

$$P_{\mathrm{v}'' \to \mathrm{v}'} = \int \psi_{\mathrm{v}'}^{*} \varepsilon \psi_{\mathrm{v}''} d\tau \tag{2.5}$$

where $\psi_{\mathrm{v}'}$ and $\psi_{\mathrm{v}''}$ are the wave functions of the v′ and v″ states (the * indicates the complex conjugate of $\psi_{\mathrm{v}'}$) and ε is the dipole moment which may be expressed as a linear function of x for small displacements about the equilibrium configuration:

$$\varepsilon = \varepsilon_0 + \left(\frac{d\varepsilon}{dx}\right)_{\mathrm{e}} x \tag{2.6}$$

where ε_0 is the dipole moment at the equilibrium internuclear distance.

The transition moment for the transition v″→v′ may be calculated by substituting the appropriate wave functions and dipole moment from Eq. (2.6) into Eq. (2.5). Transitions are allowed for a nonzero value of the transition moment. This

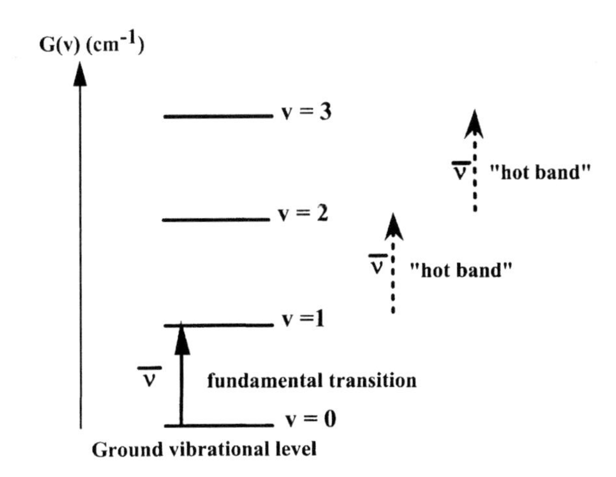

Fig. 2.2 Vibrational energy levels of the harmonic oscillator

occurs if the vibration is accompanied by a dipole-moment change; this implies that only heteronuclear, diatomic molecules will exhibit vibrational-spectral transitions. Qualitatively, we can picture the dipole moment as an oscillating dipole coupled with the electric field of the incident radiation in such a way that the energy can be exchanged between the molecule and the radiation.

In the quantum mechanical harmonic oscillator, a further restriction concerns the vibrational quantum number which can only change by one unit. Thus, transitions across more than one energy level are forbidden in the harmonic oscillator.

Therefore, transitions are allowed only if:

$$\left(\frac{d\varepsilon}{dx}\right)_{\mathrm{e}} \neq 0$$

$$\Delta \mathrm{v} = \pm 1$$

From the Boltzmann distribution, most molecules at room temperature exist at the ground vibrational level $\mathrm{v}=0$ and consequently the allowed transition $\mathrm{v}=0 \rightarrow \mathrm{v}=1$, called the fundamental transition, dominates the infrared absorption spectrum. This transition is responsible for most of the infrared absorption of interest to a chemical spectroscopist.

The other allowed transitions such as $\mathrm{v}=1 \rightarrow \mathrm{v}=2$, $\mathrm{v}=2 \rightarrow \mathrm{v}=3 \ldots$, originate from vibrationally excited levels ($\mathrm{v} \neq 0$). The corresponding bands are much weaker than the fundamental absorption. Their designation as "hot bands" comes from the fact that the excited levels have a relatively low population and increasing the temperature will increase this population and, thus, the intensity of the bands. For the harmonic oscillator, the transitions giving rise to the hot bands have the same frequency as that of the fundamental transition (Fig. 2.2).

2.2.1.2 **Anharmonic Oscillator**

Two experimental observations give evidence that molecules are not ideal oscillators. Firstly, the vibrational energy levels are not equally spaced, so the hot bands do not have exactly the same frequency as the fundamental band. Secondly, overtone transitions such as $\mathrm{v}=0$ to $\mathrm{v}=2$, 3, 4, ... are allowed. This departure from harmonic behaviour may be expressed by two effects.

The first effect, called *mechanical anharmonicity*, arises from the effect of cubic and higher terms in the potential-energy expression:

$$V = \frac{1}{2}kx^2 + k'x^3 + \ldots k' \ll k \qquad (2.7)$$

The above expression is used in the Schrödinger equation to deduce the energy levels of the allowed states of the anharmonic oscillator. The solution is obtained by an approximation or perturbation method and leads to an energy level (in cm^{-1}) that can be written as:

$$G(\mathrm{v}) = E_{\mathrm{vib}}/hc = \overline{\nu}\left(\mathrm{v}+\frac{1}{2}\right) - x_{\mathrm{e}}\overline{\nu}\left(\mathrm{v}+\frac{1}{2}\right)^2$$
$$= \overline{\nu}\left(\mathrm{v}+\frac{1}{2}\right) - X\left(\mathrm{v}+\frac{1}{2}\right)^2 \quad (2.8)$$

Here x_{e} is the anharmonicity constant and $X = x_{\mathrm{e}}\overline{\nu}$. Unlike the harmonic oscillator, energy levels are no longer equally spaced (Fig. 2.3(b)).

An empirical equation called the Morse function, illustrated in Fig. 2.1, has often been used for the anharmonic potential function:

$$V = D_{\mathrm{e}}(1 - e^{-\beta x})^2 \quad (2.9)$$

where β is a constant and D_{e} is the dissociation energy measured from the equilibrium position, which is the minimum of the curve. In terms of this potential, D_{e} is given by [34]

$$D_{\mathrm{e}} = \frac{\overline{\nu}}{4x_{\mathrm{e}}} \quad (2.10)$$

The second effect, called *electrical anharmonicity*, is responsible for the appearance of overtones corresponding to transitions between energy levels that differ by two or three vibrational quantum number units ($\Delta\mathrm{v}=+2, +3\ldots$) in the infrared spectra. The electrical anharmonicity arises from the effect of square and higher terms in the dipole-moment expression:

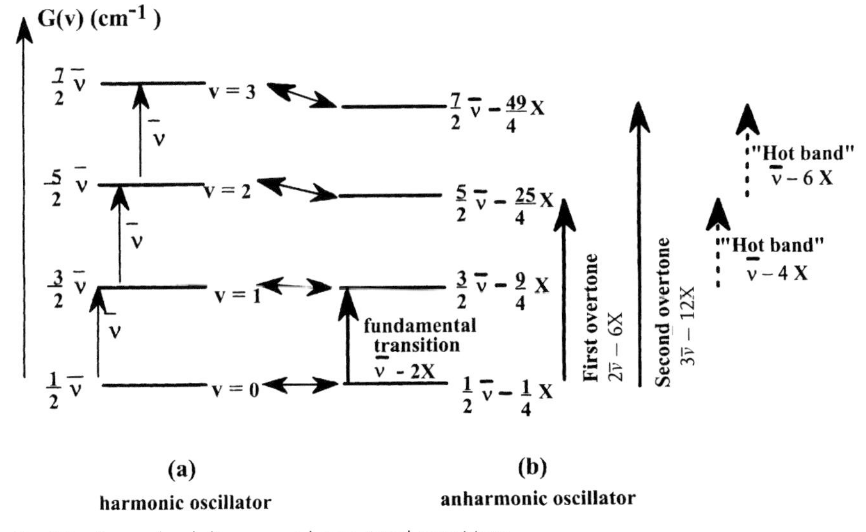

Fig. 2.3 Energy-level diagram and associated transitions

$$\varepsilon = \varepsilon_0 + \left(\frac{d\varepsilon}{dx}\right)_e x + \frac{1}{2}\left(\frac{d^2\varepsilon}{dx^2}\right)_e x^2 + \ldots \qquad (2.11)$$

The energy-level diagram, as well as the associated transitions for a diatomic molecule, are represented in Fig. 2.3(b). As can be seen, for the anharmonic oscillator, the frequencies of the overtone absorptions are not exactly 2, 3, ... times that of the fundamental absorption. On the contrary, on account of the mechanical anharmonicity, the frequency of the hot bands is less than that of the fundamental transition.

For an OH radical, from the wavenumber of the harmonic vibration:

$$\bar{\nu} = 3735.2 \text{ cm}^{-1},$$

and from the value of X=82.8 cm^{-1} the wavenumber of the fundamental transition can be calculated:

$$\bar{\nu} - 2X = 3735.2 - 2(82.8) = 3569.6 \text{ cm}^{-1}$$

It is of course less than that of the harmonic oscillator.

We can also calculate the wavenumber of the first overtone:

$$2\bar{\nu} - 6X = 2(3735.2) - 6(82.8) = 6973.6 \text{ cm}^{-1}$$

which is located in the near-infrared.

At this point, one can already understand that overtones are at the heart of NIR spectroscopy and that the key quantity which determines the occurrence and the spectral properties (frequency, intensity) of the NIR absorption bands is anharmonicity.

The intensities of the overtones also depend on anharmonicity. It has been shown [12] that those bonds with a low anharmonicity constant have much lower overtone intensities. The XH stretching transitions, which have the highest anharmonicity constants, dominate overtone spectra. On the other hand, carbonyl stretching bands are not very anharmonic and their higher overtones are found to be very low in intensity. See Tab. 2.1 for values.

Tab. 2.1 Values of the anharmonicity constant x_e for various vibrations. From [12]

x_e (ν CH)	$\sim 1.9 \cdot 10^{-2}$
x_e (ν CD)	$\sim 1.5 \cdot 10^{-2}$
x_e (ν CF)	$\sim 4 \cdot 10^{-3}$
x_e (ν CCl)	$\sim 6 \cdot 10^{-3}$
x_e (ν C=O)	$\sim 6.5 \cdot 10^{-3}$

2.2.2
The Polyatomic Molecule

A molecule containing a number, N, atoms will have ($3N$–6) vibrational degrees of freedom ($3N$–5 for linear molecules). The number of vibrational degrees of freedom gives the number of fundamental vibrational frequencies of the molecule or the number of different "normal" modes of vibration. A normal mode of vibration of a given molecule corresponds to internal atomic motions in which all atoms move in phase with the same frequency, but with different amplitudes.

2.2.2.1 Harmonic Approximation

In wave mechanics, as in classical mechanics, the vibrational molecule may be considered, in a good first approximation, as a superposition of ($3N$–6) simple harmonic motions.

Let us consider, as an example, the case of a nonlinear triatomic molecule like SO_2 which is expected to have three fundamental frequencies: ν_1, which corresponds to a symmetric stretch and which gives a band at 1151 cm^{-1}, the bending mode ν_2 with absorbance at 519 cm^{-1} and the other stretch ν_3, called the antisymmetric stretch with a wavenumber of 1361 cm^{-1} (Fig. 2.4). The three energy-level patterns are also represented in Fig. 2.4. Each energy-level pattern is that obtained

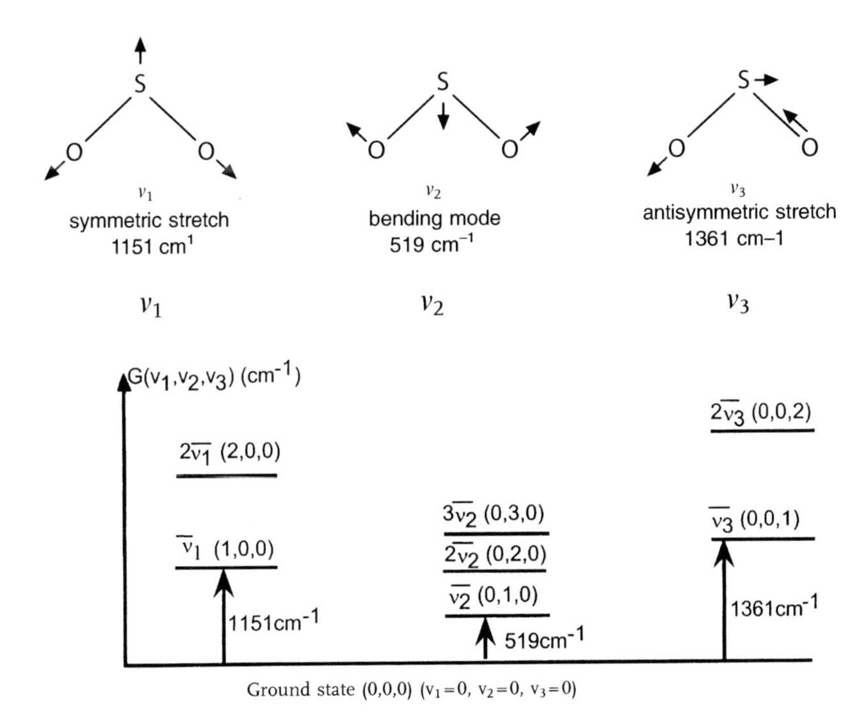

Fig. 2.4 Vibrational energy levels of sulfur dioxide, SO_2

for the harmonic diatomic oscillator. Inside each pattern, each energy level is labelled with the corresponding vibrational quantum number v_1, v_2 or v_3.

2.2.2.1.1 **Energy Levels of a Polyatomic Molecule**

Assuming harmonic oscillations, the vibrational energy of the molecule would be given by:

$$E(v_1, v_2, v_3) = h\nu_1\left(v_1 + \frac{1}{2}\right) + h\nu_2\left(v_2 + \frac{1}{2}\right) + h\nu_3\left(v_3 + \frac{1}{2}\right) \quad (2.12)$$

Again it is more convenient to express the energy in wavenumber units (cm^{-1}):

$$\begin{aligned} G(v_1, v_2, v_3) &= E(v_1, v_2, v_3)/hc \\ &= \bar{\nu}_1\left(v_1 + \frac{1}{2}\right) + \bar{\nu}_2\left(v_2 + \frac{1}{2}\right) + \bar{\nu}_3\left(v_3 + \frac{1}{2}\right) \end{aligned} \quad (2.13)$$

in which $\bar{\nu}_1$, $\bar{\nu}_2$, $\bar{\nu}_3$ are the vibrational frequencies measured in units of cm^{-1}.

For $v_1 = 0$, $v_2 = 0$, $v_3 = 0$, that is, in the lowest possible state, the vibrational energy is:

$$G(0, 0, 0) = \frac{1}{2}\bar{\nu}_1 + \frac{1}{2}\bar{\nu}_2 + \frac{1}{2}\bar{\nu}_3 \quad (2.14)$$

The energy change between G (v_1, v_2, v_3) and G (0, 0, 0) states is given by:

$$\Delta G = \bar{\nu}_1 v_1 + \bar{\nu}_2 v_2 + \bar{\nu}_3 v_3 \quad (2.15)$$

2.2.2.1.2 **Selection Rules**

If the vibrational motion were strictly a simple harmonic one, the changes in the vibrational quantum numbers for each normal vibration would be restricted to: $\Delta v_1 = 1$, $\Delta v_2 = 1$, $\Delta v_3 = 1$. In the harmonic oscillator approximation, only the fundamental modes are allowed but, in order to be "infrared active", they should also be connected with a change of dipole moment. In the harmonic oscillator model, the infrared spectrum of the SO_2 molecule should be composed of three absorption bands (Fig. 2.5).

2.2.2.2 **Influence of Anharmonicity**

2.2.2.2.1 **Overtone and Combination Bands**

As in the case of diatomic molecules, if the anharmonicity of the vibrations is taken into account, overtones (for which $\Delta v_i > 1$) and combination vibrations (for which several $\Delta v_i \neq 0$) may also occur. However, they will be much weaker than

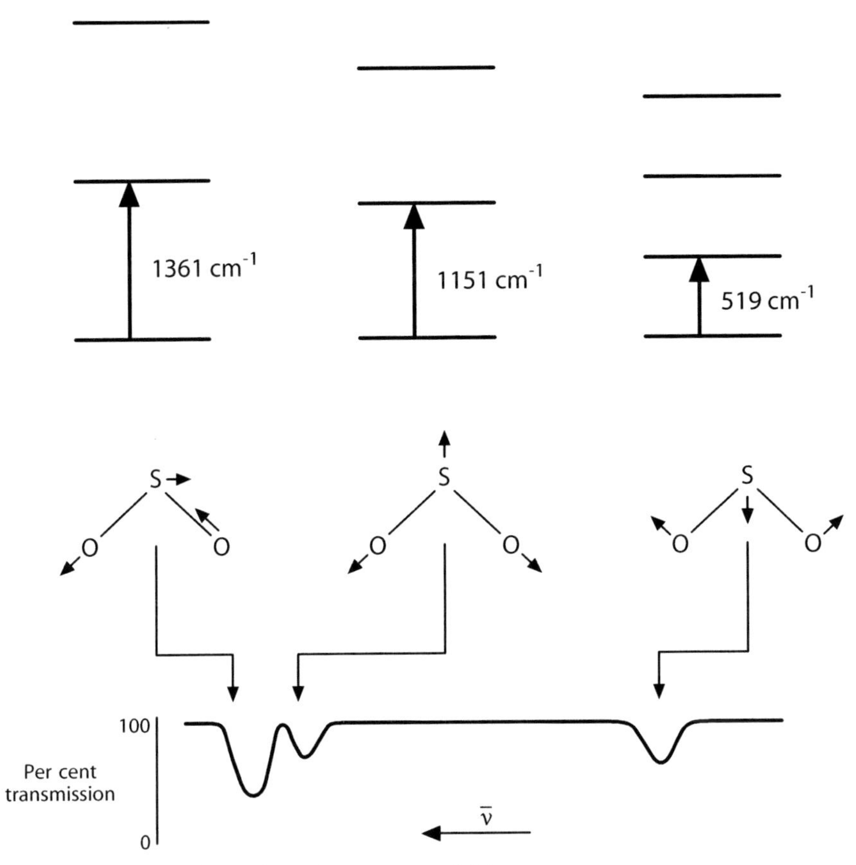

Fig. 2.5 Schematic infrared absorption spectrum of SO_2

the fundamental bands since the anharmonicities are in general slight, except for very large amplitudes of the nuclei.

Overtone and combination bands for which $\Delta v_i = 2$ or $\sum \Delta v_i = 2$, that is, transitions in which either one vibration changes by two quanta or two vibrations change by one quantum, are called binary combinations, those for which $\Delta v_i = 3$ or $\sum \Delta v_i = 3$, are called ternary combinations, and so on. Ternary combinations are weaker than binary combinations since higher approximations are involved.

To illustrate the above considerations, let us examine in Fig. 2.6 the energy-level diagram of the SO_2 molecule again. The three fundamental levels ν_1, ν_2 and ν_3, and for each normal vibration, the overtone levels are represented in the figure. The combination levels, for example the combination levels corresponding to $\nu_1+\nu_2$, $\nu_1+\nu_3$, $\nu_2+\nu_3$, are also shown.

In order to complete the interpretation of the spectrum of this molecule, it is worth mentioning that the $v_2 = 1$ level has a relatively low energy and could consequently be populated at room temperature: the ratio of the number of molecules

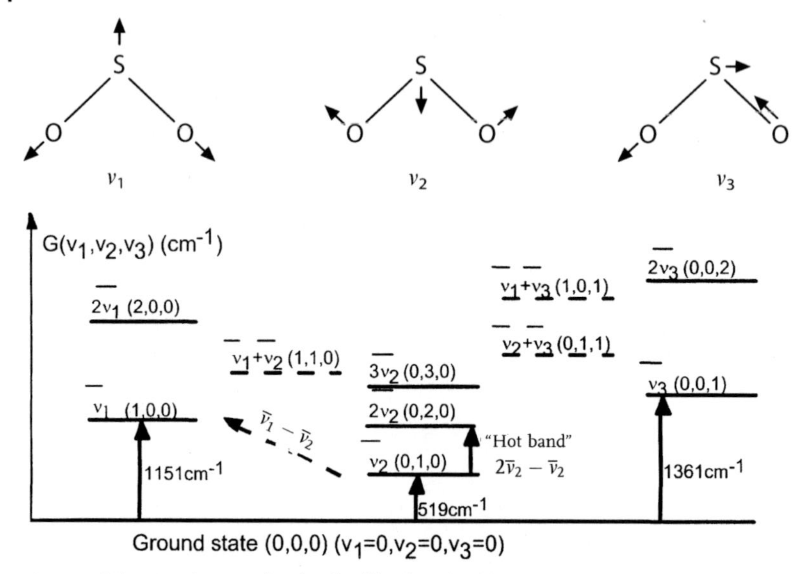

Fig. 2.6 Vibrational energy levels of sulfur dioxide, SO_2

in this level to that in the zero vibrational level at 300 K is 0.08:1. Two types of transitions from $v_2=1$ may also occur: "hot bands" ($2\bar{\nu}_2 - \bar{\nu}_2$) or difference bands ($\bar{\nu}_1 - \bar{\nu}_2$) and the probability of obtaining such bands is favoured by increasing the temperature.

As can be seen in Tab. 2.2, the number of vibrational bands that may be observed for such a simple molecule as sulfur dioxide is higher than that predicted by the harmonic oscillator model. Fundamental bands are easily identified since they are much more intense than overtone and combination bands. But we have to keep in mind that the described transitions appear in the infrared spectrum only if they are connected with a change of dipole moment. Dipole moment changes and, thus, the activity of normal modes of vibration in infrared spectra can be elegantly predicted from symmetry considerations and group theory. Without any other information than the symmetry properties of molecules, the applica-

Tab. 2.2 Assignments of some absorption bands of the SO_2 molecule

$\bar{\nu}$ *(cm^{-1})*	*Assignment*
519	ν_2
606	$\nu_1-\nu_2$
1151	ν_1
1361	ν_3
1871	$\nu_2+\nu_3$
2296	$2\nu_1$
2499	$\nu_1+\nu_3$

tion of group theory allows the determination of the number of normal modes occurring in each symmetry species as well as the activity of each mode with respect to infrared. Symmetry considerations may restrict the number of allowed transitions; on the other hand, all normal vibrations are infrared active in asymmetrical molecules.

In conclusion, in polyatomic molecules, the anharmonicity causes not only the appearance of overtones but combination bands as well. When the fundamental bands are related to C-H, N-H and O-H functional groups that occur between 4000 and 2500 cm^{-1}, the overtones and combinations make up the backbone of what is called the "near-infrared". Thus this region of the spectrum is not really so difficult and can be predicted by the rules of spectroscopy as well as any other "normal" region.

2.2.2.2.2 **Determination of the Anharmonicity Constants**

Taking into account the anharmonicity terms (higher than quadratic terms, cubic, quartic...) leads for our nonlinear triatomic model to a vibrational energy which is no longer a sum of independent terms corresponding to the different normal vibrations, but contains cross terms containing the vibrational quantum numbers of two or more normal vibrations:

$$\begin{aligned} G(\mathrm{v}_1,\mathrm{v}_2,\mathrm{v}_3) = {} & \bar{\nu}_1\left(\mathrm{v}_1+\frac{1}{2}\right)+\bar{\nu}_2\left(\mathrm{v}_2+\frac{1}{2}\right)+\bar{\nu}_3\left(\mathrm{v}_3+\frac{1}{2}\right) \\ & +X_{11}\left(\mathrm{v}_1+\frac{1}{2}\right)^2+X_{22}\left(\mathrm{v}_2+\frac{1}{2}\right)^2+X_{33}\left(\mathrm{v}_3+\frac{1}{2}\right)^2+X_{12}\left(\mathrm{v}_1+\frac{1}{2}\right)\left(\mathrm{v}_2+\frac{1}{2}\right) \\ & +X_{13}\left(\mathrm{v}_1+\frac{1}{2}\right)\left(\mathrm{v}_3+\frac{1}{2}\right)+X_{23}\left(\mathrm{v}_2+\frac{1}{2}\right)\left(\mathrm{v}_3+\frac{1}{2}\right) \end{aligned} \tag{2.16}$$

the X_{ik} are the anharmonicity constants corresponding to the $\bar{\nu}x_e$ of diatomic molecules. In the above expression, all the anharmonicity constants have negative values.

The first three terms of Eq. (2.16) are those already obtained for the harmonic oscillator, each term belonging to one of the three normal vibrations: ν_1, ν_2 or ν_3. The vibrational energy also contains six anharmonicity constants of which three belong to one of the three normal vibrations and the other three are the coupling constants.

For example, the wavenumber associated with the frequency ν_1 would be given by:

$$\overline{\nu_0^1} = \overline{\nu_1} + 2X_{11} + \frac{1}{2}X_{12} + \frac{1}{2}X_{13} \tag{2.17}$$

For a diatomic oscillator only the first two terms would appear.

The first overtone of the same vibration is, with $\mathrm{v}_1=2$, $\mathrm{v}_2=0$, $\mathrm{v}_3=0$:

$$\overline{\nu_1^{02}} = 2\overline{\nu_1} + 6X_{11} + X_{12} + X_{13} = 2\overline{\nu_1^{01}} + 2X_{11} \quad (2.18)$$

so that if we measure the overtone and the fundamental, we can compute X_{11} just as for a diatomic oscillator.

For example, in the case of sulfur dioxide, the fundamental ν_1 is located at 1151 cm^{-1} while its first overtone is observed at 2296 cm^{-1}.

We can easily deduce X_{11}:

$$X_{11} = [2296 - 2(1151)]/2 = -3\,\mathrm{cm}^{-1}$$

Polyatomic molecules also have combination bands. Let $(\nu_1+\nu_3)$ be the observed wavenumber of the vibrations ν_1 and ν_3 of the SO_2 molecule:

$$(\overline{\nu_1} + \overline{\nu_3}) = G'(1,0,1) - G''(0,0,0) = \text{observed wavenumber}$$

By replacing the appropriate values of v in the expression of the energy G', we get the following expression for the wavenumber of the combination band:

$$\begin{aligned}(\overline{\nu_1} + \overline{\nu_3}) &= \overline{\nu_1} + \overline{\nu_3} + 2X_{11} + 2X_{33} + 2X_{13} + \frac{1}{2}X_{12} + \frac{1}{2}X_{23} \\ &= \overline{\nu_1^{01}} + \overline{\nu_3^{01}} + X_{13}\end{aligned} \quad (2.19)$$

Thus, if we measure the combination band and the two fundamentals, we can compute the coupling constant X_{13}.

The combination of $\overline{\nu_1}$ (1151 cm^{-1}) and $\overline{\nu_3}$ (1361 cm^{-1}) is observed at 2499 cm^{-1}, so the anharmonicity constant X_{13} can be calculated as follows:

$$X_{13} = [2499 - (1151 + 1361)] = -13\,\mathrm{cm}^{-1}$$

In the general case of molecules that contain more than three atoms (without degenerate vibrations), Eq. (2.16) becomes:

$$G(v_1, v_2 \ldots) = \sum_i \overline{\nu_i}\left(v_i + \frac{1}{2}\right) + \sum_i \sum_{k \geq i} x_{ik}\left(v_i + \frac{1}{2}\right)\left(v_k + \frac{1}{2}\right) + \ldots \quad (2.20)$$

2.2.2.3 **Degenerate Vibrations**

The SO_2 molecule has nondegenerate vibrations; this means that all the normal vibrations of the molecule have different frequencies. The presence of threefold or higher axes of symmetry in a molecule leads to equal or degenerate frequencies.

In a doubly degenerate vibration, two vibrational modes exhibit a single frequency of vibration $\nu_a = \nu_b$; for triply degenerate vibrations, three of the vibrational frequencies have the same value $\nu_a = \nu_b = \nu_c$. The degeneracy arising from symmetry has to be distinguished from the "accidental" degeneracy where two or more of the normal frequencies happen to be approximately equal.

The phenomenon of degeneracy can be illustrated by looking at the bending vibrations of a linear triatomic molecule. It is easy to guess that these two vibrations have the same frequency. In the first mode, the atoms are moving in the plane of the molecule, in the second one, they are moving at right angles to the plane of the molecule. The two vibrations constitute a degenerate pair (Fig. 2.7).

Discarding cases of accidental degeneracy, the vibrational energy can be written as:

$$G(\mathrm{v}_1, \mathrm{v}_2, \mathrm{v}_3 \ldots) = \sum \bar{\nu}_\mathrm{i} \left(\mathrm{v}_\mathrm{i} + \frac{d_\mathrm{i}}{2} \right) \tag{2.21}$$

where d_i is the degree of degeneracy of the vibration ν_i ($d_\mathrm{i}=1$ for nondegenerate vibration).

The quantum mechanical energy levels corresponding to states in which these normal modes are excited will be degenerate. If, for example, a doubly degenerate vibration is excited by one quantum, we could have either $\nu_\mathrm{a}=1$, $\nu_\mathrm{b}=0$ or $\nu_\mathrm{a}=0$, $\nu_\mathrm{b}=1$ so the energy level is doubly degenerate. If two quanta are excited, we may have $\nu_\mathrm{a}=2$, $\nu_\mathrm{b}=0$ or $\nu_\mathrm{a}=0$, $\nu_\mathrm{b}=2$ or $\nu_\mathrm{a}=1$, $\nu_\mathrm{b}=1$ and the level is triply degenerate. Generally, the degree of degeneracy, if v_i quanta of the doubly degenerate vibration are excited, is equal to $(\mathrm{v}_\mathrm{i}+1)$; for a triply degenerate vibration, it is equal to $\frac{1}{2}(\mathrm{v}_\mathrm{i}+1)(\mathrm{v}_\mathrm{i}+2)$.

Let us specify that the general formula of the energy for the case of doubly degenerate vibration is given by the following expression if anharmonicity is taken into account:

$$G(\mathrm{v}_1, \mathrm{v}_2 \ldots) = \sum_\mathrm{i} \bar{\nu}_\mathrm{i} \left(\mathrm{v}_\mathrm{i} + \frac{d_\mathrm{i}}{2} \right) + \sum_\mathrm{i} \sum_{\mathrm{k} \geq \mathrm{i}} x_{\mathrm{ik}} \left(\mathrm{v}_\mathrm{i} + \frac{d_\mathrm{i}}{2} \right) \left(\mathrm{v}_\mathrm{k} + \frac{d_\mathrm{k}}{2} \right) + \sum_\mathrm{i} \sum_{\mathrm{k} \geq \mathrm{i}} g_{\mathrm{ik}}\, l_\mathrm{i} l_\mathrm{k} + \ldots$$

where $d_\mathrm{i}=1$ or 2 depending on whether i refers to a nondegenerate or doubly degenerate vibration and the g_{ik} terms only arise from the degenerate vibrations associated with vibrational angular momenta l_i and l_k. l_i assumes the values:

$$l_\mathrm{i} = \mathrm{v}_\mathrm{i}, \mathrm{v}_\mathrm{i} - 2, \mathrm{v}_\mathrm{i} - 4, \ldots 1 \text{ or } 0.$$

For nondegenerate vibrations $l_\mathrm{i}=0$ and $g_{\mathrm{ik}}=0$.

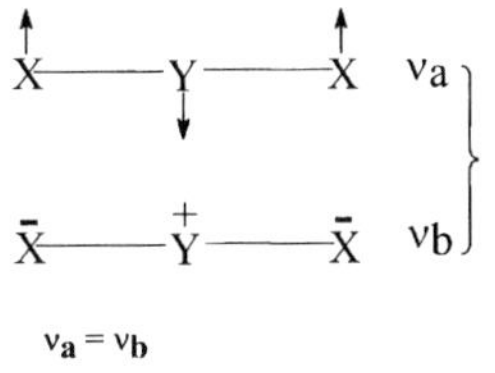

Fig. 2.7 Doubly degenerate vibrations in a linear molecule YX_2

2.2.2.4 Symmetry Considerations

A molecule may have one or more symmetry elements. Each symmetry element has a symmetry operation associated with it. The set of symmetry operations generated by the symmetry elements forms a mathematical group called a symmetry point group. The complete set of operations carried out on molecules, and which are the group elements, should satisfy four criteria in order to form a group in the mathematical sense: (a) the product of any two operations must yield another operation in the group; (b) the combination of symmetry operations must obey the associative law of multiplication (RS)P = R(SP); (c) the set possesses an identity operation (called E) such that for any element X in the group EX = XE = X; (d) every operation X possesses an inverse R^{-1} which is also an element of the group and which is defined such that $RR^{-1} = R^{-1}R = E$.

The SO_2 molecule, which has two vertical planes of symmetry ($\sigma_{(xz)}$ and $\sigma_{(yz)}$) and a twofold rotation axis of symmetry C_2, belongs to the C_{2v} point group (Fig. 2.8).

As the potential energy of the molecule is assumed to be a function of the distances between the atoms only, it remains unchanged by the symmetry operations of the equilibrium position. It follows that nondegenerate molecular vibrations, represented by displacement vectors, are either symmetric (unaltered) or antisymmetric (changed in sign) with respect to a symmetry operation of the point group of the undistorted molecule.

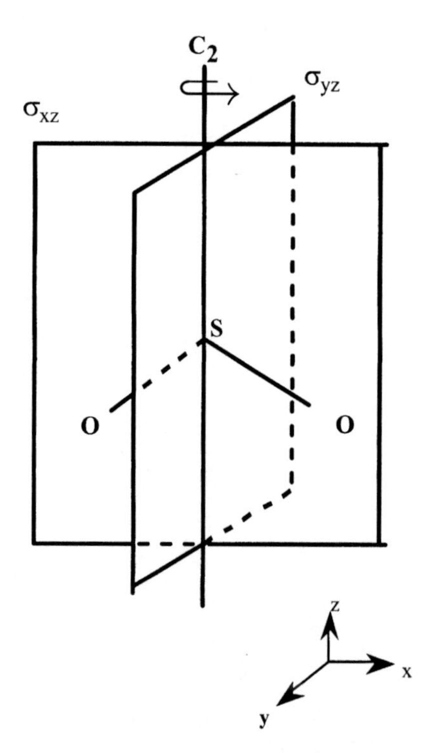

Fig. 2.8 Symmetry elements of the SO_2 molecule

The behaviour of the molecular motions with respect to symmetry operations is contained in a character table. The character table associated with any molecule belonging to the C_{2v} point group (Tab. 2.3) summarises the effect of each symmetry operation on the molecular motions, including the translational motions T_x, T_y and T_z, the rotational motions R_x, R_y and R_z and the vibrations of the atoms of the molecule.

A_1, A_2, B_1 and B_2 are said to be the irreducible representations of the C_{2v} point group. They are referred to as the symmetry species to which motions of a molecule belonging to the point group of the table can be assigned. The process by which these representations can be built up is subject to the rules of a mathematical treatment known as group theory. For a detailed study of this subject, the reader should consult some of the standard treatments of group theory [13–19]. The symmetric stretch and the bending mode of SO_2 are symmetric with respect to all the symmetry operations (Tab. 2.4) (+1 represents symmetric behaviour, there is no change in direction when the symmetry operation acts on the displacement coordinate). The irreducible representation A_1 reflects the symmetry behaviour of these two vibrational modes.

Tab. 2.3 Character table for the C_{2v} point group

C_{2v}	E	$C_{2(z)}$	$\sigma_{(xz)}$	$\sigma'_{(yz)}$	*Rotations*	*Translations*
A_1	1	1	1	1		T_z
A_2	1	1	–1	–1	R_z	
B_1	1	–1	1	–1	R_y	T_x
B_2	1	–1	–1	1	R_x	T_y

Tab. 2.4 Symmetry properties of the fundamental modes of the SO_2 molecule

	E	C_2	$\sigma_{(xz)}$	$\sigma_{(yz)}$	
A_1	1	1	1	1	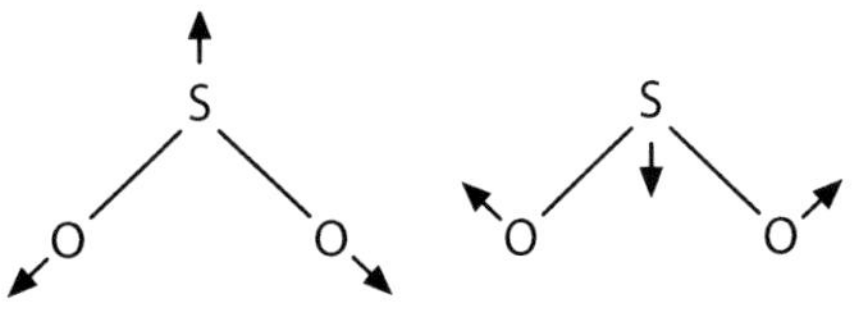 ν_1 symmetric stretch; ν_2 bending mode
B_1	1	–1	1	–1	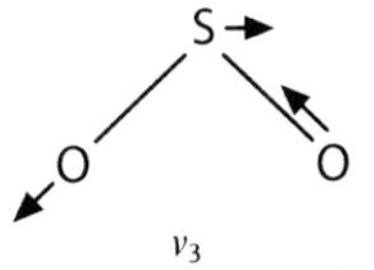 ν_3 antisymmetric stretch

The antisymmetric stretch is antisymmetric with respect to the axis and also antisymmetric with respect to the plane $\sigma_{(yz)}$. Its behaviour is given by the irreducible representation B_1.

2.2.2.4.1 Selection Rules

As already mentioned, a fundamental vibrational transition from v=0 to v=1 can occur if one of the transition moment components:

$$\int_{-\infty}^{+\infty} \psi^*_{v=1}\, \varepsilon_x \psi_{v=0}\, d\tau$$

$$\int_{-\infty}^{+\infty} \psi^*_{v=1}\, \varepsilon_y \psi_{v=0}\, d\tau$$

$$\int_{-\infty}^{+\infty} \psi^*_{v=1}\, \varepsilon_z \psi_{v=0}\, d\tau$$

is nonzero. The quantities ε_x, ε_y and ε_z are the components of the dipole moment. The values of these integrals must be unchanged for all the symmetry operations applied to the molecule. The dipole components will be transformed under the various symmetry operations in exactly the same way as are the displacement arrows representing T_x, T_y and T_z.

Considerations of the symmetry properties of the components of the $\psi_{v=0}$ and $\psi_{v=1}$ wavefunctions and the fact that the symmetry properties of the dipole components are the same as those of the component of translation, leads to the following selection rule:

"a normal mode of vibration is infrared active if it belongs to the same symmetry species as at least one component of translation".

For sulfur dioxide, the symmetric stretching and the bending modes, which belong to the same symmetry as T_z, are infrared active, as is the antisymmetric stretch which belongs to the same symmetry as T_x.

An alternative method for obtaining vibrational spectra is by Raman spectroscopy. A Raman spectrum of a sample is observed as a shift in the frequency of monochromatic radiation when this is scattered by the sample. This shift is associated with a change between two vibrational energy levels. A fundamental transition is Raman active if an integral of the type:

$$\int_{-\infty}^{+\infty} \psi^*_{v=1}\, \alpha_{ij} \psi_{v=0}\, d\tau$$

has a nonzero value. α_{ij} is one component of the polarisability and i and j are x, y or z. It can be shown that a fundamental transition is Raman active if it belongs to the same symmetry species as at least one component of the polarisability tensor.

2.2.2.4.2 Spectroscopic Activity of Overtones and Combination Bands

The near-infrared region is also concerned with symmetry aspects since group theory can predict the activity of the overtones and the combination bands.

The symmetry of any combination band may be determined by combining the symmetry properties of its components: the symmetry species of the combination is the same as that of the direct product of the characters of the vibrational symmetry species of the two normal modes, ν_i and ν_j, involved. The selection rule for combination is the following: the product representation $\Gamma_i\Gamma_j$ must transform in the same way as T_x, T_y or T_z. For example, the spectral activity of the combination $\nu_1(A_1)+\nu_3(B_1)$ is obtained by the product: $A_1 \times B_1$. It is infrared active since B_1 and T_z share the same symmetry species.

For nondegenerate modes, the even overtones will belong to the totally symmetric species while the odd overtones will have the same species as the fundamental vibration. For the SO_2 molecule, which belongs to the C_{2v} group, the symmetry species of the overtones of the $\nu_3(B_1)$ vibration can be summarised as:

$$B_1^n = \begin{cases} n & \text{even} & \rightarrow & A_1 \\ n & \text{odd} & \rightarrow & B_1 \end{cases}$$

From the analysis of overtones and combination bands, NIR spectroscopy can make information on forbidden fundamental vibrations available. For example, among the six vibrations of the methylene group represented in Fig. 2.9, only the twisting mode A_2 is infrared active. It is possible to reach the wavenumber of this mode in the CH_2X compounds from its overtone and from combinations where it is involved [20].

The spectroscopic activity of overtones of degenerate vibrations involve a more complex analysis contained in books specialised in molecular vibrations [13, 14].

2.2.2.5 Fermi and Darling-Dennison Resonances

Although the described concepts make a NIR spectrum comprehensible, some other characteristic features that could be of importance in the near-infrared region have to be mentioned. The first feature concerns the resonances which could play a role at high energies.

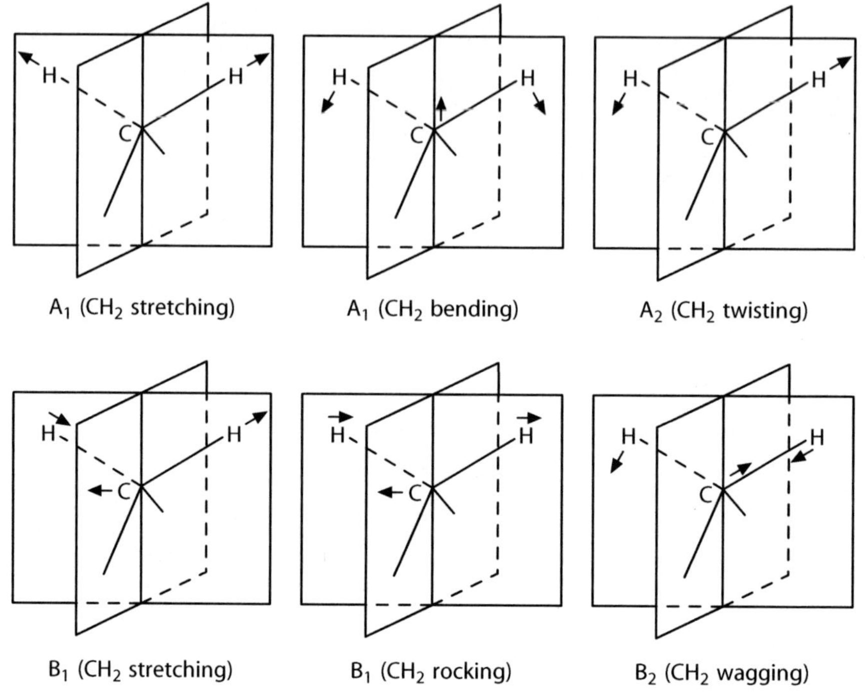

Fig. 2.9 Vibrations of the CH_2 group

2.2.2.5.1 **Fermi Resonance**

A resonance that leads to a perturbation of the energy levels can occur if two vibrational levels belong to the same symmetry species and have similar energy. Such a resonance may occur between an overtone or a combination band which happens to have the same symmetry and nearly the same frequency as that of a fundamental vibration. This accidental degeneracy, called Fermi resonance, leads to two relatively strong bands rather than one strong band for the fundamental. The two bands are observed at somewhat higher and lower frequencies than the expected unperturbed positions of the fundamental and overtone (or combination). An example of this effect can be found in the CO_2 spectrum. This molecule has three fundamental frequencies at 667, 1337 and 2349 cm^{-1}, of which the one at 667 cm^{-1} is a doubly degenerate bending frequency. The overtone level 2×667 would lie quite near to the fundamental at 1337 cm^{-1} and one component of this overtone is of the same symmetry as the fundamental in question. Resonance occurs between the two levels, pushing them to 1285 and 1388 cm^{-1} and causing each actual level to contain a contribution from each component of the interaction.

Another example can be seen in the spectra of many aldehydes where the overtone of the C-H in-plane bending vibration expected near 2800 cm^{-1} is close to the frequency of the C-H stretching vibration. The interaction is revealed by the presence of two bands of approximately equal intensity, symmetrically displaced about the expected position for the overtone.

The reader who desires to study the subject more thoroughly should consult some of the standard references [13, 14, 19, 21, 22].

2.2.2.5.2 Darling-Dennison Resonance

A second-order resonance that can be of importance in NIR spectra has been found to occur in the water molecule. This kind of perturbation, discussed by Darling and Dennison [23, 24], can also occur in other molecules containing symmetrically equivalent XH bonds. Like SO_2, H_2O is a nonlinear symmetrical molecule exhibiting three normal modes of vibration: v_1 (symmetrical stretch at 3657 cm^{-1}), ν_2 (bending vibration at 1595 cm^{-1}) and ν_3 (antisymmetric stretch at 3756 cm^{-1}). The two stretching modes ν_1 and ν_3 have similar wavenumbers but belong to two different symmetry species (A_1 and B_1) and therefore cannot interact directly. However, levels associated with the vibrational quantum numbers v_1, v_2, v_3 and (v_1–2), v_2, (v_3+2) can interact by resonance since they belong to identical symmetry species and have similar energies. The hypothesis of such resonance is supported by the fact that overtone bands occurring in the near-infrared range with appreciable intensity appear in pairs (Tab. 2.5). For example, those with quantum numbers (2, 0, 1) and (0, 0, 3) are located at 10613.12 and 11032.36 cm^{-1}, respectively, while (2, 1, 1) and (0, 1, 3) states give rise to the pair at 12151.22 and 12565.01 cm^{-1} (Tab. 2.5). While the energy levels for which no perturbation occurs are given in Eq. (2.22), a perturbation constant γ, which measures the effect of the resonance, has to be introduced in order to account for the observed energy levels.

Tab. 2.5 Some excited states for H_2O (from [23])

$v_1v_2v_3$	*Symmetry species*	*Wavenumbers* (cm^{-1})
201	B_1	10613.12
003	B_1	11032.36
211	B_1	12151.22
013	B_1	12565.01
301	B_1	13830.92
103	B_1	14318.77
311	B_1	15347.91
113	B_1	15832.47
401	B_1	16899.01
203	B_1	17495.48

2.2.2.5.3 **Local Mode Model**

The local mode model has been used extensively for the treatment of high energy (>8000 cm^{-1}) C-H stretching overtone spectra. While the treatment based on normal modes, which has been described above, is generally used in the frequency region characteristic of fundamental vibrations and low quanta overtone and combination modes, the local mode model has proved to be appropriate for a description of high energy levels [25–37].

Molecules containing symmetrically equivalent sets of stretching vibrations such as H_2O, SO_2, NH_3 or CH_4 ... can exhibit "near-normal" or "near-local" behaviour. In order to understand which parameter governs the switch from normal to local mode character, we can consider the case of molecules containing two equivalent bond oscillators XY_2 like H_2O or SO_2.

The coupling between the two oscillators causes them to form the familiar in-phase (symmetric stretch, ν_s) and out-of-phase (antisymmetric stretch, ν_a) normal vibrations given by these expressions:

$$\overline{\nu_s} = \overline{\nu_M} - C$$

$$\overline{\nu_a} = \overline{\nu_M} + C$$

where $\overline{\nu_M}$ is the wavenumber associated with either of the bonds regarded as an independent Morse oscillator and represents the anharmonic frequency of a single bond stretching (the subscript M has been added to emphasise the relation to the Morse function of a single bond stretching). It can be calculated as follows:

$$\overline{\nu_M} = [\overline{\nu_a} + \overline{\nu_s}]/2$$

C, which represents the interbond coupling, can easily be calculated by the following equation:

$$C = [\overline{\nu_a} - \overline{\nu_s}]/2$$

It is interesting to note that the SO_2 molecule exhibits a particularly large coupling parameter $C = 105$ cm^{-1}.

The main idea of the local mode model is to treat a molecule as if it were made up of a set of equivalent diatomic oscillators and the reason for local-mode behaviour at high energy may be understood qualitatively as follows. As the stretching vibrations become more highly excited, the anharmonicity term $X = x_e\overline{\nu}$ tends, in certain cases, to dominate the effect of the interbond coupling: the vibrations become uncoupled vibrations and occur as "local modes". The cause of this decoupling is the strong bond anharmonicity which can, in certain cases, quench any interbond coupling term. The overtone spectrum can thus be interpreted in the same way as that of an anharmonic diatomic molecule. This is why it is said that overtone spectra get simpler at high energy. Experimentally, it is found that a switch from normal to local mode character occurs for high energy transitions corresponding to $\Delta v \geqslant 3$.

Tab. 2.6 gives the various parameters for SO_2 and H_2O, as well as the bond anharmonicities derived by Child and Lawton [27]. $\bar{\nu}_M$ and X_M are the anharmonic wavenumber and anharmonicity constant associated with either of the bonds considered as decoupled. The harmonic frequency of a single bond stretching, $\bar{\nu}$, is the wavenumber of the anharmonic vibration plus the anharmonicity term:

$$\bar{\nu} = \bar{\nu}_M + 2X = \left[\frac{\bar{\nu}_a + \bar{\nu}_s}{2}\right] + 2X$$

The important parameter which gives rise to "near-normal" or "near-local" behaviour is the ratio of coupling strength to bond anharmonicity, C/X. This latter point is demonstrated by the case of the two molecules H_2O and SO_2 which illustrate the two extremes of behaviour. For the SO_2 molecule, the very large coupling parameter dominates the very small anharmonicity; this leads to a normal-mode behaviour: the overtone and combination states can be described according to the familiar normal-mode behaviour (Tab. 2.7).

At the other end of the scale, the local mode picture was used to analyse the vibrational states of the water molecule in the fundamental region and at up to several quanta of excitation. The effect is illustrated in Tab. 2.8, which shows the

Tab. 2.6 Specific parameters for the H_2O and SO_2 molecules

	H_2O	SO_2
$\bar{\nu}_a$ (cm^{-1})	3756	1361
$\bar{\nu}_s$ (cm^{-1})	3657	1151
$\bar{\nu}_M$ (cm^{-1})	3706.5	1256
$\bar{\nu}$ (cm^{-1})	3875.3	1271
X_M (cm^{-1})	84.4	7.5
C (cm^{-1})	49.5	105

Tab. 2.7 Excited states for SO_2

$\bar{\nu}$ (cm^{-1}) (experimental)	Assignment
1151	ν_1
1361	ν_3
2296	$2\nu_1$
2499	$\nu_1+\nu_3$
2715	$2\nu_3$
3431	$3\nu_1$
3630	$2\nu_1+\nu_3$
4054	$3\nu_3$
4751	$3\nu_1+\nu_3$
5166	$\nu_1+3\nu_3$

Tab. 2.8 Excited states for H_2O (energies given in cm^{-1}) in the $v_2=0$ bending state (from [35])

	(v_1, v_2, v_3)	**$\bar{\nu}$ (experimental)**	***Assignment***
	1 0 0	3657	ν_1
v=1			
	0 0 1	3756	ν_3
	2 0 0	7201	$2\nu_1$
v=2	1 0 1	7250	$\nu_1+\nu_3$
	0 0 2	7445	$2\nu_3$
	[n, m]		
		10600	
	[3,0]±		
v=3		10613	
	[2,1]+	10869	
	[2,1]–	11032	
		13830	
	[4,0]±		
		13828	
v=4	[3,1]+	14221	
	[3,1]–	14319	
	[2,2]	14537	
		16898.4	
	[5,0]±		
		16898.8	
v=5	[4,1]+	17458.2	
	[4,1]–	17495.5	
	[3,2]+	17748	
	[3,2]–	17971	

stretching vibrational energy levels of H_2O as determined experimentally. The data and assignments of the excited vibrational states are those reported by Mills [35]. The results refer to the lowest bending state, $v_2=0$. The lower excited states ($\Delta v \leq 2$) are treated on the normal mode basis and are labelled by the conventional normal quantum numbers (v_1, v_3); they correspond to the states $2\nu_1$, $2\nu_3$, $\nu_1+\nu_3$. The higher excited states are described in terms of local modes denoted $[n,m]\pm$ where n represents the number of quanta in one bond and m in the other and ± designates the symmetry with respect to permutation of the two bonds.

The data show an irregular spacing of levels at high v and the lowest energy states for each v are a closely spaced pair for which the excitation is localised in one bond only. The lowest doublet for each manifold occurs at almost the energy of the corresponding OH diatomic vibrator.

2.3 Assignments of NIR Bands

In order to be used for both qualitative and quantitative analyses, NIR spectra need to be interpreted in a manner analogous to the interpretation of MIR spectra. Thus the customary methods used to assign the bands in the mid-infrared can be applied in the near-infrared range. We can mention essentially:

- group frequencies,
- deuteration
- polarisation measurements
- two-dimensional spectroscopy.

2.3.1 Group Frequencies

The reader will have understood that overtones and combination bands are at the heart of NIR spectroscopy and that the key quantity is anharmonicity, which determines the occurrence and the spectral properties (frequency, intensity) of the NIR bands.

As shown previously, the bonds with the highest anharmonicity are those involving hydrogen, the lightest atom. These bonds, which vibrate at high energy and with large amplitude when undergoing stretching motions, carry the most intensity. Thus the NIR spectral region is dominated by absorption associated with XH_n functional groups. These absorptions arise from the overtones of fundamental stretching bands or combinations involving the stretching and bending modes of vibration of such groups. Some overtones and combination bands occur with sufficient regularity to characterise molecular groups in the same manner as in the fundamental bands in the mid-IR region.

Spectral identification and analytical applications of the NIR region have been discussed in several comprehensive reviews [4, 38–42]. A chart established by Goddu and Delker contains the major spectra-structure correlations [43]. However, it is interesting to review some specific absorptions of various XH groups.

2.3.1.1 C-H Absorptions

In *aliphatic hydrocarbons*, the first set of combination bands occurs between 2000 and 2400 nm, the first overtones between 1600 and 1800 nm and the second overtones between 1000 and 1200 nm (note that an absorption band is located by its wavenumber $\bar{\nu}$ in cm^{-1} in a MIR spectrum and usually by its wavelength λ in nm in a NIR spectrum).

Olefinic CH groups give rise to specific NIR absorption bands (1620 and 2100 nm for the vinyl group: 1180, 1680, 2150 and 2190 nm for *cis* olefines) while *aromatic CH bonds* have first and second overtones located at 1685 and 1143 nm [4], respectively.

In a study carried out by Kelly and Callis [44], NIR spectroscopy was used for the simultaneous estimation of the major classes of hydrocarbon constituents (aliphatic, aromatic and olefinic) of finished gasolines. Two multivariate calibration approaches have been employed in order to correlate the volume percents of the three hydrocarbon classes with NIR spectra. It was shown that the wavelengths selected statistically are consistent with real absorbance features of the CH bonds of different functional groups within the hydrocarbon molecules that make up the gasoline mixture.

The CH absorption bands in *oxygenated compounds* like epoxides have proved to be useful analytical bands for qualitative and quantitative evaluation of the curing process of epoxy resins [45–54].

On the other hand, recent advances in the development of optical-fibre FTIR and the availability of silica-type optical fibres for NIR radiation, make real-time remote sensing of reactive systems possible. Terminal epoxides display a specific band at 2207 nm, which is ascribed to a combination band of epoxy stretching and bending vibrations whose absorption intensity decreases systematically during reaction and can therefore be used in kinetic studies. Additionally, the CH stretch in the epoxy group appears at 1650 nm.

2.3.1.2 **O-H Absorptions**

The first and second overtones of the O-H stretch vibrations in *alcohols* and *phenols* are located around 1400 and 1000 nm, respectively, while the stretching-bending combination occurs near 2000 nm.

The OH stretch and its overtones are well known to be very sensitive to the environment [55]. In a study of specific interactions in miscible blends, Ghebremeskel et al. [56] showed that the information obtained from the OH-stretch overtone absorption in the NIR region was consistent with that obtained from the OH-stretch fundamental in the MIR region. Their analysis of a model system, isopropanol in a mixed solvent (heptane/ethyl acetate), reveals NIR spectral shifts due to interaction between the hydroxyl of the isopropanol and the ester functionality of the ethyl acetate. In heptane, the unassociated OH bond gives rise to a sharp absorption centred around 1410 nm, while the NIR spectrum in ethyl acetate is dominated by a broad absorption centred at 1435 nm that is ascribed to the associated species. Similarly, the NIR spectra of blends of polycaprolactone (PCL) and phenoxy [poly(hydroxypropyl)ether of bisphenol-A] show that the peak maximum of the OH stretch overtone decreases in wavelength from 1431 to 1425 nm as the fraction of PCL increases [56].

In connection with OH absorptions, one of the major applications of NIR spectroscopy needs to be mentioned, that is, the determination of moisture in food analysis [2, 57, 58]. Water has specific bands at 1940 nm (combination) and 1440 nm (first overtone of the OH stretch) which have been very useful in the study of the state of water in various samples (see Chapter 11).

The surface analysis of silicas offers another example of the application of NIR spectroscopy. A silica surface is generally characterised by the number of silanol groups and their distribution; this can influence their reactivity to some extent [59]. Three types of surface hydroxyls have been identified: isolated, vicinal (on adjacent silicon atoms) and geminal (two hydroxyl groups on the same silicon atom). Adsorption of water is governed by the hydroxyl configurations on the silica surface. Free Si-OH groups exhibit a band at about 2220 nm, which is attributed to a combination of the stretching and bending modes, and a band at about 1385 nm, which corresponds to the first overtone of the O-H stretch. Although the different types of silanols have been tentatively assigned in the MIR region, NIR spectroscopy has proved successful in determining the bonding environments of hydrogen-bearing species [60].

2.3.1.3 N-H Absorptions

Overtones and combination bands of amines have been shown to be suitable for quantitative analyses since they are well resolved and do not overlap. The NIR bands at about 1500 and 2000 nm ascribed, respectively, to the first overtone and to a combination of the stretching and bending modes, have been used to determine the change in primary amine concentration during epoxy-amine reactions [45–49, 51–54].

Aromatic amines have been extensively studied by Whetsel, Roberson and Krell who have tried to correlate variations in band intensity and frequency with the electronic nature and position of the substituents on the ring. Concerning primary aromatic amines, they have mentioned a combination band at about 1972 nm, first overtone asymmetric and symmetric stretching bands near 1446 and 1492 nm and the second overtone symmetric at about 1020 nm [61, 62].

2.3.2 Deuteration

The shift in frequency obtained by deuterating XH groups aids in the assignment of bands simply by observing which bands are altered. If we assume that the force constants are unaltered by isotope substitution, the shift in observed frequency can be attributed principally to mass effects. The wavenumber ratio of the normal and isotope species is given by:

$$\frac{\overline{\nu}_{XH}}{\overline{\nu}_{XD}} = \sqrt{\frac{\mu_{XD}}{\mu_{XH}}} \cdot \sqrt{2} \, .$$

2.3.3 Polarisation Measurements

Polarisation measurements can be used to reveal information about the transition moments of vibrational modes in solid oriented compounds, such as uniaxially stretched polymers.

The parameters commonly used to characterise the degree of molecular orientation are the dichroic ratio R or the dichroic difference ΔA defined by:

$$R = A_{\|}/A_{\perp}$$

$$\Delta A = A_{\|} - A_{\perp}$$

where $A_{\|}$ and $A_{\perp}$ are the absorbencies of the selected absorption band measured with radiation polarised parallel and perpendicular to the stretching direction, respectively (Fig. 2.10).

Analysis of the dichroic behaviour of the absorption bands can be of help in making band assignments. On the other hand, polarisation measurements can allow the determination of molecular orientation in stretched polymers.

It is well known that the mechanical properties of polymers are strongly influenced by the molecular orientation that occurs during stretching or during various forming processes, and measurement of this orientation is of particular importance for a better understanding of the molecular mechanisms involved in polymer deformation.

Fig. 2.11(a) gives an example of the mid-infrared spectrum of a film of silicone– which is a poly(dimethylsiloxane). This film, which is about 100 μm thick, exhibits very strong absorption bands associated with fundamental modes. This example points out one practical problem often met in the mid-infrared spectra. The problem arises from the requirement of band absorbance to be sufficiently low to permit use of the Beer-Lambert law. Thin films, often less than 50 μm, must then be used. This problem can now be overcome by using NIR spectroscopy [63].

Fig. 2.11(b) shows the NIR spectrum of a silicone film of about 2 mm thick. The reduced intensity of the NIR absorptions makes a wide range of the spectrum available for the evaluation of the dichroic effects.

The dichroic behaviour of the spectral pattern between 4000 and 4500 cm^{-1} is given in Fig. 2.12. The observed bands are assigned to combinations of the type $\nu+\delta$: in this case ν and δ are related to a stretching and a bending mode of the

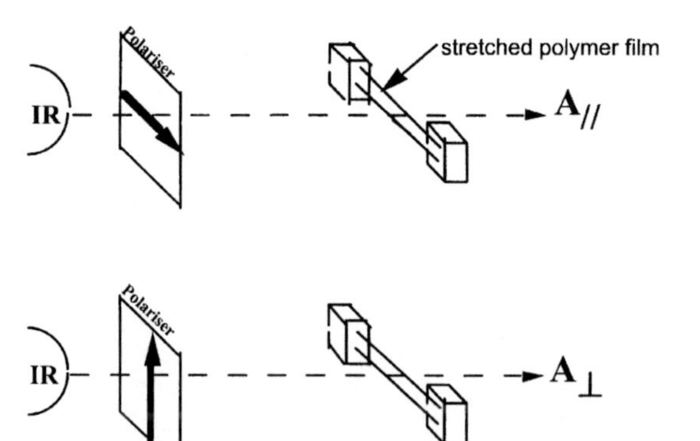

Fig. 2.10 Polarisation measurements

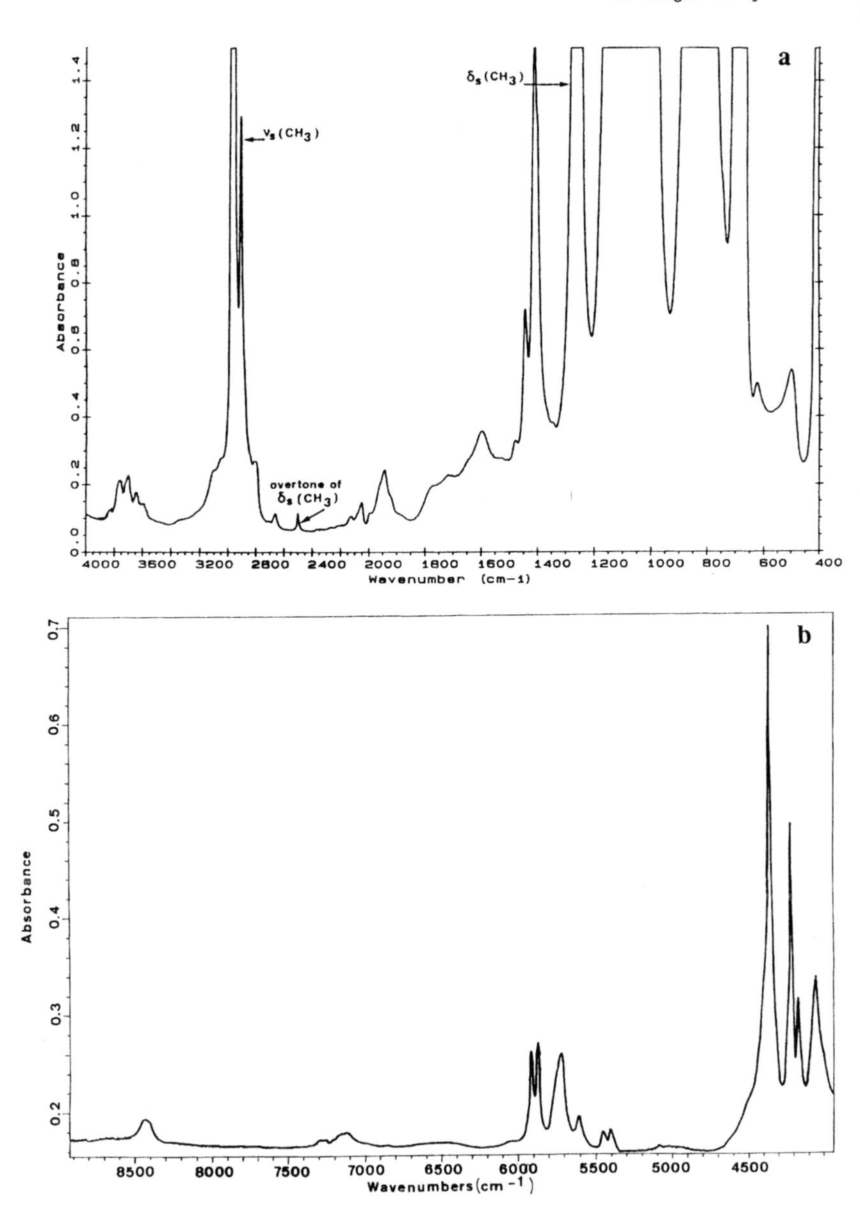

Fig. 2.11 Mid- (a) and near- (b) infrared spectra of a silicone film

methyl group, respectively. For each band, the dichroic difference and, consequently, the anisotropy of the sample increases with the deformation α of the polymer (α represents the ratio of the final length of the sample to its initial length before deformation).

The values of the orientation derived from the dichroic difference ΔA are plotted against deformation in Fig. 2.13. The measurements are carried out in the

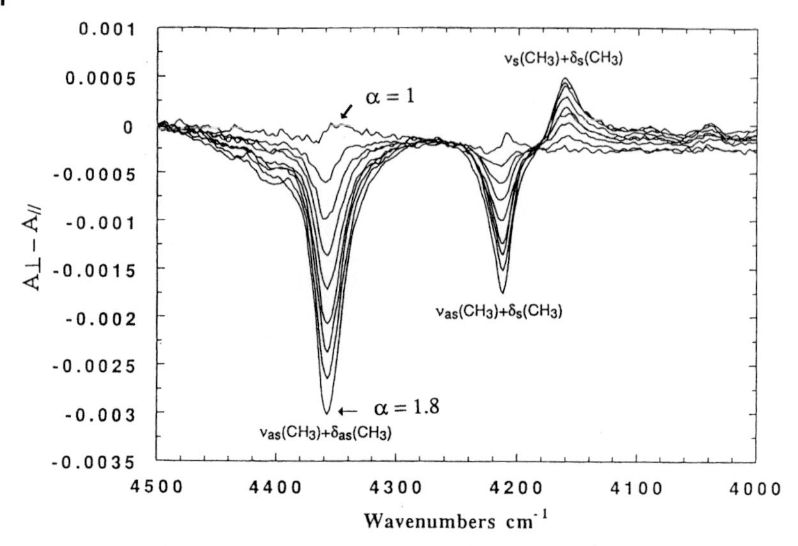

Fig. 2.12 Strain dependence of the dichroic difference for the bands located between 4000 and 4500 cm^{-1}

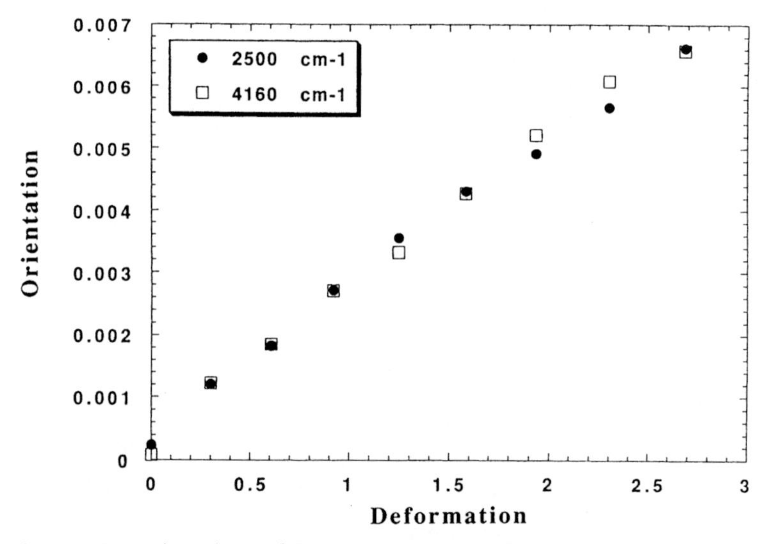

Fig. 2.13 Strain dependence of the orientation derived from experiments carried out in the mid-infrared (2500 cm^{-1}) and near-infrared region (4160 cm^{-1})

mid- and near-infrared range. We demonstrate here that NIR spectroscopy can give similar information to mid-infrared spectroscopy but with the significant advantage that specimens several mm thick can be examined.

2.3.4 Two-Dimensional Correlation Spectroscopy

Two-dimensional vibrational spectroscopy can also provide new opportunities for analysing vibrational spectra. In the 2D approach, samples are subjected to various external perturbations, such as temperature, pressure or stress. One advantage of this technique is to enhance the spectral resolution of highly overlapped NIR bands. Generalised 2D correlation analysis has proved to be a powerful tool in clarifying NIR spectra that appear rather complicated owing to the large number of overtones and combination bands that overlap with each other [64, 65]. But the possibility of correlating MIR and NIR spectral ranges could be exploited more in order to get unambiguous assignments of the NIR absorptions.

2.4 Conclusion

The purpose of this chapter was to show the potential usefulness of NIR spectroscopy for solving analytical problems. By emphasising the vibrational origin of NIR absorptions, this work is intended to bridge the gap between statistical approaches of NIR data analysis and traditional vibrational spectroscopy in order to convince chemometricians that an absorbance or a group of various absorbances that appear in chemometric models could be correlated to specific absorbers of the molecule. This could provide a better understanding of the analytical information present in NIR spectra.

2.5 References

[1] C. A. Watson, *Anal. Chem.* **1977**, *49(9)*, 835 A.

[2] B. G. Osborne, T. Fearn, P. H. Hindle, *Practical NIR Spectroscopy with Applications in Food and Beverage Analysis.* Longman Scientific & Technical **1993**.

[3] J. D. Kirsch, J.K. Drennen, *Appl. Spectrosc. Rev.* **1995**, *30(3)*, 139.

[4] L. G. Weyer, *Appl. Spectrosc. Rev.* **1985**, *21(1/2)*, 1.

[5] H. W. Siesler, *Makromol. Chem. Macromol. Symp.* **1991**, *52*, 113.

[6] C. Kradjel in *Handbook of Near-Infrared Analysis* (Eds. D.A. Burns, E.W. Ciurczak), Marcel Dekker, **2001**, pp. 659–701.

[7] K.A. Bunding-Lee, *Appl. Spectrosc. Rev.* **1993**, *28(3)*, 231.

[8] B. Descales, I. Cermelli, J. R. Llinas, G. Margail, A. Martens, *Analusis Magazine* **1993**, *21(9)*, M25.

[9] H. Martens, T. Naens in *Multivariate Calibration* (Wiley), Chichester, U.K., **1989**.

[10] H. Mark in *Handbook of Near-Infrared Analysis* (Eds. D.A. Burns, E.W. Ciurczak), Marcel Dekker, New York, **2001**, pp. 129–184.
[11] H.-R. Bjorsvik, H. Martens. in *Handbook of Near-Infrared Analysis* (Eds. D.A. Burns, E.W. Ciurczak), Marcel Dekker, New York, **2001**, pp. 185–207.
[12] W. Groh, *Makromol. Chem.* **1988**, *189*, 2861.
[13] G. Herzberg in *Molecular Spectra and Molecular Structure, Vol. II. Infrared and Raman Spectra of Polyatomic Molecules*, Van Nostrand Reinhold, **1945**.
[14] E. B. Wilson, J. C. Decius, P. C. Cross in *Molecular Vibrations: The Theory of Infrared and Raman Vibrational Spectra*, McGraw-Hill, **1955**.
[15] F. A. Cotton in *Chemical Applications of Group Theory*, Wiley, **1963**.
[16] D. S. Schonland in *An introduction to Group Theory and its uses in Chemistry*, Van Nostrand, **1965**.
[17] L. H. Hall in *Group Theory and Symmetry in Chemistry*, McGraw-Hill, **1969**.
[18] J. M. Hollas in *Symmetry in Molecules*, Chapman and Hall, **1972**.
[19] N. B. Colthup, L.H. Daly, S.E. Wiberley in *Introduction to Infrared and Raman Spectroscopy*, Academic Press, **1975**.
[20] R. Iwamoto (private communication).
[21] G. Herzberg in *Molecular Spectra and Molecular Structure, Vol I. Spectra of Diatomic Molecules* Van Nostrand Reinhold, **1950**.
[22] G. M. Barrow in *Introduction to Molecular Spectroscopy*, McGraw-Hill, **1962**.
[23] B. T. Darling, D. M. Dennison, *Phys. Rev.* **1940**, *57*, 128.
[24] D. M. Dennison, *Rev. Mod. Phys.* **1940**, *12*, 175.
[25] B. R. Henry, W. Siebrand, *J. Chem. Phys.* **1968**, *49*, 5369.
[26] B. R. Henry, *Acc. Chem. Res.* **1977**, *10*, 207.
[27] M. S. Child, R.T. Lawton, *J. Chem. Soc. Faraday Trans.* **1981**, *1971*, 273.
[28] O. S. Mortensen, B. R. Henry, M. A. Mohammadi, *J. Chem. Phys.* **1981**, *75*, 4800.
[29] M. S. Child, L. Halonen, *Adv. Chem. Phys.* **1984**, *57*, 1984.
[30] I. M. Mills, A.G. Robiette, *Mol. Phys.* **1985**, *56*, 743.
[31] I. M. Mills, F. J. Mompean, *Chem. Phys. Lett.* **1986**, *124*, 425.
[32] B. R. Henry, *Acc. Chem. Res.* **1987**, *20*, 429.
[33] I. M. Mills, *Mol. Phys.* **1987**, *61*, 711.
[34] J. L. Duncan, *Spectrochim. Acta* **1991**, *47A*, 1.
[35] I. M. Mills in *8th International Conference on Fourier Transform Spectroscopy* (Eds. H. M. Heise, E. H. Korte, H. W. Siesler) 96 SPIE, Lübeck-Travemünde, **1991**.
[36] B. I. Niefer, H. G. Kjaergaard, B. R. Henry, *J. Chem. Phys.* **1993**, *99*, 5682.
[37] H. J. Kjaergaard, D. M. Turnbull, B. R. Henry, *J. Chem. Phys.* **1993**, *99*, 9438.
[38] W. Kaye, *Spectrochim. Acta* **1954**, *6*, 257.
[39] W. Kaye, *Spectrochim. Acta* **1955**, *7*, 181.
[40] O. H. Wheeler, *Chem. Rev.* **1959**, *59*, 629.
[41] R. F. Goddu, *Adv. Anal. Chem. Instrum.* **1960**, *1*, 347.
[42] K. B. Whetsel, *Appl. Spectrosc. Rev.* **1968**, *2*, 1.
[43] R. F. Goddu, D. A. Delker, *Anal. Chem.* **1960**, *32*, 140.
[44] J. J. Kelly, J. B. Callis, *Anal. Chem.* **1990**, *62*, 1444.
[45] G. A. George, P. Cole-Clarke, N. S. John, G. Friend, *J. Appl. Polym. Sci.* **1991**, *42*, 643.
[46] N. A. St-John, G.A. George, *Polymer* **1992**, *33(13)*, 2679.
[47] C. J. DeBakker, G. A. George, N. A. S. John, *Spectrochim. Acta* **1993**, *49A (5/6)*, 739.
[48] C. J. DeBakker, N. A. S. John, G. A. George, *Polymer* **1993**, *34(4)*, 716.
[49] B. G. Min, Z. H. Stachursi, J. H. Hodgkin, G. R. Heath, *Polymer* **1993**, *34(17)*, 3620.
[50] V. Strehmel, T. Scherzer, *Eur. Polym. J.* **1994**, *30(3)*, 361.
[51] J. Mijovic, S. Andjelic, *Macromolecules* **1995**, *28*, 2787.
[52] J. Mijovic, S. Andjelic, C. F. W. Yee, F. Bellucci, L. Nicolais, *Macromolecules* **1995**, *28*, 2797.
[53] J. Mijovic, S. Andjelic, *Polymer* **1995**, *36(19)*, 3783.

[54] R. J. Varley, G. R. Heath, D. G. Hawthorne, J. H. Hodgkin, G. P. Simon, *Polymer* **1995**, *36(7)*, 1347.
[55] C. Sandorfy, *Bull. Polish Acad. Sci. Chem.* **1995**, *43*, 7.
[56] Y. Ghebremeskel, J. Fields, A. Garton, *J. Polym. Sci. B* **1994**, *32*, 383.
[57] J. B. Reeves III, *Appl. Spectrosc.* **1995**, *49*, 181.
[58] J. B. Reeves III, *Appl. Spectrosc.* **1995**, *49*, 295.
[59] M. P. Wagner, *Rubber Chem. Technol.* **1976**, *49*, 703.
[60] T. Uchino, T. Sakka, M. Iwasaki, *J. Am. Ceram. Soc.* **1991**, *74*, 306.
[61] K. B. Whetsel, W. E. Robertson, M. W. Krell, *Anal. Chem.* **1957**, *29*, 1006.
[62] K. B. Whetsel, W. E. Robertson, M. W. Krell, *Anal. Chem.* **1958**, *30*, 1598.
[63] T. Buffeteau, B. Desbat, L. Bokobza, *Polymer* **1995**, *32(22)*, 4339.
[64] Y. Ozaki, Y. Liu, I. Noda, *Macromol. Symp.* **1997**, *119*, 49.
[65] Y. Ozaki, I. Noda, *J. Near Infrared Spectrosc.* **1996**, *4*, 85.

3
Instrumentation for Near-Infrared Spectroscopy

Satoshi Kawata

In terms of instrumentation, near-infrared (NIR) spectroscopy is not distinctively different from either visible or infrared spectroscopy, although it may sound unique. Conventional instrumentation used for the visible or infrared regions can also be used for the NIR region, depending on purpose or on environment. In the short-wavelength region ($<2\ \mu m$) of NIR, visible or an ultraviolet (UV) spectrometers are often used. Likewise, in the long-wavelength region ($>2\ \mu m$) of NIR, infrared spectrometers can be used if some optical elements are replaced.

NIR spectrometers are distinctive not only because of their applied wavelength region but also because of their specialised applications. In general, not such a high resolving power is required for NIR spectroscopy compared to visible, UV or infrared spectroscopy because NIR spectra appear as multiples of combination tones and overtones of absorption lines. The measurement of NIR spectra is often made under difficult conditions, such as in the field or on the factory line. In-process monitoring is a typical use of NIR spectroscopy in industry, while in laboratories it is usually used for in vivo or in situ measurements. Samples to be analysed by NIR spectroscopy are often heterogeneous and rough, so that the measurement of absorption of the sample can be by diffuse reflectance. A large throughput and a fast light collection method are necessary for NIR measurement.

Interference-filter spectroscopy, multichannel Fourier-transform spectroscopy and acousto-optic tuneable filter spectroscopy are particularly well known as suitable for NIR spectroscopy. Special cells and integrating spheres are often used for NIR spectroscopy for improving the collection efficiency of diffuse reflected light.

The purpose of this chapter is to describe the instrumentation for NIR spectroscopy and to make a comparison between a variety of the instruments used.

3.1
Configuration of Near-Infrared Spectrometers

Sample information in the near-infrared region is usually collected as an absorption spectrum through transmission measurements or diffuse-reflectance measurements with a NIR spectrometer. The absorption spectrometers are generally categorised into two configurations shown in Fig. 3.1 (a) and (b). The difference between these

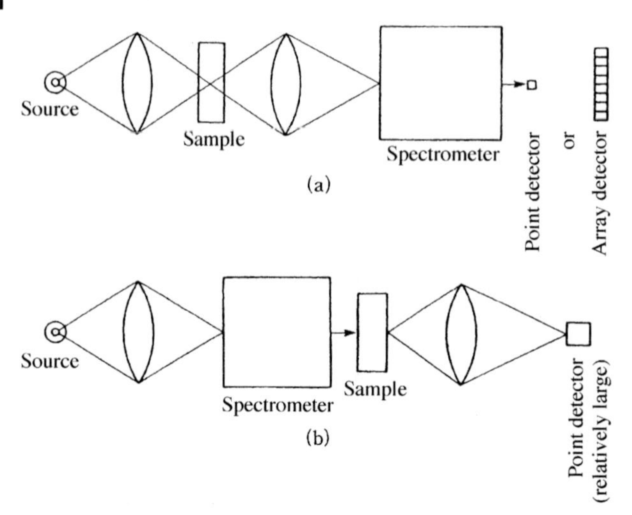

Fig. 3.1 (a) Arrangement of source-sample-spectrometer-detector. This configuration is used for IR, fluorescence, Raman spectroscopy, and spectroscopy using a multichannel detector. When light is irradiated to the sample, the sample may be damaged, or heated up. (b) Arrangement of source-spectrometer-sample-detector. This configuration is used for visible, UV spectroscopy, IR Fourier-transform spectroscopy. The transmission direction of light may be deviated after the light emerged from the sample

two is simply the position of the sample relative to the spectrometer. In the configuration shown in Fig. 3.1 (a), light with a broad-band spectrum that is emitted from a thermal source is absorbed by the sample and is then introduced into the spectrometer to obtain the spectrum. On the other hand, in the configuration shown in Fig. 3.1 (b), the light is first monochromated by a monochromater, or first interferes with the predivided beam in the interferometer, and then illuminates the sample to obtain the intensity of absorbed light. The spectrum is obtained by scanning the wavelength of the monochromater, or by scanning the path difference between the two beams in the interferometer in the case of a Fourier-transform spectrometer.

When a sample is set before the spectrometer, as shown in Fig. 3.1 (a), both the light component of the wavelength of interest and the component of the unmeasured wavelength region irradiate the sample together; this may heat up the sample and change its physical characteristics. For this reason, configuration (a) is not recommended for measurements in the visible and UV ranges. However, this configuration is very often used in infrared absorption spectroscopy, since the temperature of infrared sources is low enough to make this negative effect negligible.

If the sample is set after the spectrometer, as shown in Fig. 3.1 (b), the beam from the sample may deviate during propagation in the sample. This depends on the position of the sample, the sample cell and optical components and, with a small sized detector, results in unstable or inefficient detection. However, the detector should be made as small as possible to reduce thermal noise. Fourier-transform spectrometers for the infrared region are, however, made in configuration (b).

In NIR measurements, diffuse reflection or scattering of light at the sample is very commonly detected. In both cases, the surface of the sample is extensive and the light is scattered at or in the sample into a wide range of angles, so that introduction of all the light into the spectrometer is difficult. Hence, in NIR spectroscopy, configuration (b) is commonly preferred. In the case of interference-filter spectroscopy, configuration (b) is better because unnecessary components of the light are cut off before the sample position. In the case of grating spectrometers, either configuration (a) or (b) is chosen depending on the experimental conditions.

Configuration (a) has to be used for multichannel spectroscopy because a wide-range spectrum is being simultaneously measured with a multichannel detector. Nevertheless, heating is not a problem because multichannel detection saves on measurement time. The combination of Fourier-transform spectroscopy methods with multichannel detection could further reduce the measurement time owing to the possibility of high throughput of the optical system for a scattering extended sample without the loss of spectral resolution. The sample is also set before the spectrometer in the case of fluorescence spectroscopy and Raman spectroscopy including FT-Raman spectroscopy in NIR region.

In the next section, a variety of instrumentation for NIR spectroscopy will be discussed.

3.2 Interference-Filter Spectroscopy

The most popularly used optical element for NIR spectroscopy is an interference filter. As shown in Fig. 3.2, an interference filter has a simple configuration made of a transparent dielectric thin film sandwiched between two films. The bandwidth is approximately 10 to 20 nm, which is not very narrow but reasonable for NIR applications because NIR spectra are composed of rather broad peaks as combinations of vibration modes in the infrared.

Since an interference filter is wide and thin, it fits the practical applications of NIR spectroscopy well. Because of the compact size and its low cost, interference filters are widely used as an optical element of NIR spectroscopy.

3.2.1 Principle of Interference-Filter Spectroscopy

The principle of spectroscopy with an interference filter can be explained as follows. When a collimated beam is incident on a dielectric film having two parallel boundaries on both sides, the light travels a multiple number of round passes in the film by partially reflecting/transmitting at the two sides of the film, as shown in Fig. 3.2. The intensity of transmitted light through this film at wavelength λ_p is at a maximum when the path length ($2nd$) of the round trip within the film is exactly the same as the wavelength λ_p of light selected, multiplied by an integer number m,

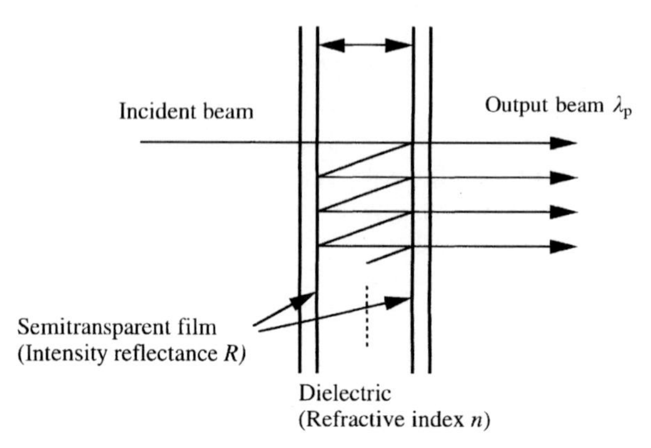

Fig. 3.2 A schematic diagram of an interference filter

$$m\lambda_p = 2nd \qquad m = 1, 2, 3, \ldots \tag{3.1}$$

where d and n are the thickness and the refractive index of the film, respectively. The intensity maximum is the result of the phase matching of light transmitted straight through the film to that after reflections at two sides of the film making a round trip. It matches those of light transmitted after any number of multiple reflections between two sides of the film. The integer number m is called the order. For most cases of near-infrared interference-filters, $m = 1$ or 2 is chosen for the length d in Eq. 3.1.

At the wavelength that has a phase difference between the transmitted light and reflections, the intensity through the film becomes zero, and 100% of the light is reflected back to the incident side. The summation of higher order components (multiple number of reflections) makes the light intensity small if there is a phase difference between them.

Mathematically, the transmittance spectrum $T(\lambda)$ of a film is given by summing the amplitude of transmission beams of different numbers of reflections infinitely. By using the intensity reflectance R of the film surface and incident angle θ inside the film, it is reduced to

$$T(\lambda) = \frac{(1-R)^2}{(1-R)^2 + 4R\sin^2(2\pi nd\cos\theta/\lambda)} \tag{3.2}$$

Fig. 3.3 (a) and (b) show the transmittance spectra of an interference filter as a function of wavenumber and as a function of wavelength for different values of reflectance R, respectively. In this calculation, the optical thickness nd of the film is set to be 1 µm. It is seen in the figures that as the reflectance R of the film surface increases from 10% to 90%, the transmittance band narrows. The ratio F of peak interval to peak width, which is called the "Finesse", is used as a value to cal-

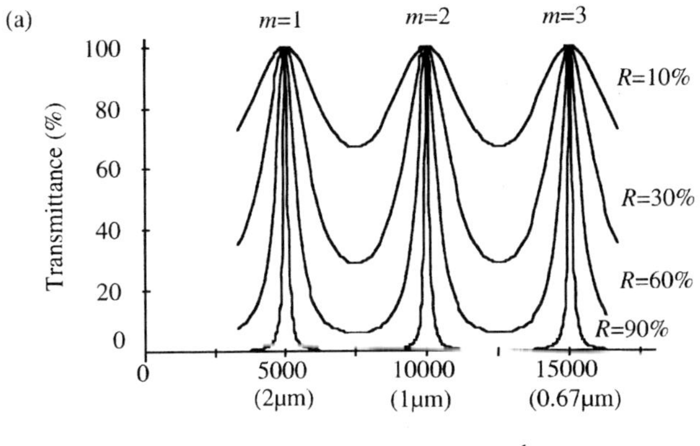

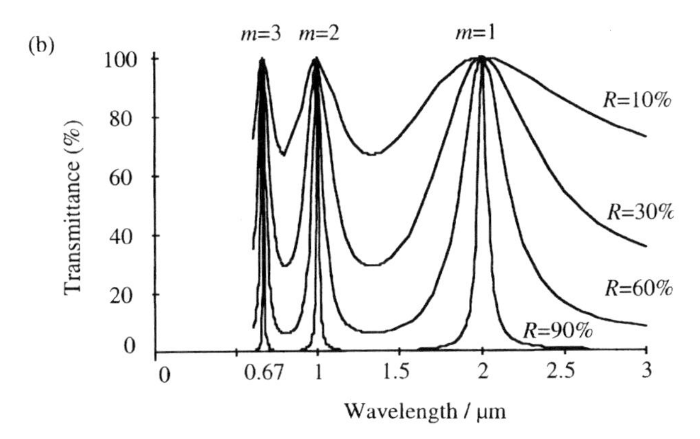

Fig. 3.3 Transmittance of interference filters as a function of wavenumber (a), and as a function of wavelength (b) ($m=1$, $\lambda_p=2$ μm)

culate the resolving power of interference filters for spectral band selection. Finesse can be expressed by using only the reflectance R of film, such that

$$F = \frac{\pi\sqrt{R}}{1-R} \tag{3.3}$$

Although, theoretically, an extremely high R provides extremely high spectral resolution, the degree of parallelism and flatness of the film surfaces and the degree of collimation of the beam incident on the film, limit the resolution in practice. Fortunately, in the practical application of NIR spectroscopy, very high resolution is not required.

The light of wavelengths of unselected orders (in Fig. 3.3, 1 μm for $m=2$, 0.67 μm for $m=3$) is not detected because such wavelengths are outside the spec-

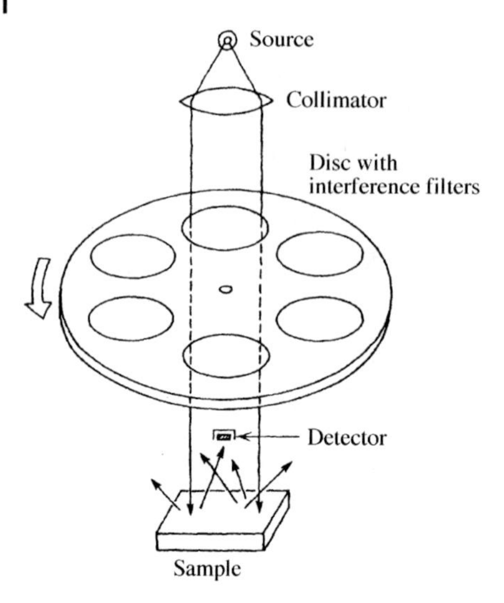

Fig. 3.4 Wavelength selection with an interference-filter analyzer. By rotating the disc, a different filter is selected, that is, a different wavelength is selected in turn

tral range sensed by the detector, or transparent to other optical elements. If necessary, a colour filter is added to block such regions.

3.2.2 Wavelength Scanning

In interference-filter spectroscopy, the wavelength can be scanned by changing the refractive index n and the thickness d of the thin film. This is known as the Fabry-Pérot interferometer and is commonly used in high-resolution spectroscopy. The Fabry-Pérot interferometer is large and expensive, and has several restrictions as to the environment in which it can be used. Another type of interference-filter device more commonly used for near-infrared spectroscopy is a set of filters with different thicknesses. The selected wavelengths are successively changed by turning a disc on which several interference filters of different thicknesses are arranged along the circumference. Fig. 3.4 shows an example of such a device.

Interference filters with the surface slanting along the circumference or linearly towards the other end such that the thickness gradually changes (Fig. 3.5 (a)) are also used for sweeping wavelengths, and are known as variable-interference filters. When light enters part of such a filter, the filter scans by making one rotation (or one linear movement) to sweep the selected wavelengths.

Fig. 3.6 (a) shows a method of wavelength scanning used for near-infrared spectrophotometers [3] in which the angle of light incident on the filters changes to vary the optical path length. This method is used for finely adjusting the wavelength. However the spectral broadening becomes too large as the angle increases (as seen in Fig. 3.6 (b)), so that the method is used for fine adjustment near the peak wavelengths by turning several interference filters of different ranges as shown in Fig. 3.7.

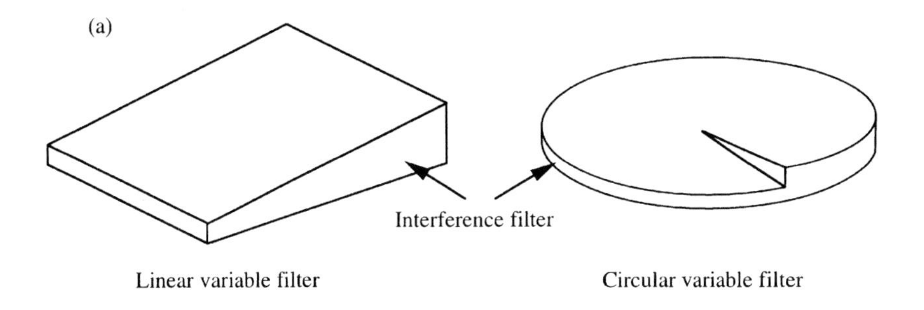

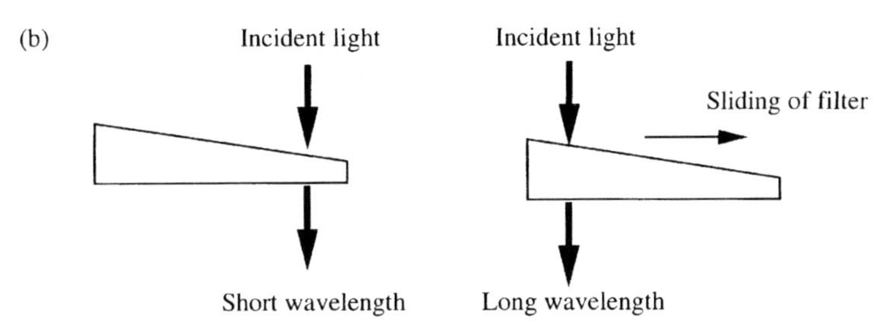

Fig. 3.5 (a) Variable interference-filter. (b) Wavelength scanning with a variable interference-filter

3.3 Diffraction-Grating Spectroscopy

Diffraction-grating spectroscopy, which is most generally used for the regions ranging from ultraviolet to far-infrared, is also used as a standard method for the near-infrared region. Unlike an interference filter, a diffraction grating can obtain the full range of the near-infrared spectrum with high resolution (around 0.1 to 1 nm, for example). With this method, however, the luminous fluxes from the light source or the sample are significantly cut due to the narrow entrance slit of the spectroscope. Due to this inefficient use of light, diffraction-grating spectroscopy requires a measurement time from as long as 30 seconds to several tens of minutes. Many commercially available near-infrared spectrophotometers with diffraction gratings are made by slightly extending the wavelength regions of visible and ultraviolet diffraction-grating spectrophotometers.

3.3.1 Principle of Diffraction-Grating Spectroscopy

Fig. 3.8 shows the principle of spectroscopy by using a diffraction-grating. The state shown in this figure is when light of wavelength λ_p is incident on an opaque screen (metal) with several slits (apertures) at regular intervals (interval d). In accordance with the principle of the diffraction of light, the light incident on

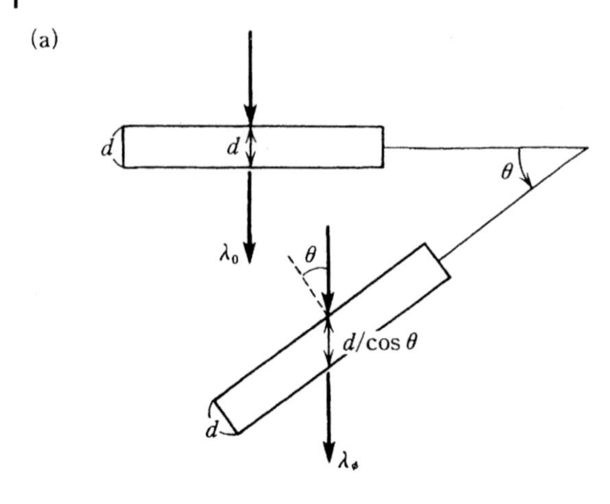

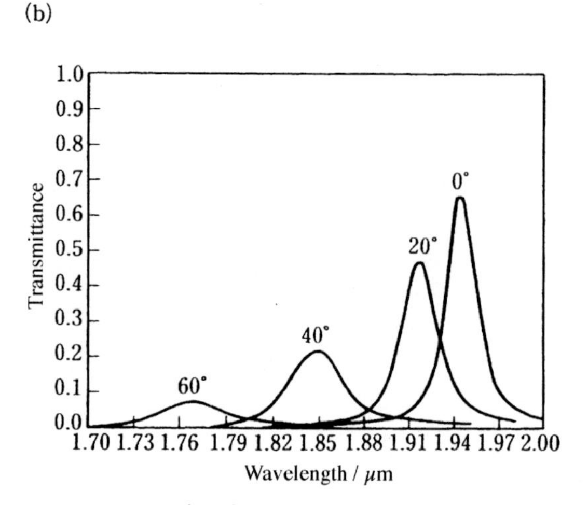

Fig. 3.6 (a) Wavelength scanning by rotating an interference-filter. (b) Transmittance spectra for different angles. Spectral broadening to visible for a large angle

the screen is diffracted at the slits, then emitted divergently. If the incident light is assumed to be collimating (the ongoing direction is fixed without broadening of angle) at this point, the light is strongly diffracted only in a specific direction. This angle of diffraction β is related to the angle of incidence α as follows:

$$d(\sin\alpha + \sin\beta) = m\lambda_p, \quad m = \pm1, \pm2 \tag{3.4}$$

This equation shows that the path difference $d(\sin\alpha + \sin\beta)$ of the in-phase diffraction of light between the light paths through the neighbouring slits is an integral multiple (m) of the wavelength λ_p. Specifically, the phases of each light beam that goes out through the slits become equivalent at an angle that satisfies this

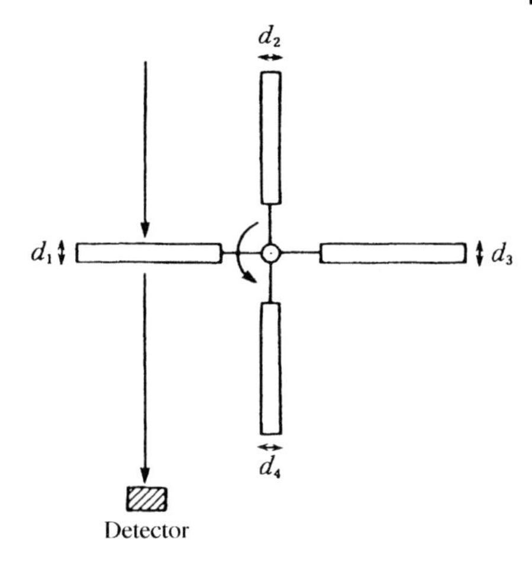

Fig. 3.7 Wavelength scanning by rotating a set of different interference-filters. By rotating this filter set, a wavelength band is selected. Fine tuning is realized by tilting an individual filter

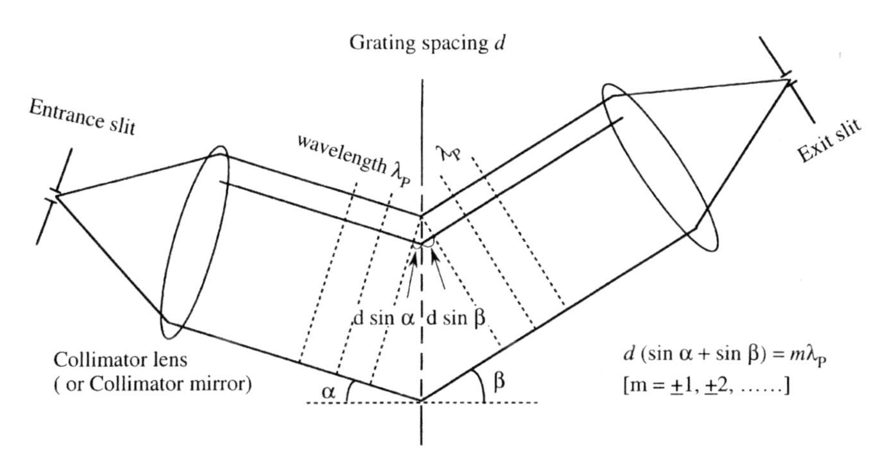

Fig. 3.8 Principle of grating spectroscopy

condition. The integral number m is called the order of diffraction. $m=0$ is called 0-th order light and corresponds to light that transmits rectilinearly.

Thus, spectroscopy with diffraction gratings is quite similar to spectroscopy with interference filters, as described previously, except that the grating method is based on the interference of multiple beams through slits at different positions. The resolution improves as the number of slits increases. Fig. 3.9 shows the characteristics of spectral transmittance as a function of wavenumber when the angles of incidence and observation are fixed at α and β, respectively; this corresponds to Fig. 3.3 (a) for interference filters.

If either the angle of incidence or the angle of observation broadens, the spectral peak to be selected becomes blurred. To collimate the incident and diffracted

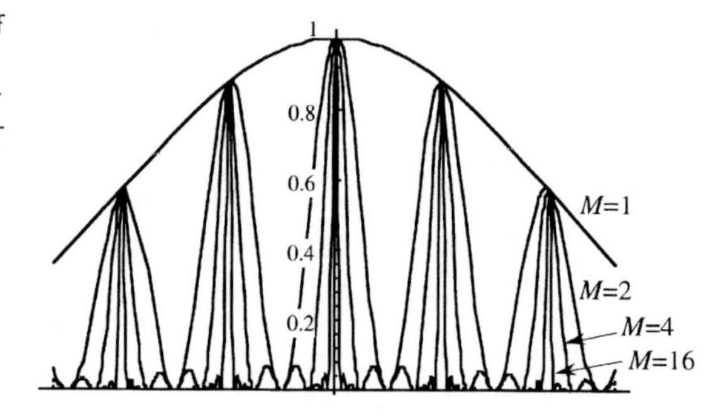

Fig. 3.9 Transmittance spectra of transmission gratings (as a function of wavenumber) for fixed values of incident angle α and observation angle β; M: the number of slits

light, slits with collimator (paralleling) lenses, or concave mirrors, are installed on both ends of the spectrometer to restrict broadening of the angle, as shown in Fig. 3.8. The opening of the slits is, for example, around 1 to 10 μm. Light producing the spectral broadening from the light source or sample is cut by the slits and, thus, is eliminated from spectroscopic measurement. Moreover, the broadening of the angle of radiation from a light source or a sample is naturally restricted by the sizes of the diffraction gratings, collimator lenses and concave mirrors. Although the image of a light source (or a sample) can be reduced to fit inside the slits by a lens or a mirror, the distribution of radiation angles widens, and hence most of the light is unable to reach the collimators and diffraction gratings.

Fig. 3.10 (a) shows the principle of diffraction spectroscopy by using reflection gratings with numerous metal wires (indicated by small circles in the figure) that are equivalent to slits for the transmission type shown in Fig. 3.8. Fig. 3.10 (b) shows a diffraction grating when both the angle of incidence and the angle of diffraction are on the same side as the angle of specular reflection. Fig. 3.10 (a) is called the Czerny-Turner mount [4], Fig. 3.10 (b) the Littrow mount [5, 6] and Fig. 3.10 (c) the Ebert-Fastie mount [7, 8]. The relationship given in Eq. 3.4 also applies to these reflection-type diffraction gratings.

Fig. 3.11 shows a diffraction grating called a blazed grating. This diffraction grating consists of a group of inclined reflection surfaces, rather than slits or wires, and, thus, generates a intense diffracted light only within a specific range of angles. With such a blazed grating, 100% efficiency of a certain order can be obtained by selecting the blaze angle θ_B such that the angles of incidence and diffraction of a certain order are symmetric to the perpendicular angle of the reflection surface at a selected wavelength. In practice, the blaze angle is made smaller than the optimum angle, and an intermediate blaze shape between a precise waveform of blazing and a sine waveform is selected. Otherwise, the blazed grating can be too sensitive to the angle for high efficiency of diffraction.

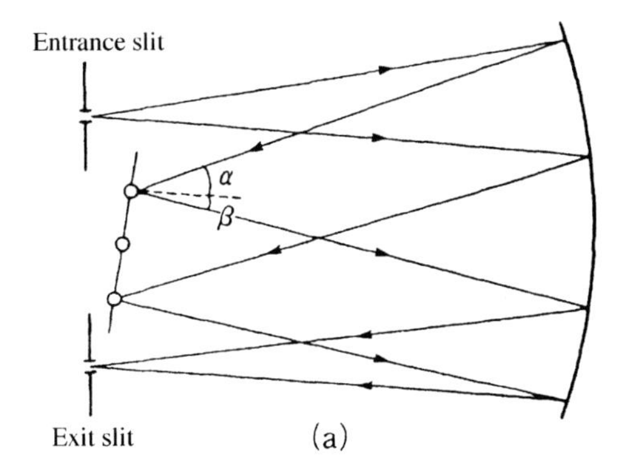

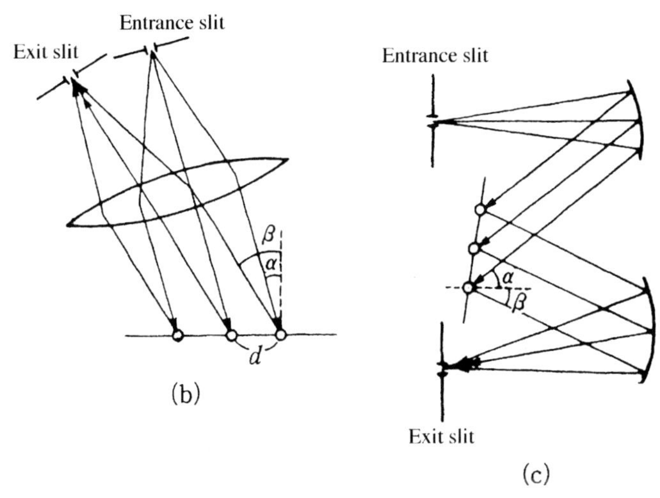

Fig. 3.10 Three types of reflection gratings. (a) Czerny-Turner mounting. (b) Littrow mounting. (c) Ebert-Fastic mounting

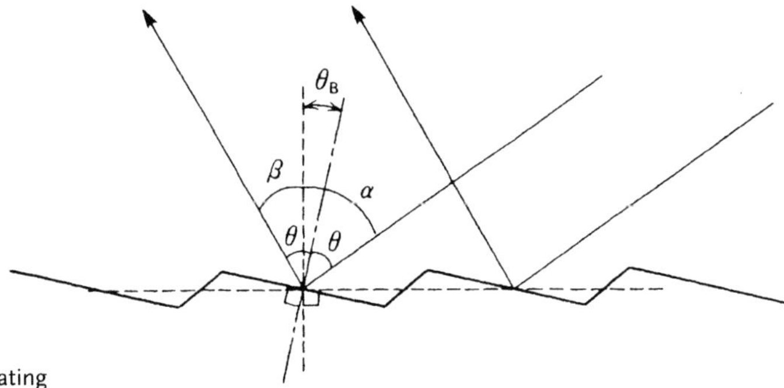

Fig. 3.11 Blazed grating

3.3.2
Wavelength Scanning for Grating Spectroscopy

Fig. 3.12 shows how wavelength sweeping can be conducted easily and continuously with diffraction-grating spectroscopy by turning a diffraction grating. When the angle of incidence and the angle of observation are fixed at θ_i and θ_d, respectively, and the surface of the diffraction grating is turned by θ from the reference surface, the wavelength λ_m of generated diffracted light of the m-th order at θ_d is given as

$$\lambda_m = \frac{d}{m}\{\sin(\theta_d - \theta) - \sin(\theta_i - \theta)\} \tag{3.5}$$

in which m is the order of diffraction. In Fig. 3.12, the angle of observation λ_d coincides with the angle of the m-th order diffraction for light with wavelength λ_m. At a fixed angle λ_d, the wavelength λ to be measured can be continuously scanned by turning the diffraction grating.

The problem with this method is the influence of diffracted-light components of orders other than the target order that is being measured. For instance, when diffracted light of the first order with a wavelength of 1 μm is measured, diffracted light of the second order with a wavelength of 2 μm overlaps the angle of

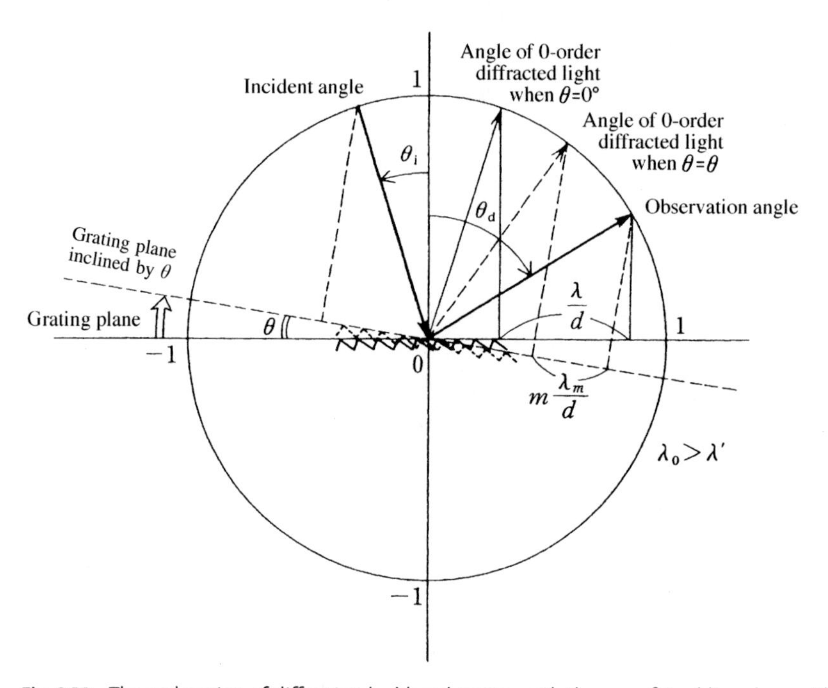

Fig. 3.12 The explanation of diffraction by blazed grating with the use of Ewald's sphere with wavenumber

observation. Thus, with diffraction-grating spectroscopy, diffraction of other orders must be cut off in advance to prevent their overlapping the diffracted light of the target wavelength. In near-infrared spectroscopy, overlapping of the second-order diffraction with a wavelength double that of the measured light of the first-order diffraction does actually occur. This problem can be resolved by, for example, using two spectrometers interconnected with each other, and sweeping the wavelengths continuously. A prism or a diffraction grating can be installed on the first spectrometer as a wavelength dispersion element. However, an infrared cut filter is preferred because of its small size and low cost.

Another problem caused by the diffraction of nontarget orders is a phenomenon called "anomaly". This consists of a combination of two phenomena as follows: Firstly, when the angle of diffraction β of an unmeasured order of diffraction becomes exactly 90°, the diffracted light of this order loses intensity, that is, diffraction becomes impossible; this results in redistribution of the light intensity of this order to the diffracted light of the order being measured. Hence, the light intensity of diffraction being measured increases. Secondly, when the angle of incidence α increases to a certain angle, the wavenumber of the diffracted light of an order that has become nonirradiating matches the wavenumber of the collective oscillation of electrons on the metal surface, thus causing resonance in the surface plasma oscillation. As a result, the energy of nonirradiating diffracted light is absorbed by the collective oscillation of the electrons (thermal absorption), and the intensity of the diffracted light of the order being measured decreases abnormally. In the near-infrared region, anomalies due to the diffracted light of the +1 order occur. For example, under the conditions in which the angle of incidence $\alpha=0°$, the diffraction order being observed $m=-1$, the interval of the diffraction grating $d=1.25\ \mu m$ (800 lines/mm), the amplitude of the diffraction grating is 0.125 μm and the diffraction grating is made of aluminum, an anomaly of the second category is seen at $\lambda=1.0\ \mu m$. This anomaly is shown graphically in Fig. 3.13.

3.3.3
Multichannel Spectroscopy with a Polychromator

A concave diffraction grating, which functions both as a diffraction grating and as a converging mirror, is often used for near-infrared spectrophotometers. This grating is advantageous because it does not require two collimator concave mirrors, so the device can be compact and such a grating can remove the aberrations when used as a polychromator.

As shown in Fig. 3.14, a polychromator uses a multichannel detector for simultaneously detecting spectra that are dispersed by a diffraction grating. It requires no slit for outgoing light because each detector element of the multichannel detector plays the role of a slit. This type of spectrometer has been used historically in visible and ultraviolet spectroscopy as a spectrograph for recording a spectrum on a photographic plate. Recently, it has again become widely used with a solid-state imaging device that consists of photoelectric detector arrays aligned linearly at

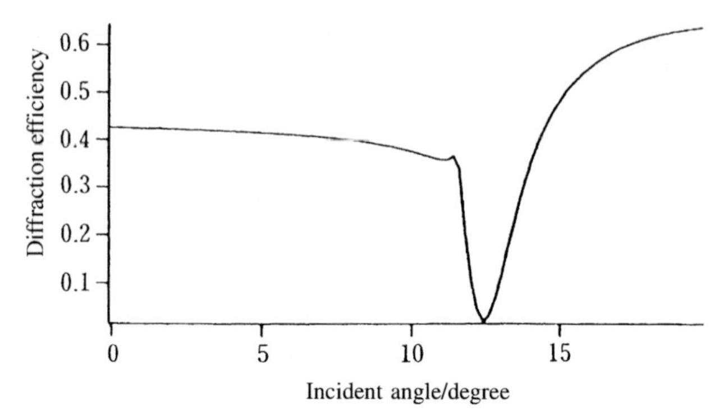

Fig. 3.13 Example of anomaly that is observed in an NIR spectrum. Diffraction efficiency of −1 order diffraction light as a function of incident angle for the wavelength 1.0 μm. Grating is made of aluminium (refractive index $n = 1.45 + 8.0i$) with 800 grooves/mm, corresponding to the picture of 0.125 μm. A small projection shown around 11° is a redistribution phenomenon anomaly due to the passing off of the 1st order diffraction light. The sharp dip shown around 12° indicates anomaly caused by surface plasmon polariton excited on the aluminum surface

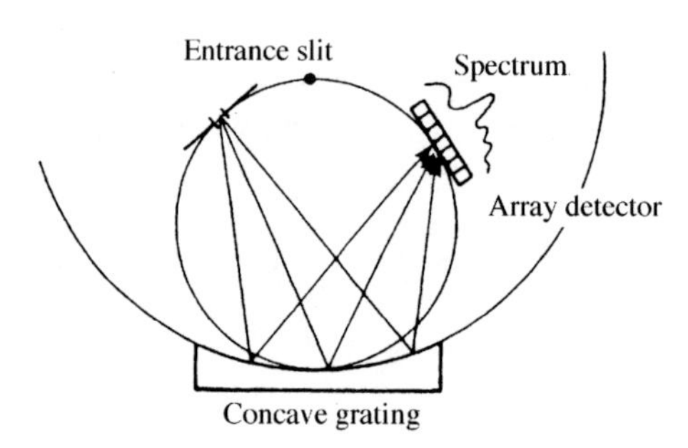

Fig. 3.14 Polychromator. The incident light is dispersed by the concave grating, and all the spectral components of diffraction are simultaneously detected by the multichannel array detector

high density, such as a CCD (charge-coupled device). Spectrophotometers that use such multichannel detectors are suitable for high-speed measurement, since they do not require scanning mechanics for turning the diffraction grating. In addition, they efficiently use the light for detection because a multichannel detector detects all the ranges all the time, simultaneously. Thus, multichannel spectroscopy can be performed with a simple and handy spectrophotometer for the visible and ultraviolet regions. In the near-infrared region as well, multichannel detectors are about to enter practical use. Although not yet complete as products, near-infrared multichannel spectrophotometers have the potential to be used soon as near-infrared spectroscopes.

3.3.4
Production Methods of Reflection-Type Diffraction Gratings

There are two methods of producing reflection-type concave diffraction gratings. In the conventional method, lines are engraved one by one with a diamond cutter by using a ruling engine on the metal deposition on the glass substrate. With the more recent method, a photoresist is coated homogeneously onto a glass substrate by using a spinner, onto which interference fringes are registered holographically by using laser beams, then the entire diffraction grating is produced in one go by developing the fringes. Such a diffraction grating is called a holographic grating. The diffraction grating produced in this manner is replicated for actual use, although it is sometimes used as it is by depositing metal onto it by vacuum evaporation.

The method of engraving lines with a ruling engine has the merit that the blaze angle can be accurately selected by changing the shapes of the addendum of the diamond cutter; however, it can produce a spurious spectrum called "ghosts" due to the accumulation of errors of engraving position as the number of lines increases.

Holographic gratings are free from this problem. In addition, by suitably arranging the optical system, it is comparatively easy to produce irregularly arranged lines with which spectra can focus on a flat field where all the elements of multichannel detectors are aligned linearly. The blaze angle can be set by irradiating ion beams at a blaze angle after holographic recording of the grating.

Currently, holographic gratings are generally used for near-infrared spectrophotometers, and it is hardly necessary to produce diffraction gratings with a ruling engine any more.

3.4
Spectroscopy with Acousto-Optical Diffraction Gratings

Eq. 3.4, a relational expression of the angle of diffraction, shows that the selected wavelength λ can be swept by making the grating interval d variable and by keeping the angle of incidence α and the angle of observation β constant. This idea is the same as that of Fabry-Pérot spectroscopy, in which wavelengths are swept by making the optical length nd variable as in Eq. 3.2, as discussed in the previous section on interference-filter spectroscopy. The question is how to produce diffraction gratings that can make the grating interval variable. One answer is by the use of sound. The pitch (wavelength) of sound waves in a substance changes when the sound frequency changes. A sound wave forms a series of low and high density regions, by which light is diffracted. This phenomenon is called the acousto-optical effect. Diffraction-grating spectroscopy by using this effect was proposed in the 1960s for high-speed visible spectroscopy [10]. It can also be a practical method in the near-infrared region, depending on the purpose.

3.4.1
Schematics of Acousto-Optical Diffraction Gratings

Fig. 3.15 shows the principle of spectroscopy by using an acousto-optical diffraction grating. A piezoelectric element (piezo transducer) is adhered to a material that is transparent in the near-infrared region and onto which an AC electric signal is applied; a compression wave (i.e., sound wave) is then generated inside the material. The pitch (wavelength) d of the compression wave is obtained by dividing the propagation velocity v of the sound wave inside the material by the AC frequency f. Since the frequency f can be varied electrically, the grating interval d can easily be changed by changing the frequency f. In the case of tellurium dioxide (TeO_2), for example, which is a commonly used material, $v=616\ ms^{-1}$ (transverse waves), and $d=8.2\ \mu m$ for $f=75$ MHz. Sound waves of such a high frequency are not audible to the human ear and are called ultrasonic waves.

An ultrasonic wave, as a compression wave, works as a refraction-index grating for light. As shown in Fig. 3.15, when the phase of the light reflected from a certain reflection surface of an ultrasound grating becomes identical to the phase of the adjacent reflection surface, the diffraction reaches a maximum intensity because the light beams interfere with each other in phase.

Diffraction by using such a thick diffraction grating, called Bragg diffraction, is well known in the structural analysis of crystals by X-ray and electron beam. As will be discussed later, Bragg diffraction is different from the diffraction caused by a thin diffraction grating and shown in Fig. 3.8 (Raman-Nath diffraction). When the variation of refractive index of an acoustic wave grating is small, it is regarded as Raman-Nath diffraction regardless of its thickness.

3.4.2
Characteristics of Spectroscopy by Bragg Diffraction

As shown in Fig. 3.16 (a), Bragg diffraction is generated only when the wavenumber vector k_d (its length is the same as that of the incident light with the direction of progression of the diffracted waves) of diffraction coincides with the synthesis

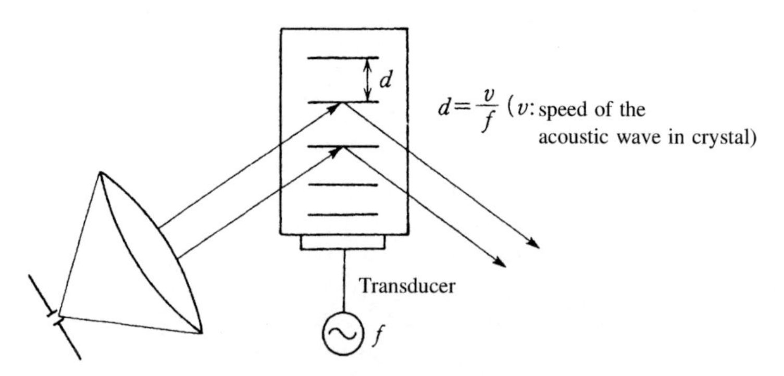

Fig. 3.15 A geometry of diffraction by an acousto-optic grating

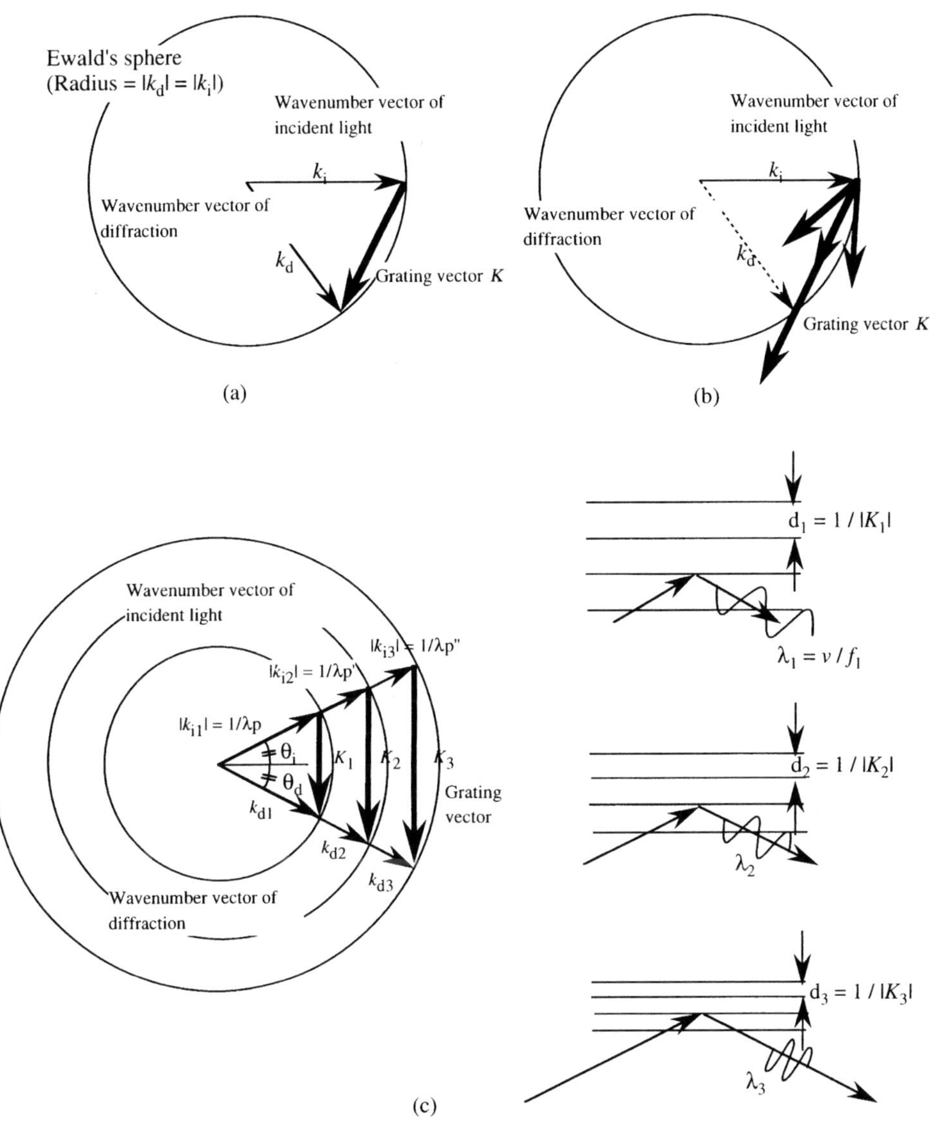

Fig. 3.16 The principle of Bragg diffraction. (a) The cases when the Bragg diffraction condition is satisfied. (b) The cases when the Bragg diffraction is not satisfied. The product of wavenumber vector of incident light + grating vector does not exist on the Ewald sphere, resulting is that the light passes through the grating without diffraction. (c) Bragg diffraction for different wavelengths (and grating vectors)

of the wavenumber vector k_i (the direction of light incidence) of incident wave and the grating vector K (its length is the inverse number of grating interval) of the diffraction grating. Since the wavelength $|k|$ or the wavelength λ of light is never changed by diffraction at a grating, the ends of the grating vectors of both the incident light and the diffracted light exist on the same surface of a sphere (Ewald's sphere as shown in Fig. 3.16 (a)). Thus, only a diffraction grating which meets the ends of these vectors can diffract the light. Diffraction is not generated under other conditions (See Fig. 3.16 (b)). This is the Bragg's diffraction condition, which corresponds to the state in which this is equivalent to an ordinary diffraction with a blazed grating of an appropriate blazing angle.

In ordinary, isotropic Bragg diffraction, the angle of incidence and the angle of Bragg diffraction relative to the diffraction grating are identical and their signs are always opposite, as shown in Fig. 3.16 (c). The grating constant K of the grating created acousto-optically is varied by changing the sound frequency f. Light of a certain wavelength is obtained that corresponds to the sound frequency as an output, by fixing both the angle of incident light and the angle of observation of diffraction (See Fig. 3.16 (c)).

The biggest merit in creating a diffraction grating acousto-optically is that wavelengths can be swept at extremely high speed. Since wavelengths can be swept by changing the frequency of the AC signal generator, and without any mechanical operations, diffraction wavelengths can be changed in microseconds. This wavelength-sweeping method makes it possible to both vary the frequency continuously and to obtain several wavelengths selectively. This characteristic is particularly suitable for practical near-infrared spectroscopic analyses.

When light is made to pass through an anisotropic material and the direction of polarisation of the diffracted light differs from that of the incident beam, the wavenumber of the diffracted light is not identical to that of the incident beam (attributable to the change of refractive index) and thereby fails to meet Bragg's diffraction condition. Fig. 3.17 (a) shows this case. Now, we assume that the incident beam is an ordinary ray and the diffraction is an extraordinary ray. Since the refractive index for an extraordinary ray varies depending on the direction of light transmission, its wavenumber vector is on an ellipse. Case A in Fig. 3.17 (a) easily fails Bragg's diffraction condition, when wavenumber matching is disturbed by changing the grating vector K, even a little, by scanning the acoustic wave frequency. This is thus superior in the selectivity of the wavelength of diffracted light and, hence, can be used as a wavelength selection filter. On the other hand, Case B does not so easily fail to meet Bragg's diffraction condition, even when the acoustic grating vector is changed, so that the diffraction direction can be changed without changing the wavelength. This configuration is thus suitable as an acousto-optical deflector of a laser-beam.

As described so far, the performance of anisotropic Bragg diffraction is better than isotropic Bragg diffraction when used as a laser-beam deflector or a wavelength filter in spectroscopy.

It can also be seen from the figure that diffraction of nontarget orders is not generated, neither does any anomaly occur. This is another merit of spectroscopy by using Bragg diffraction.

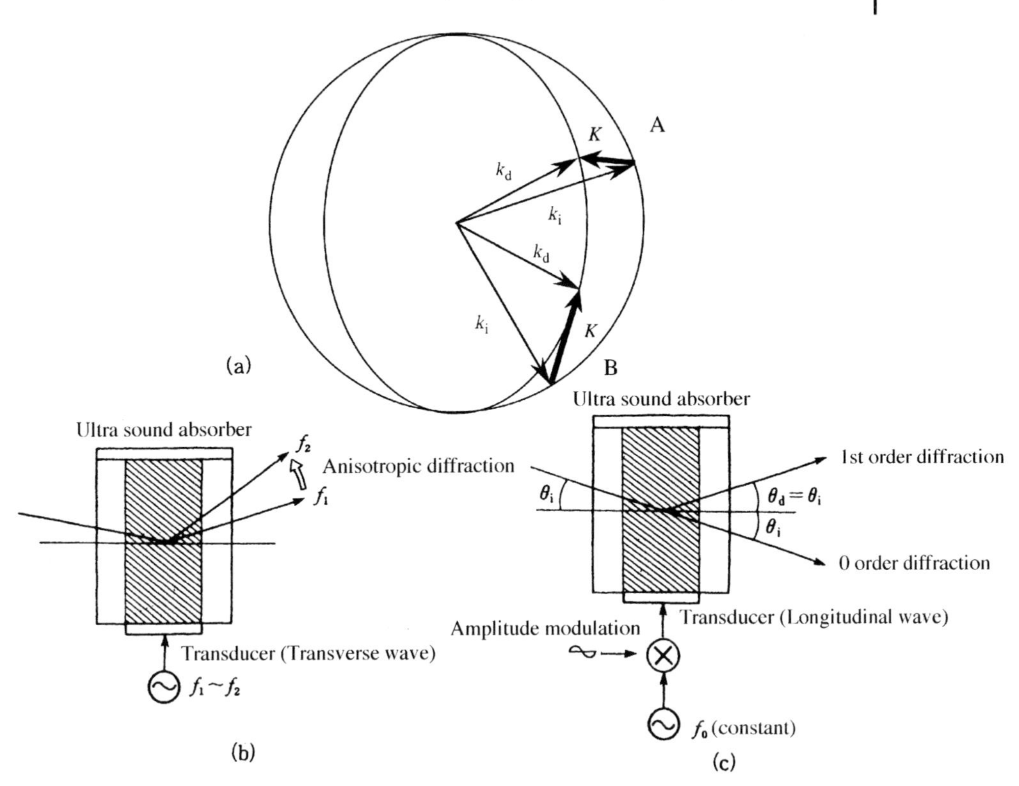

Fig. 3.17 (a) Anisotropic Bragg diffraction. (b) Acousto-optic deflector (AOD), and (c) Acousto-optic modulator (AOM) are the examples of this phenomenon

3.4.3 Application and Materials of Acousto-Optical Elements

Acousto-optical elements have been widely used in the field of laser optics, such as acousto-optical deflectors (AOD) and acousto-optical modulators (AOM). AOD is used for changing the beam angle by changing the wavelength of sound and is actually used in laser scanning microscopes. AOM is used for modulating the light intensity by modulating the amplitude of the sound signal and for high-speed laser-beam intensity modulators and laser beam shutters. Both of these are characterised by their ability to control the laser beam nonmechanically.

Although it was a long time ago when an acousto-optical element was first applied to spectroscopy [11], it has not, in practice, been used in general spectroscopy, except for mode analyses of semiconductor lasers and high-speed wavelength sweeping of dye lasers [12, 13]. This is because of its extremely low resolving power (defined as the number of channels to be resolved into the wavenumber region being measured, or band B divided by wavenumber broadening $\delta\nu$), attributable to the fact that the number of ultrasound gratings is far smaller than that of ordinary diffraction gratings. Although a large grating with many slots is needed to achieve high resolving power, the aperture of a practical acousto-optical element is only a few millimetres.

An acousto-optical material should be transparent with a high figure of merits $M_e = n^6 p^2/\rho v^3$ (n, refractive index; p, photoelasticity; ρ, density and v, sound velocity in the medium) in the target wavelength region. Materials used in the near-infrared region include TeO_2 (350 to 5000 nm), $LiNbO_3$ (400 to 4500 nm), $PbMoO_4$ (420 to 5500 nm), and SiO_2 (210 to 2400 nm). Despite the unsatisfactory quality of acousto-optical elements as a diffraction grating compared with that of an ordinary diffraction grating made with high precision, the quality of diffracted light is within an allowable range thanks to Bragg diffraction, which is due to the thickness of the volume.

The biggest problem is the spectral distortion, which occurs when the power of the incoming light is too large to make the high power of the light beam which raises the temperature locally due to the thermal absorption in the area where the light is incident, resulting in a change in refractive index and hence displacing and broadening the angle of diffraction. Other problems are the convergence of the diffracted beam when the speed of wavelength scanning is too fast, and the wavelength shift of diffracted light (Doppler effect) by the velocity of ultrasound waves (a positive use of this spectral shift for spectroscopy has been proposed [12]). While these problems could be insurmountable in laser physics and engineering, they are almost negligible in the application to near-infrared spectroscopy.

3.5 Fourier-Transform Spectroscopy

Fourier-transform spectroscopy has been commonly used for spectroscopic analyses in the mid-infrared region; diffraction-grating spectroscopy is rarely used except for special cases. This is because of the following advantages with Fourier-transform spectroscopy: 1) the collection efficiency of photon fluxes is high (Jacquinot's advantage or the advantage of optical throughput) because light from the light source or the sample with a wide area and a wide angle of radiation can be guided into the spectroscope efficiently; 2) the detection efficiency of signals is high (Fellgett's advantage or the advantage of multiplexing) because all the wavelengths are detected simultaneously and 3) high resolution can be obtained because its wavenumber precision is high (Connes' advantage).

Since these merits are significantly important for the near-infrared region, the present near-infrared spectrophotometers that use diffraction gratings are likely to be replaced by Fourier-transform near-infrared spectrometers (FT-NIR) in the future.

3.5.1
Principle of Fourier-Transform Spectroscopy

As shown in Fig. 3.18, Fourier-transform spectroscopy uses a two-beam interferometer (in the figure, a Michelson interferometer). Light from the light source is collimated by a lens (or a concave mirror) and is divided into two beams by a beam splitter. These beams make a round trip of different distances. Upon returning to the beam splitter again, they are superimposed (interfered) by the beam splitter and form an image of the light source on the sample. Then, the light absorbed by the sample passes through the lens (or the convex mirror) again and forms an image of the light source on the detector, which is photodetected.

When one of the two reflection mirrors of the interferometer is shifted along the optical axis, the optical path lengths of the two beams change. Two beams intensify each other when the difference d of the optical-path lengths (twice the distance because an optical path is a round trip) satisfies

$$d = m\lambda + \lambda/2, \quad m = \ldots, -2, -1, 0, 1, 2, ., \tag{3.6}$$

The same applies to both Eq. 3.1 for interference-filter spectroscopy and Eq. 3.4 for diffraction grating spectroscopy. However, unlike interference-filter or diffraction-grating spectroscopy, which are based on the multibeam interferometer, Four-

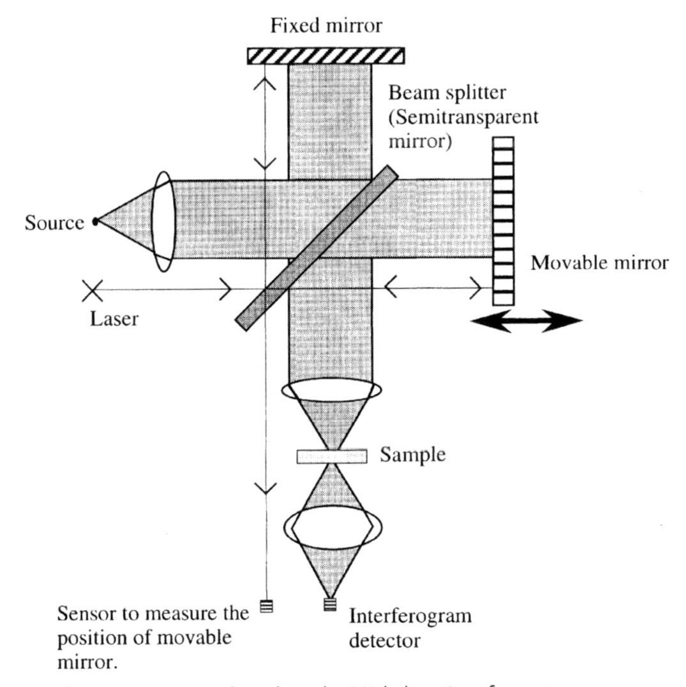

Fig. 3.18 The Fourier-transform spectrometer based on the Michelson interferometer

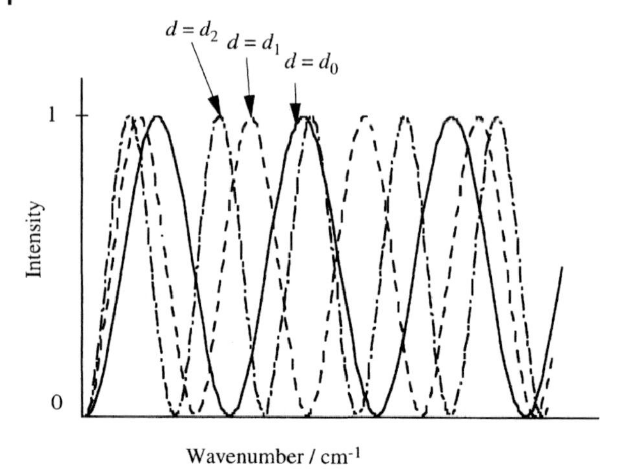

Fig. 3.19 Intensity spectra obtained by a two-beam interferometer as a function of wavenumber for different optical path differences (d)

ier-transform spectroscopy is based on the two-beam interferometer. The spectral characteristics do not have sharp peaks as shown in Figs. 3.3 and 3.9, but have a sine-wave characteristic relative to d as shown in Fig. 3.19. The reason why $\lambda/2$ is added in Eq. 3.6 is because it cancels the phase shift of π (or $\lambda/2$) which is generated when the light reflects outside the surface of the beam splitter. This happens only to one of the two beams. If the intensity of interference is detected at the source-side position, this phase bias does not occur.

To conduct spectral measurements, the optical path length d of the two beams is changed (i.e., one of the reflection mirrors is moved), and the intensity of interference is measured at the detector. The intensity $i(d)$ is called an interferogram and is given by the following equation in which the spectrum for measurement is expressed by $I(\nu)$:

$$i(d) = \int I(\nu)(1 - \cos 2\pi \nu d) \cdot d\nu \qquad (3.7)$$

and ν is the wavenumber (= $1/\lambda$).

Eq. 3.7 indicates that the detection intensity of light can be given by the difference between the noninterference light component $\int I(\nu)d\nu$ and the interference light component $\int I(\nu) \cos 2\pi\nu d \cdot d\nu$. When Eq. 3.7 is cosine-transformed relative to d, the intensity spectrum $I(\nu)$ of the light can be obtained. Here we neglected AD bias $I(0)$, which is actually added due to the noninterference light component, because this information is not necessary. Eq. 3.7 shows that this spectroscopy is cosine-transform spectroscopy rather than Fourier-transform spectroscopy.

3.5.2 Characteristics of Fourier-Transform Spectroscopy

In this section, the three advantages of Fourier-transform spectroscopy mentioned before are discussed in more detail.

3.5.2.1 Optical Throughput Advantage

As shown in Fig. 3.18, a slit is not necessary for Fourier-transform spectroscopy at either the entrance or the exit of the spectrometer. The slit is the most important element for diffraction-grating spectroscopy, including spectroscopy with an acousto-optical element to pick up the component of a certain wavelength. Thus, a Fourier-transform spectrophotometer collects a much larger number of photons into the optical system than a diffraction-grating spectrophotometer. This is known as the advantage of optical throughput. The area of light sources (or samples) is, however, still subject to restrictions. When the light source has an extent, the component of light emission incident obliquely upon the optical system and the component incident in parallel with the optical axis become different in optical path length; this results in the blurring of the interferogram. The maximum difference in the optical path length between the two beams determines the resolution of Fourier-transform spectroscopy. While the size of the aperture degrades the resolution, in practice, Fourier-transform spectrophotometers are, nevertheless, brighter than diffraction-grating spectrophotometers by at least one order of magnitude. Since near-infrared spectroscopic analyses do not require very high resolution in comparison with mid-infrared analyses, Fourier-transform spectroscopy becomes more advantageous by opening the aperture wider.

3.5.2.2 Multiplexing Advantage

With Fourier-transform spectroscopy, a single photodetector continuously detects light of all wavelengths. With diffraction grating and interference-filter spectroscopy, on the other hand, only a specific wavelength is detected at a certain period of measurement, and the light of other wavelengths is discarded. Accordingly, the efficiency of using signals of Fourier-transform spectroscopy is higher than that of other spectroscopic methods. As a result, Fourier-transform spectroscopy is superior in terms of the signal-to-noise ratio (S/N-ratio) when the thermal noise of the detector is a predominant noise. Diffraction-grating spectroscopy is superior when the intensity of the light source varies while the interferogram is being measured or when photon noise is more dominant than thermal noise [14].

3.5.2.3 Resolution

The spectral resolution of Fourier-transform spectroscopy is determined by the maximum difference between the optical path lengths of two beams (i.e., the length of interferogram), while the resolving power is determined by the number of sampling points (length/sampling interval) of the interferogram. The interferogram can be sampled with high precision by introducing a stabilised single-mode laser beam of a short wavelength into the interferometer, and sampling the signal at the crossing point of the interference fringe with the threshold during the mirror scanning. High wavenumber precision (high resolution) is also one of the advantages of Fourier-transform spectroscopy compared with other spectroscopic methods. Although there are other issues regarding Fourier-transform spectroscopy, such as phase correction, apodisation and zero-filling, they are not discussed here [15, 16].

3.5.3 Various Types of the Michelson Interferometer

With Fourier-transform spectroscopy, the interferogram is obtained by scanning the difference in the optical path lengths of two beams of a two-arm interferometer. The Michelson interferometer, shown in Fig. 3.18, is a typical two-beam interferometer, but there are many other interferometers that can be used for Fourier-transform infrared spectroscopy. The Genzel interferometer is constructed such that two beams reach the same reflection mirror from opposite directions, hence the difference in optical path lengths obtained relative to the amount of movement of the reflection mirror is twice that obtained by the Michelson interferometer. With the Genzel interferometer, a beam splitter that is small in size can be made because it is on the image plane of the light source. It is an optical system in which the combined mechanism of the beam splitter and the reflection mirror rotates for scanning the optical-path difference.

The Transept interferometer scans the optical-path difference by inserting a wedge-type prism of a high refractive index into the optical path. The sampling precision is high and the optical system is easily adjustable in this interferometer because the change of optical-path difference is smaller than the amount of prism movement. In addition, a beam splitter with a sufficient thickness can be turned to scan the difference between the optical paths, which changes the optical path lengths of the two beams.

Other interferometers used for far-infrared spectrophotometers include an interferometer that uses lamella gratings and a Martin-Puplett interferometer (polarisation-divided interferometer), but these are unlikely to be used for near-infrared spectroscopy.

3.5.4 Polarisation Interferometer

A polarisation interferometer was first proposed in the 1960s by Mertz as a new device for Fourier-transform spectroscopy for the visible and near-infrared regions [17] and was put to practical use as one type of Fourier-transform near-infrared spectrophotometer in the 1990s [18]. Fig. 3.20 shows its optical system. In this optical system, a triangular prism of high refractive index is moved perpendicularly to the optical axis to change the optical distances of the two polarising components of the incident light that pass through the crystal. Since a birefringent crys-

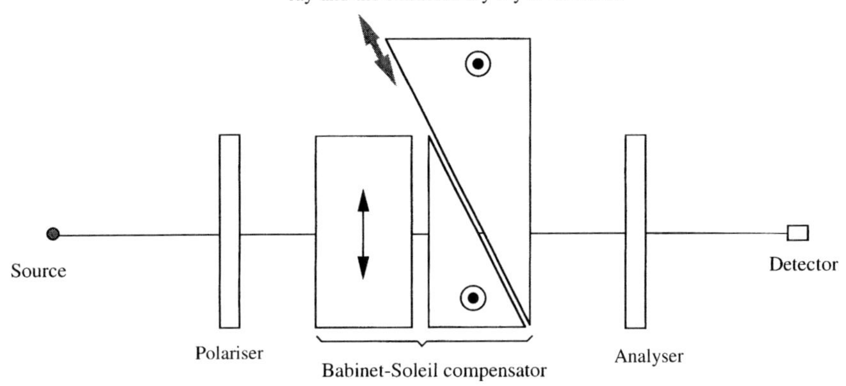

Fig. 3.20 Polarisation interferometer for Fourier-transform spectroscopy

tal has two different refractive indexes for the two orthogonal polarising components, the difference between the optical distances relative to the respective polarising components is varied by changing the physical thickness of the crystal. To make the intensity ratio of the two polarisations 1:1, a polariser is inserted in front of the interferometer at an angle of 45°. Behind the interferometer is another polariser at an angle of 45° for interfering with the two polarised light beams.

This interferometer system has advantages for field-use spectrophotometers because: 1) it is not greatly affected by vibration since the two beams pass through the same optical path; and 2) the device can be easily operated because the difference between the optical path lengths is scanned by sliding the triangular birefringent prism and not by adjusting the angle of reflection precisely.

3.5.5
FT-NIR Raman Spectroscopy

For Raman scattering spectroscopy, a high-resolution diffraction grating is used with a short-wavelength laser such as an argon laser (488 nm), as the light source for excitation. A short wavelength is desirable for excitation for Raman scattering because the cross section of Raman scattering is inversely proportional to the fourth power of the wavelength and because of the availability of highly sensitive detectors in the short wavelength region. However, recently, near-infrared solid lasers, such as the semiconductor laser and the Nd:YAG laser, have become available as light sources for excitation and, hence, a Raman spectrum is often measured with such near-infrared lasers. The major merits of Raman scattering spectroscopy in the near-infrared region are 1) there is no interference from fluorescence in the near infrared region; 2) small, inexpensive and powerful lasers are available for use in the near infrared region and 3) FTIR can be used in the near-infrared region by replacing some of the optical elements. The three advantages of Fourier-transform spectroscopy described above are also desirable merits for near-infrared Raman spectroscopy.

3.6
Multichannel Fourier-Transform Spectroscopy

The advantages of Fourier-transform spectroscopy have been mentioned repeatedly. However, for multichannel diffraction grating spectroscopy, these advantages may not be sufficient when compared with a polychromator (Section 3.3), which uses a multichannel detector with many photodetector elements that are aligned with high density. With a multichannel detector, the band being measured is continually read by the detector elements of the respective channels without wavelength scanning; thus, a benefit similar to that of multiplexing of Fourier-transform spectroscopy can be obtained. With Fourier-transform spectroscopy, the spectral precision deteriorates markedly when the light source or the measurement environment (for instance, the refraction index in optical paths) changes during the measurement. This is known as the multiplexing disadvantage [14]. This problem does not happen with polychromator spectroscopy when using a multichannel detector. Nevertheless, a polychromater with a multichannel detector is inferior to Fourier-transform spectroscopy in terms of optical throughput, since a multichannel detector requires an entrance slit for the incoming light.

The characteristics and superiority of diffraction-grating spectroscopy with a multichannel detector and those of Fourier-transform spectroscopy with a single detector are therefore complementary, and the choice of the method depends on the purpose.

Multichannel Fourier-transform (MCFT) spectroscopy is a new spectroscopic method, in which a multichannel detector is installed on a Fourier-transform spectrophotometer to utilise the advantages of both. Although this method was proposed around 1980 [1, 19], it was not put to practical use immediately. The results of practical implementation were finally reported in the 1990s, as multichannel detectors with sensitivity for both the infrared and near-infrared regions were developed [20, 21].

With this method, the interferogram is formed in space but not in time and is simultaneously detected by a multichannel detector. Accordingly, there is no mechanically movable scanner, unlike ordinary Fourier-transform spectroscopy, and, hence, a stable interferometer can be constructed. This method can give time-resolved spectroscopy of varying or transitional phenomena with high precision at high speed and is strong against the fluctuation of light source. A triangular optical-path interferometer and a Savart plate birefringent polarisation interferometer (discussed later) can be used as the interferometer with no restrictions on the size of the light source (sample). Thus, an extremely high collection efficiency of photons is realised for diffuse reflection samples in comparison with the optics of conventional Fourier-transform spectroscopy. This merit is of great importance because many samples for near-infrared spectral analysis are light-scattering or diffuse reflection samples.

3.6.1 Principle of Multichannel Fourier-Transform Spectroscopy

Fig. 3.21 shows a schematic diagram of a multichannel Fourier-transform spectrophotometer that is based on the Michelson interferometer. To produce an interferogram spatially with this interferometer, two reflection mirrors (or one of two) of two arms in the interferometer two-beams are inclined reverse to the optical axis. Then, the virtual position M2′ of the mirror M2 in the position of mirror M1, is displaced from the position of M1 by θx (θ corresponds, as a result, to the angle made by the two beams), being proportional to the distance x from the optical axis. When this mirror position is observed by using lens L2 (with angle magnification α), interference fringes, generated due to the difference between the optical path lengths of the two beams, appear in the image plane. By placing a self-scanning multichannel array detector (such as a CCD) in this image plane, the optical-path difference can be scanned electronically; this is done by mechanically operating mirrors with ordinary Fourier-transform spectroscopy. This is equivalent to observation of Young's interference fringes from two pinholes S1 and S2 (corresponding to the two real images of the light source created by lenses L1 and L2). When the light source S to be measured is not small, the plane where fringes are formed (relative to the optical axis direction) is localised on the image plane of mirrors M1 and M2 focused by lens L2; thus, the image sensor must be placed exactly in this plane.

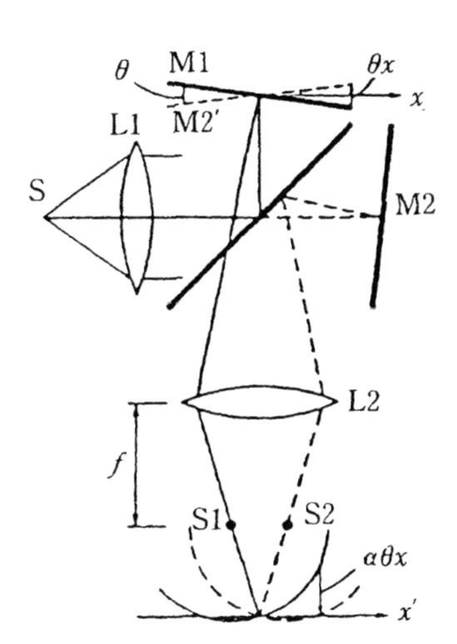

Fig. 3.21 Multichannel Fourier-transform spectrometer based on the Michelson interferometer

3.6.2 Multichannel Fourier-Transform Spectroscopy with a Polarising Interferometer with a Savart Plate

Besides the Michelson interferometer, many other interference optical systems have been proposed for multichannel Fourier-transform spectroscopy [1]. In particular, a common path birefringent interferometer that uses a birefringent prism called the Savart plate has excellent potential for practical application in near-infrared spectroscopy. This interferometer boosts the optical throughput of Fourier-transform spectroscopy because the size of the light source does not restrict the resolution at all [20, 21]. Fig. 3.22 shows a schematic diagram of a multichannel Fourier-transform spectrophotometer that uses this interferometer. The Savart plate is made by combining two birefringent crystals with which linear polarised light, normally incident on the crystal, is decomposed into two components, each displaced laterally and emitted as two parallel beams. These two beams are then made to interfere on the detector by a lens; this gives rise to an interferogram. Two polarisers of 45° are installed one before and one after the Savart plate to interfere with the two beams with an intensity ratio of 1:1.

Since this interferometer has an in-line structure, this does not cause a vignetting effect by the limit of view in optics. The light source (or sample) can be placed sufficiently near the lens because no slit is required. In addition, since the system has an in-line structure with only one lens, the system is extremely com-

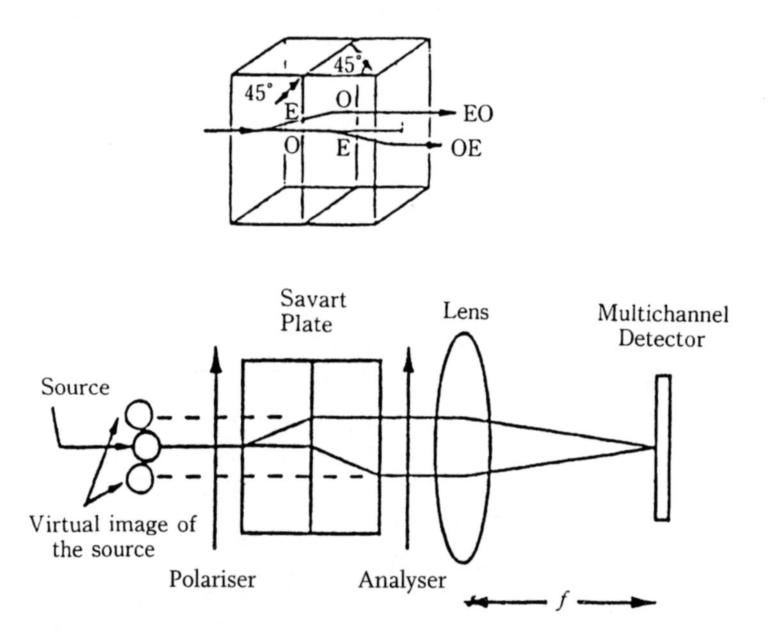

Fig. 3.22 Multichannel Fourier-transform spectrometer based on the polarisation interferometer with a Savart plate

Tab. 3.1 Comparison between spectroscoic methods in performance

	Resolution in wave-number (spectral width)	*The number of channels (resolving power in wavenumber)*	*Flatness of spectrum*	*Removal contamination of high-order diffraction*	*Optical throughput (slits)*	*Real-time measurement*	*Scanning speed*	*Dimension*	*Remarks*
Interference-filter spectroscopy	△	×	○	○	○	×	×	△~○	small
Variable interference filter	×	△	×	○	△	×(△)	○	◎	small and easy in operation
(Fabry-Pérot)	◎	○	○	○	○	×	○	△	
Grating spectroscopy	○	◎	○	×	×	××	×	×	
(Hadamard spectroscopy)	△	△	○	×	×	○	×~△	×	
Polychromator	△	△	△	××	×	○	◎	△	transient phenomena, no mechanically moved parts
Acousto-optic element	×	△	×	○	×	×	◎	◎	Measurements in a narrow small size band
Fourier-transform spectroscopy	○	◎	○	○	○	○	○	×	Multiplex advantage
Multichannel Fourier transform-spectroscopy	△	△	○	○	◎	○	◎	◎	compact, transient phenomena

◎ Excellent, ○ superior, △ good, × inferior

pact and easily adjustable. Furthermore, this interferometer is effective for measuring diffuse reflection samples and multiple scattering samples because a large size of sample does not restrict the resolution.

A rutile crystal is most commonly used for the Savart plate, with a metal wire grid or a polarising film for the polariser and the analyser, although other materials can also be used depending on the wavelength region.

3.7 Comparison of Spectrometers

Tab. 3.1 summarises the various types of spectroscopy discussed. Each of them has both advantages and disadvantages, so that the best type of spectroscopy depends on the purpose. Considering that ease of operation is required for field-use and in-process near-infrared measurements, diffraction-grating spectroscopy with an acousto-optical element and multichannel Fourier-transform spectroscopy with an array detector have good future potential. Interference-filter spectroscopy that has been used conventionally is also a reasonable spectroscopy and, thus, will continue to be used. Fourier-transform spectrophotometers are considered to be optimal as spectrophotometers with high precision for laboratory use.

3.8 References

[1] S. Kawata, *J. Spectrosc. Soc. Jpn.* **1989**, *38*, 415.

[2] For example, BRAN+LUEBBE, InfraAlyzer Systems, Norderstedt, Hamburg.

[3] *Near-Infrared Technology in the Agricultural and Food Industries*, (Eds. P. Williams, K. Norris), American Association of Cereal Chemists, St. Paul, MN, **1987**.

[4] F. K. Kneubühl, J. F. Moser, H. Steffen, *J. Opt. Soc. Am.* **1966**, *56*, 760.

[5] T. K. Cubbin, W. M. Sinton, *J. Opt. Soc. Am.* **1952**, *42*, 113.

[6] R. A. Oetjen, W. H. Haynie, W. M. Ward, R. L. Hansler, H. E. Shauwecker, E. E. Bell, *J. Opt. Soc. Am.* **1952**, *42*, 559.

[7] W. G. Fastie, *J. Opt. Soc. Am.* **1952**, *42*, 641.

[8] H. Ebert, *Wied. Ann.* **1889**, *38*, 489.

[9] H. Noda, M. Koike, *J. Spectrosc. Soc. Jpn.* **1989**, *38(3)*, 174.

[10] S. E. Harris, R. W. Wallamce ((Q2A)), *J. Opt. Soc. Am.* **1969**, *59*, 744.

[11] T. Yano, A. Watanabe, *Appl. Phys. Lett.* **1974**, *24*, 256.

[12] L. D. Hutcheson, R. S. Hughes, *Appl. Opt.* **1974**, *13*, 1395.

[13] D. J. Taylor, S. E. Harris, S. T. K. Nieh, *Appl. Phys. Lett.* **1971**, *19*, 269.

[14] S. Minami, *Fourier Transform Infrared Spectroscopy* (Ed. J. Hiraishi), Gakkai Shuppan Center, **1982**, p. 28.

[15] G. A. Vanasse, H. Sakai, *Progress in Optics, Vol. 6*, North-Holland, Amsterdam, **1967**.

[16] S. Kawata, *Waveform Data Analysis and Spectral Analysis*, Workshop of Infrared-

Raman Spectroscopy Division of SIJ, **1992**.

[17] L. Mertz, *Transformations in Optics*, Wiley, New York, **1965**, p. 53.

[18] BRAN+LUEBBE, Infra Prover, Norderstedt, Hamburg.

[19] T. Okamoto, S. Kawata, S. Minami, *Appl. Opt.* **1984**, *23(2)*, 269.

[20] S. Kawata, Y. Inouye, S. Minami, *Compact Multichannel FTIR-Sensor with a Savart-Plate Interferometer*, Proc. SPIE Vol. 1145, Fourier Transform Spectroscopy, Conference, **1989**, p. 567.

[21] M. Hashiomoto, S. Kawata, *Appl. Opt.* **1992**, *31(28)*, 6096.

4
New Techniques in Near-Infrared Spectroscopy

Satoshi Kawata

4.1
Near-Infrared Light Sources

In practical applications of near-infrared spectroscopy, spectra to be measured are mostly overtones and combination bands of several functional groups. Therefore, light sources with broad spectra are used and measurement of a single line is not particularly important. Measurements of absorption at a specific wavelength, including those of primary and secondary overtones of molecular vibration transitions, do not require very high wavenumber resolution. Accordingly, thermal radiation with continuous spectral radiation is often used for the light source in near-infrared spectroscopy. There is no light source that has been developed exclusively for near-infrared spectroscopy, rather different light sources are used for infrared spectroscopy or visible and ultraviolet spectroscopy depending on the purpose.

On the other hand, a monochromatic light source is required for devices that need to sense a specific target (such as in vivo haemoglobin concentrations in blood) and for near-infrared Raman measurement. Strong, monochromatic near-infrared light sources actually in use include semiconductor lasers, solid lasers, gas lasers and light emitting diodes (LED).

4.1.1
Thermal Radiation

A radiant divergence spectrum $W(\lambda)$ of an ideal radiation source (black body radiation) from a thermal radiation body (black body furnace) is given by the following equation based on Planck's law of radiation:

$$W(\lambda) = c_1 \lambda^{-5} / \{\exp(c_2/\lambda T) - 1\} \tag{4.1}$$

where T is the temperature (K) of the black body; λ, the wavelength (µm); $c_1 = 1.91 \times 10^{-12}$ Wcm^2; and $c_2 = 1.438$ cm deg. Fig. 4.1 shows the spectral distribution. This figure shows that the peak wavelength of black body radiation with a temperature of 2000 K appears at around 1.5 µm, while that of black body radiation with a temperature of 1000 K appears at around 2.9 µm.

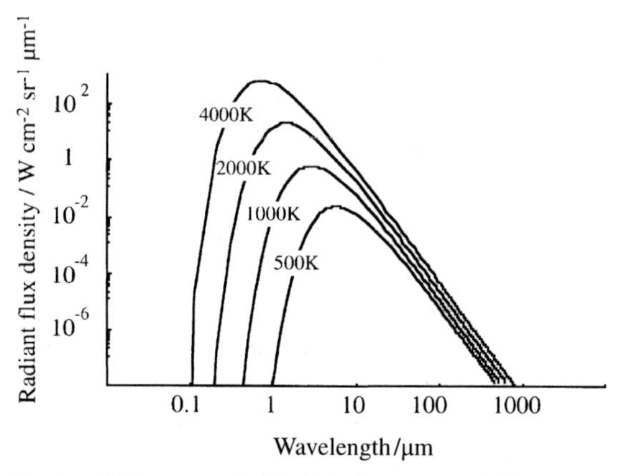

Fig. 4.1 NIR spectra of a black body source at different temperatures

Since actual lamps are not complete black bodies, the radiant divergence is lowered due to reflection on the surface of the radiator. Surface treatment is therefore applied to minimise the reflectance ρ (thus maximising the emissivity, $\varepsilon(=1-\rho)$). Nevertheless, the true temperature of the light source is still higher than the apparent temperature of radiation.

4.1.1.1 **Tungsten Halogen Lamp**

Tungsten halogen lamps used as visible and ultraviolet light sources (wavelength ~2.5 μm) are also the most typically used as near-infrared continuous spectral light sources. Inexpensive, ordinary tungsten lamps that do not contain halogen gas can also be used as a near-infrared light source as long as their valve material is transparent in the near-infrared region. However, halogen lamps have longer life, higher temperature and greater stability. Since the power supply for the light source should be stable, stabilised DC power sources are used rather than AC power sources (particularly those designed for commercial frequency are not recommended). Fig. 4.2 shows an example of the structure of a tungsten halogen lamp.

In addition to halogen lamps, other light sources for visible and ultraviolet regions can also be used for near-infrared spectroscopy. Duplex lamps (deuterium and tungsten combination lamps) have radiation between 0.2 and 2.5 μm, and xenon lamps between 0.2 and 2.0 μm. Nevertheless, special attention is required for xenon lamps, because the spectral distribution of these lamps is not smooth in the near-infrared region.

4.1.1.2 **Nichrome Heater and Globar**

Black body radiation for infrared spectroscopy can also be used as a continuous spectral light source for the near-infrared region. Globar (or siliconite: 1.1 to 100 μm) is

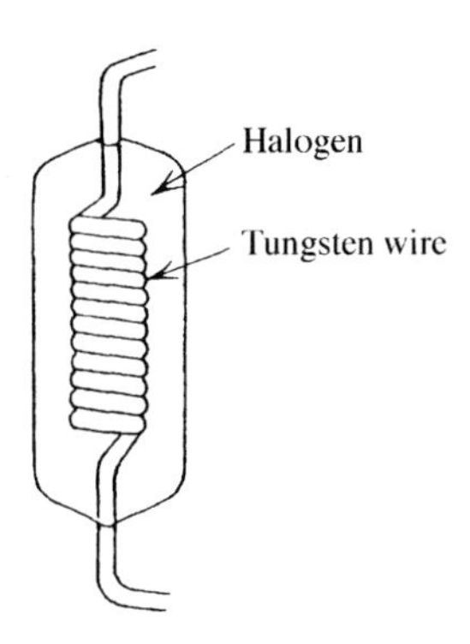

Fig. 4.2 A schematic diagram of a tungsten halogen lamp

generally used for infrared light sources; it is made from silicon carbide (SiC) mixed with additives, shaped into a bar and sintered at high temperature. This thermal light source has high emissivity and can emit spectra similar to black body radiation. When used for infrared spectroscopy, an electric current of around 5 to 10 A (50 to 100 W) is applied to make it exothermic up to a temperature of around 1400 K. When used for the near-infrared region, a still higher temperature is desired. A Nichrome wire heater [1], which is also known to be an infrared light source, can be a practical near-infrared light source. It has a simple structure of a ceramic rod with a Nichrome wire coated with a silicon carbide or an oxide film and has a long life. It is used at a temperature of around 1000 K in the infrared region.

Other infrared light sources can also be used as near-infrared light sources; however, their radiant quantities are small for the near-infrared region because their radiation peaks are at wavelengths longer than 2 μm.

4.1.2 Laser and Light Emitting Diode

4.1.2.1 Light Emitting Diode and Semiconductor Laser

A light emitting diode (LED) made of gallium arsenide (GaAs) has a peak wavelength at 940 nm and can be used for measuring the quantities of fats and oils, for example [2]. An LED made of GaAlAs (aluminum is added to gallium in an adequate ratio) covers a wavelength region of 650 to 900 nm, while an LED made of InGaAs covers 1.0 to 1.6 μm. An InGaAs LED can measure the moisture content from the absorption of the OH-overtone at 1.45 μm.

The use of laser beams with high convergence is essential in the measurement of the respiration of animals. The transmission of near-infrared light through the heads of cats or humans is detected after weak absorption of near-infrared light [3]. It is expected that in vivo near-infrared measurements will be practical in the near future by combining semiconductor lasers in the region of 700 to 1200 nm (absorption bands of haemoglobin, myoglobin and cytochrome oxidase) under picosecond-pulsed operation with TOF (time-of-flight) measurement, or by using frequency heterodyne detection with current injection control.

Sharpening the gain curve of an LED and injecting a large current result in "lasing". This is the mechanism of a semiconductor laser or a laser diode (LD). The broadening of the peak of the LD is 0.1 to 2 nm in wavelength, which is extremely narrow in comparison with the 20 to 100 nm of LEDs.

It must be noted that in terms of measurement, the spectral peak should be not narrower than necessary, since the influence of coherent noise (interference, speckle noise, etc.) becomes more conspicuous as the spectral peak narrows.

4.1.2.2 **Other Lasers**

Among the numerous solid-state lasers for the near-infrared region, the Nd:YAG laser is widely used. In addition to the well-known region at 1.064 μm, the Nd:YAG laser has ten emission lines in the region between 1.052 and 1.123 μm and nine lines in the region between 1.319 and 1.144 μm. A titanium sapphire (Ti:sapphire) laser, known as a solid-state tunable laser, can scan the wavelengths in the region between 700 and 1000 nm without discontinuity. Usually, an argon ion laser of several watts is used to excite a titanium sapphire laser. Dye lasers (from 400 nm to 1 μm in general, although the region differs depending on the dye) are also used as tunable lasers. However, they will probably be replaced by solid-state lasers such as Ti:sapphire lasers due to the lack of a good dye, in particular for the near-infrared region.

Wavelength-tunable lasers [4–7] that use parametric oscillation of non-linear optical materials, such as BBO (Barium borate, BaB_2O_4), are also starting to enter practical use. Although they cover the wavelength region between 0.35 and 3.65 μm, their emissions involve a lot of non-linear optical effects because they use pulse oscillations.

4.2
Near-Infrared Detectors

Photo-multipliers (PMT) and silicon photo diodes, both used for detecting visible light, are not used for near-infrared spectroscopy, except for the regions of wavelength shorter than 1 μm, since their sensitivity in the near-infrared region is low. Golay cells and thermocouples, which are in practical use in the mid-infrared region, as well as thermo-detectors that use pyroelectric crystals made of TGS (Triglycine sulfate) and $LiTaO_3$, have flat spectral properties in the near-infrared region. However, they are not commonly used as detectors for near-infrared spectroscopy due to their low sensitivity in comparison with quantum detectors, which will be discussed later.

Although called "infrared," an infrared photographic film only covers the very near-infrared region as it has a peak at 750–800 nm and its wavelength sensitivity is up to around 1 μm. A silicon photo-diode also has a wavelength sensitivity of up to around 1 μm. Some infrared sensors and infrared cameras are used in this region.

4.2.1
Photoconduction Effect

Lead sulfide (PbS) detects photons over a wide range of the near-infrared with high sensitivity. This element is a semiconductor in which resistance decreases when infrared rays penetrate due to photoconduction. It covers the range of 1 to 2.5 μm at ordinary temperatures and up to 3.5 μm when thermoelectrically cooled to −77 °C by using a Peltier element. Lead selenide (PbSe) is not very sensitive, although it has sensitivity from 1.5 to 4.5 μm at ordinary temperatures and up to around 5.8 μm when cooled. Mercury cadmium telluride (HgCdTe: usually called MCT) can be made to cover the sensitive range from 2 to 5 μm by choosing the composition ratios of HgTe and CdTe.

Fig. 4.3 shows the characteristics of spectroscopic sensitivity of these elements. The axis of ordinate D^* is given by the following equation:

$$D^* = \{(S/N)/P_D\}(f/A)^{1/2} \tag{4.2}$$

or by the following equation by using noise equivalent power (NEP):

$$D^* = (1/NEP)(f/A)^{1/2} \tag{4.3}$$

here f is the width of the frequency band being measured, A the effective area of element, P_D the effective value of the energy of incident light, S the effective value of signal voltage and N the effective value of noise voltage. D^* expresses the relative detection sensitivity per unit area of detector, per unit chopping frequency.

When a photoconductive detector is used, the signal from the sample being measured may be changed by the surrounding temperature, because dark resistance, sensitivity and response characteristics vary with temperature. The lock-in detection method is used to counter this problem. In the lock-in method, the light irradiating the sample is modulated or chopped (on and off) at certain frequencies, and only the components between these chopping frequencies are extracted from the detected signals. With Fourier-transform spectrophotometers, the interferogram being detected is automatically modulated at the mean frequency of the spectral band of light as an interferogram, so that the signal can be obtained as an AC signal without a mechanical chopper to modulate the light.

4.2.2
The Photovoltaic Effect

Like a silicon photodiode, a germanium (Ge) photodiode with a p-n junction shows a photovoltaic effect with sensitivity in the near-infrared region (0.8 to 1.9 μm). Although this can be used at ordinary temperatures, it is better to cool it down thermoelectrically to reduce dark currents. Its sensitivity can be remarkably improved when cooled to the temperature of liquid nitrogen.

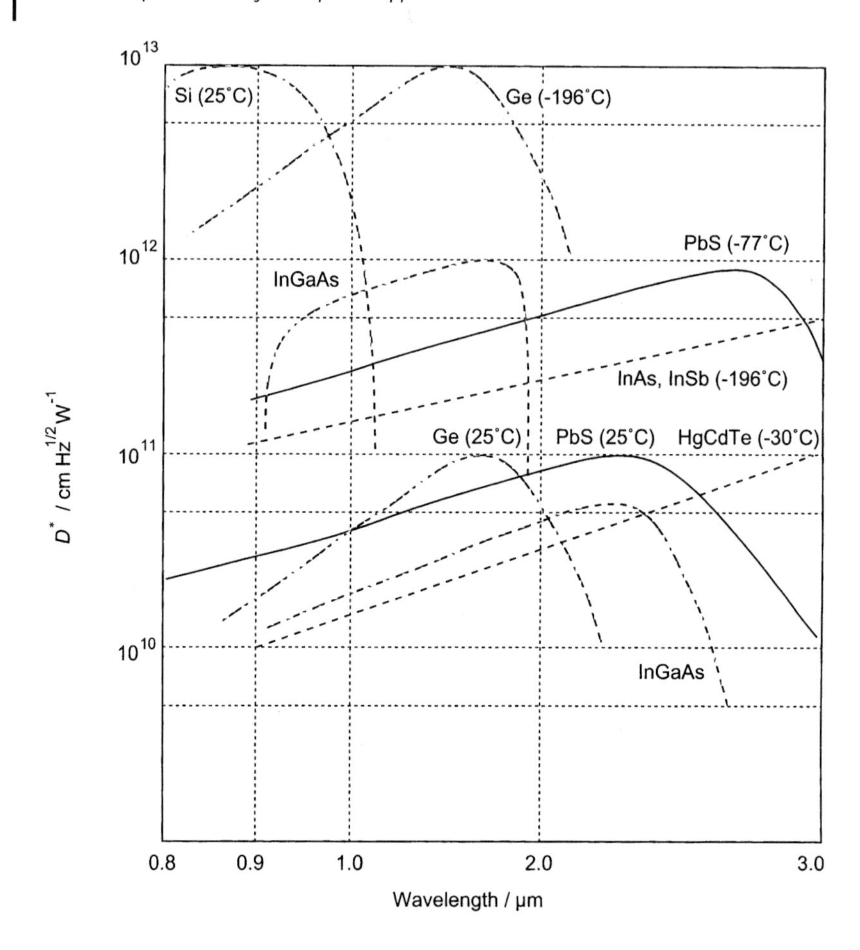

Photoconductive type

PbS PbSe HgCdTe

Photovoltaic type

Ge InAs InSb InGaAs HgCdTe

Fig. 4.3 Sensitivity spectra in NIR for various photoconductive detectors (*normal lines*) and photovoltaic detectors (*broken lines*), respectively

Other photovoltaic elements with a p-n junction include indium arsenide (InAs) and indium antimonide (InSb). InAs has sensitivity at 1 to 3.1 μm, and InSb at 1 to 5 μm; thus, they cover a wide range of the near-infrared region. Fig. 4.3 shows the characteristics of spectroscopic sensitivity of these elements in broken lines.

Silicon can also be used for near-infrared spectroscopy in the wavelength region shorter than 1 μm; it has detection sensitivity up to around 1.1 μm.

4.2.3
Multi-Channel Detectors

A device in which infrared photo-detection elements are aligned one- or two-dimensionally is called a near-infrared multichannel detector, and such a device is used in infrared security cameras. This detector can detect spatially distributed near-infrared images, including near-infrared spectra, without mechanically scanning.

A vidicon tube with a photoelectric surface to sense near-infrared photons up to around 2 μm (Fig. 4.4) has been conventionally used as a multichannel detector. Recently, however, solid image sensors, such as CCDs, are becoming widely used as the semiconductor integration technology progresses. The length and width of each element of a near-infrared image sensor are around 10 to 100 μm.

Near-infrared multichannel detectors made of 256 elements of PbS or MCT are commercially available. It may be difficult, however, to increase the number of elements with present technologies. In addition, CCDs cannot be used for the scanning section, since the output signals are read by synchronously detecting the respective channels along with chopping of the incident light.

Near-infrared multichannel detectors that use the photovoltaic effect are made of germanium or InSb. Although it used to be difficult to produce a high-density

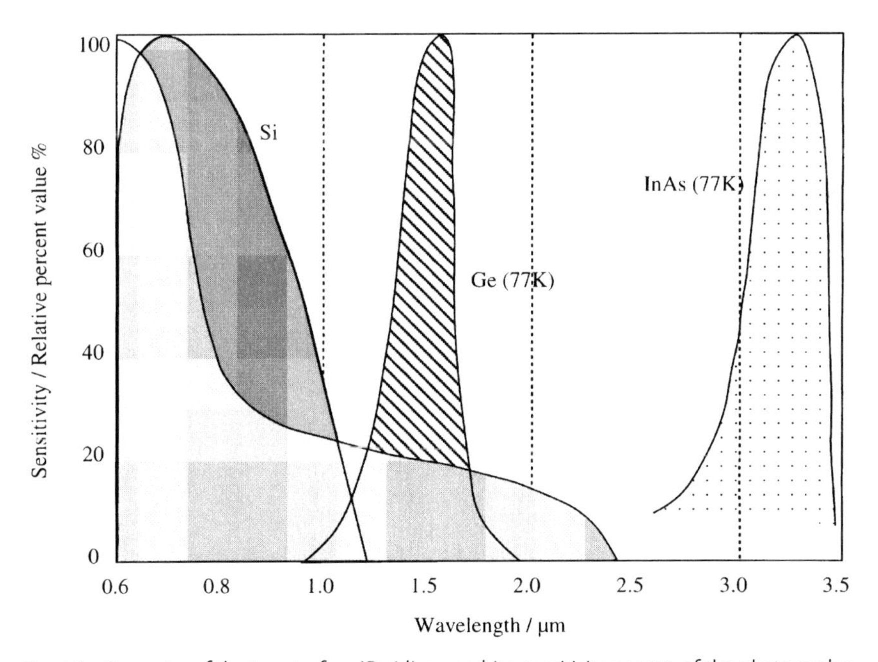

Fig. 4.4 Geometry of the target of an IR vidicon and its sensitivity spectra of the photoconductive thin film of NIR vidicon

InSb array, commercial products of an array composed of 512×512 elements are now available.

Silicon detectors do not sense the range longer than 1.1 μm because the band gap of silicon is about 1.1 eV. However, by coating the silicon substrate with a very thin layer of metal to form a siliconised film (silicide), photons with energy higher than the height of the Schottky barrier are absorbed to form photocurrents at the interface between the siliconised metal film and the silicon. If the Schottky barrier is lower than the band gap of silicon, near-infrared photons with wavelengths longer than 1 μm can be detected. The cut-off wavelength of a Schottky barrier detector made of palladium silicide (Pd_2Si) and p-type silicon is up to 3.5 μm; that of platinum silicide (PtSi), up to 6 μm and that of iridium silicide (IrSi) up to 10 μm [8]. The light for measurements is usually guided into the detector from the silicon side, so that the shortest wavelength limit is 1.1 μm or the absorption edge of silicon.

Since such Schottky-barrier detectors use silicon for the substrate, the technology of the silicon integration-circuit process can be used to fabricate high-density multi-channel detectors with many pixels. In particular, CCDs with up to 4096 elements that use PtSi have already been produced [9] and are used in infrared spectroscopy [10].

The quantum efficiency of Schottky barrier near-infrared detectors is dependent upon the wavelength (energy), and the sensitivity drops drastically near the cut-off wavelength. Sufficient sensitivity can be obtained by lowering the barrier, although this will increase the dark current. Therefore, the elements need to be cooled to the temperature of liquid nitrogen. Recently, the Stirling engine with reverse cycle has been used to cool them to this temperature (–193 °C/80 K). By separating the cooling section from the compressor (engine) by installing pipes, the vibration noise can be reduced. Although this has made the cooling section compact, problems still remain in terms of life at the photosensing section. A mini cooler that uses the Joule-Thomson effect can also be compact and brings the temperature down to that of liquid nitrogen [11]. A cooler that works on the Peltier effect can be much more compact and easily operated, although the temperature is around –120 °C (153 K).

4.3 Optical Elements for the Near-Infrared Region

Glasses and crystals used for the visible region can also be used as transparent window materials and lenses in the near-infrared region. However, many of them absorb secondary overtones of the OH-group at 1.4 μm.

Fig. 4.5 shows the near-infrared transmission spectra of various optical materials. CaF_2 and MgF_2 are known to be transparent materials for the entire near-infrared region and are used as lenses, windows and the substrate of filters.

Optical fibers are often used for guiding near-infrared rays into the spectroscope, and materials for fibers must be also transparent in the near-infrared

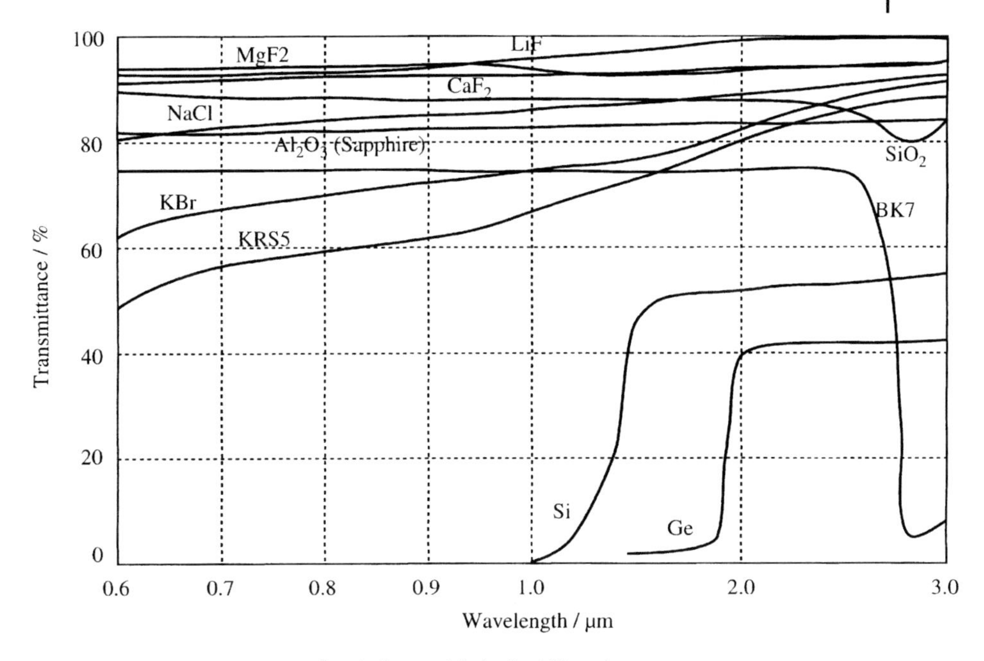

Fig. 4.5 Transmittance spectra of optical materials in the NIR region

region. The main issue is the production of fibers with minimum OH-content to reduce the absorption at 1.4 μm and 1.9 μm.

As a birefringent material, calcite can be used in the short wavelength region (up to μm). Rutile (TiO_2) has a higher birefringence in the entire near-infrared region. Although inferior to rutile, MgF_2 crystals also have a reasonable birefringence.

Metal can also be used for coating the mirrors for image formation and collimation instead of lenses and the inner surface of the integrating spheres. The metals used have optical characteristics in the near-infrared region roughly equivalent to (or better than) those in the visible region.

4.4 References

[1] M. Nishimoto, *J. Spectrosc. Soc. Jpn.* **1985**, *38*(5), 327.

[2] D. R. Massie, *Quality Detection in Foods* (Ed. J. J. Gaffney), ((Q2A)) **1976**.

[3] F. F. Jöbsis, *Science* **1977**, *198*, 1264.

[4] J. A. Giordmaine, R. C. Miller, *Appl. Phys. Lett.* **1966**, *9*, 298.

[5] R. W. Wallace, *Appl. Phys. Lett.* **1970**, *17*, 497.

[6] L. S. Goldberg, *Appl. Phys. Lett.* **1970**, *17*, 489.

[7] R. G. Smith, J. E. Geusic, H. J. Levinstein, J. J. Rubin, S. Singh, L. G. van Uitert, *Appl. Phys. Lett.* **1968**, *12*, 308.

[8] M. Kimata, M. Denda, N. Yutani, S. Iwade, N. Tsubouchi, *Proc. of SPIE* **1988**, *930*, 11.

[9] M. Denda, M. Kimata, S. Iwade, N. Yutani, T. Kondo, N. Tsubouchi, *Proc. of SPIE* **1987**, *819*, 279.

[10] M. Hashiomoto, S. Kawata, *Appl. Opt.* **1992**, *31*(28), 6096.

[11] N. Watanabe, *J. Jpn. Soc. Precision Eng.* **1990**, *56*, 1980.

5
Near-Infrared FT-Raman Spectroscopy

Yukio Furukawa

5.1
Introduction

Vibrational spectroscopy is a useful method for the structural analysis of materials, because it gives us information about their configuration, conformation, the nature of their chemical bonds, etc. Vibrational spectra can be obtained by infrared and Raman spectroscopy, which are to be regarded as complementary, rather than alternative, methods of obtaining the vibrational spectra.

Raman spectra are measured with visible laser lines. It was generally believed that near-infrared (NIR) light was not suitable for obtaining high-quality Raman spectra, because the cross section of Raman scattering becomes smaller with increasing excitation wavelength. However, Chantry et al. [1] suggested the possibility of Raman measurements with an NIR excitation source and an interferometer. Since Hirschfeld and Chase [2] reported successful NIR Fourier transform (FT) Raman measurements in 1986 using a spectrophotometer with a Michelson interferometer and the 1.064 μm laser line, a large number of studies have been made. The progress of NIR FT-Raman spectrometry has been reviewed in several articles [3–10]. It has been demonstrated that NIR FT-Raman spectrometry is a powerful tool for obtaining Raman spectra of various materials.

NIR FT-Raman spectrometry with a Michelson interferometer has various advantages in measuring Raman spectra. In most cases, when a visible laser line is used as an excitation source, a strong luminescence background is observed. The strong background prevents us from detecting Raman scattering, because the Raman intensity is much weaker than that of the luminescence. The luminescence is emitted because the sample itself, or an impurity in the sample, is pumped to an electronically excited state by the visible light. On the other hand, when an NIR laser line is used, luminescence is not emitted. This is because the sample or the impurity is not excited to the electronically excited state. Although the cross section of Raman scattering is small with NIR excitation, it is expected that the throughput advantage (Jacquinot's advantage) and the multiplex advantage (Fellgett's advantage) in FT spectrometry [11] ensure high-quality spectra. NIR FT-Raman spectrometry is the most probable method for overcoming the fluorescence problem and is a useful method for the application of Raman spectroscopy. High wavenumber accuracy (Connes' advantage) in FT spectrometry releases us from troublesome wavenumber calibrations in dispersive spectrometry and allows for a long accumulation time and for taking difference spectra. These merits are useful beyond our expectations in spectroscopic analysis. Double modulation measurements and pulsed-laser excited measurements have been developed recently. These methods have potential in new applications of FT-Raman spectrometry. In this chapter, we describe the principles, the instrumentation, the applications and the recent developments of NIR FT-Raman spectrometry.

5.2 Principles of FT-Raman Spectrometry

5.2.1 Raman Scattering

When systems like gases, liquids and solids are irradiated by monochromatic light of wavenumber $\tilde{\nu}_0$, light of wavenumber $\tilde{\nu}_0 \pm \tilde{\nu}_i$ as well as that of wavenumber $\tilde{\nu}_0$ is scattered. The former is called *Raman* scattering and the latter *Rayleigh* scattering. The lower wavenumber components of the Raman bands ($\tilde{\nu}_0 - \tilde{\nu}_i$) are called the *Stokes* lines and the higher wavenumber components ($\tilde{\nu}_0 + \tilde{\nu}_i$) are called the *anti-Stokes* lines. The Raman spectrum of liquid carbon disulfide in is shown in Fig. 5.1 as an example. Note that the light of wavenumber $\tilde{\nu}$ is regarded as photons whose energy E is given by

$$E = h\nu = \frac{hc}{\lambda} = hc\tilde{\nu} \tag{5.1}$$

where ν is the frequency, c is the velocity of light and h is Planck's constant. Raman scattering originates from the interaction between light and matter. Schematic vibrational energy level diagrams associated with Raman scattering are shown in Fig. 5.2. When the system interacts with incident light of wavenumber $\tilde{\nu}_0$, it may make a transition from a lower energy level E_1 to an upper energy level E_2. It must acquire the

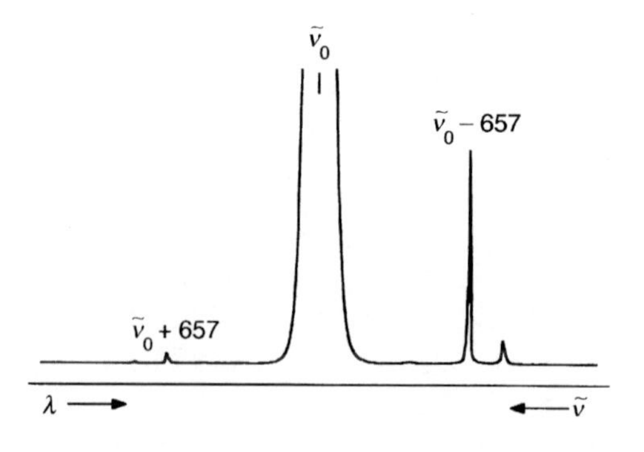

Fig. 5.1 Rayleigh and Raman spectra of carbon disulfide (liquid) excited with the 514.5-nm laser line. $\tilde{\nu}_0$, incident wavenumber

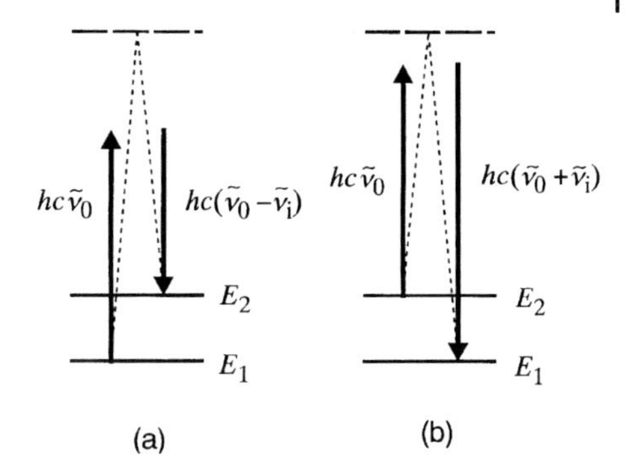

Fig. 5.2 Schematic energy level diagrams associated with Raman scattering: (a) Stokes line; (b) anti-Stokes line

necessary energy $\Delta E = E_2 - E_1 = hc\tilde{\nu}_i$ from the incident light. This energy is provided by the annihilation of one photon of the incident light of energy $hc\tilde{\nu}_0$ and the simultaneous creation of one photon of smaller energy $hc\ (\tilde{\nu}_0 - \tilde{\nu}_i)$, so that scattering of the light of lower wavenumber $\tilde{\nu}_0 - \tilde{\nu}_i$ occurs. This is a Stokes line. Alternatively, the interaction of the light with the system may cause a transition from the upper energy level E_2 to the lower energy level E_1 if the system is in the upper excited level. In this case it makes available energy $E_2 - E_1 = hc\tilde{\nu}_i$. Again one photon of the incident light of energy $hc\tilde{\nu}_0$ is annihilated, but in this instance one photon of higher energy $hc(\tilde{\nu}_0 + \tilde{\nu}_i)$ is simultaneously created, so that scattering of light of a higher wavenumber $\tilde{\nu}_0 + \tilde{\nu}_i$ occurs. This is an anti-Stokes line. Raman bands are characterised not by their absolute wavenumber $\tilde{\nu}_0 \pm \tilde{\nu}_i$ but by the magnitude of their wavenumber shifts $\Delta\tilde{\nu} = \pm\tilde{\nu}_i$. Such wavenumber shifts are called the *Raman shifts.* When it is necessary to distinguish between Stokes and anti-Stokes Raman bands, we define the sign of $\Delta\tilde{\nu}$ to be positive for Stokes bands and negative for anti-Stokes bands. Thus we can obtain vibrational spectra by measuring Raman scattering.

Stokes Raman scattering is measured in most studies since the intensity of anti-Stokes relative to Stokes Raman scattering decreases rapidly with increasing Raman shift. This is because anti-Stokes Raman scattering involves the transition from a populated upper energy state to a lower energy state. The thermal population of such a higher state decreases exponentially as its energy $hc\tilde{\nu}_i$ above the lower state increases.

According to the classical theory of Raman scattering [12], the condition for Raman activity is that at least one component of the polarisability tensor derivatives with respect to a normal coordinate, taken at the equilibrium position, should be non-zero. On the other hand, the corresponding condition for infrared absorption is that at least one component of the dipole moment derivatives with respect to a normal coordinate, taken at the equilibrium position, should be non-zero. The selection rule for Raman scattering is different from that for infrared absorption. Accordingly, a complete knowledge of the vibrational energy levels of a system is likely to require the study of both Raman scattering and infrared absorption. Thus infrared and Raman spectroscopies are to be regarded as complementary methods of investigating vibrational spectra.

A resonance Raman effect is expected when the wavenumber of excitation light is close to or coincident with that of an electronic transition of the system. In resonance Raman spectra the intensities of certain Raman bands are greatly enhanced relative to their off-resonance values. Thus the resonance effect greatly enhances the sensitivity of Raman spectroscopy. It is solely vibrational modes associated with the chromophore in a large molecule that can be enhanced at resonance. Thus, information relating to the chemically significant parts of macromolecules can be studied by resonance Raman spectroscopy. There have been several excellent reviews [13–16] of resonance Raman spectroscopy.

5.2.2
FT-Raman Measurement

We will briefly mention the basic concepts of FT-Raman spectrometry with a Michelson interferometer, which consists mainly of a beam splitter, a fixed mirror and a movable mirror, as shown schematically in Fig. 5.3. Laser light is incident on a sample and scattered light is collected and passed through optical filters to eliminate only the Rayleigh scattering. After the light has passed through an aperture, it is introduced into a Michelson interferometer. The Michelson interferometer is a device that can divide a beam of light into two paths and then recombine the two beams which now have a path difference. Let us consider a beam of monochromatic light of wavenumber $\tilde{\nu}$ being incident on the beam splitter. Half of the light is transmitted to the fixed mirror FM and half is reflected to the movable mirror MM. After reflection at FM and MM, the two beams are recombined at the beam splitter. Half of the light is transmitted to reach a detector D. The optical path difference x between the two beams is called the *retardation*. The intensity of the light reaching the detector becomes

$$E(x) = 2B(\tilde{\nu})(1 + \cos 2\pi\tilde{\nu}x) \qquad (5.2)$$

where $B(\tilde{\nu})$ is the observed spectrum. It consists of a constant (dc) component (the first term of Eq. (5.2)) and a modulated (ac) component (the second term of Eq. (5.2)). Only the ac component is important in spectroscopic measurements.

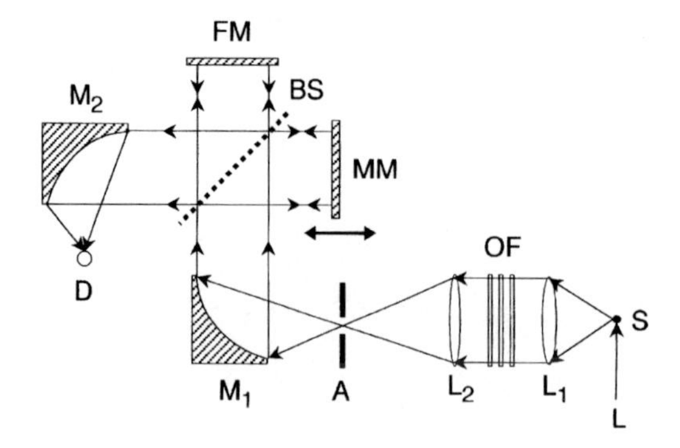

Fig. 5.3 Schematic illustration of a Michelson interferometer: L, laser light; S, sample; L_1 and L_2, lenses; OF, optical filters; A, aperture; M_1 and M_2, mirrors; BS, beam splitter; MM, movable mirror; FM, fixed mirror; D, detector

This ac component is referred to as the *interferogram* $F(x)$. When the source is a continuum, the interferogram can be represented by the following integral:

$$F(x) = \int_{0}^{\infty} 2B(\tilde{\nu}) \cos 2\pi\tilde{\nu}x d\tilde{\nu} \tag{5.3}$$

by supposing that $B(\tilde{\nu})$ is an even function: $B(-\tilde{\nu}) = B(\tilde{\nu})$. Thus, Eq. (5.3) can be expressed as follows:

$$F(x) = \int_{-\infty}^{\infty} B(\tilde{\nu}) \cos 2\pi\tilde{\nu}x d\tilde{\nu} \tag{5.4}$$

This equation is one half of a cosine Fourier transform (FT) pair; the other being

$$B(\tilde{\nu}) = \int_{-\infty}^{\infty} F(x) \cos 2\pi\tilde{\nu}x dx \tag{5.5}$$

(The properties of Fourier transforms are described in detail in references [17] and [18]). This equation means that the Raman spectrum can be obtained from the FT of the interferogram. Every point in the interferogram contains information about every wavenumber in the spectrum. This is the origin of the multiplex advantage (Fellgett's advantage) in FT spectrometry. The intensity of an observed spectrum is affected by the wavenumber-dependent response of the spectrophotometer. The latter is a composite of a number of factors: the optical filters, the beam splitter, the detector, the electronics, etc. Thus, the true intensity of the Raman spectrum $I(\tilde{\nu})$ can be obtained by correcting the intensity of the observed spectrum $B(\tilde{\nu})$ from the following equation:

$$I(\tilde{\nu}) = \frac{B(\tilde{\nu})}{C(\tilde{\nu})} \tag{5.6}$$

where $C(\tilde{\nu})$ is the instrument response function of the spectrophotometer.

The procedures for obtaining Raman spectra are schematically shown in Fig. 5.4. The bare spectrum is first obtained by the FT of the observed interferogram. We then make an intensity calibration and convert the wavenumber of the x axis to Raman shift in cm^{-1}. It is recommended [19] that the Raman shift increases from right to left. In this representation, it is easy for us to compare Raman spectra to infrared spectra. The ordinate of a Raman spectrum denotes the relative intensity of Raman scattering because it is difficult to determine the absolute intensity of Raman scattering.

In most FT-Raman interferometers, the movable mirror is scanned continuously with a velocity of v cm s^{-1}. The retardation can then be expressed as:

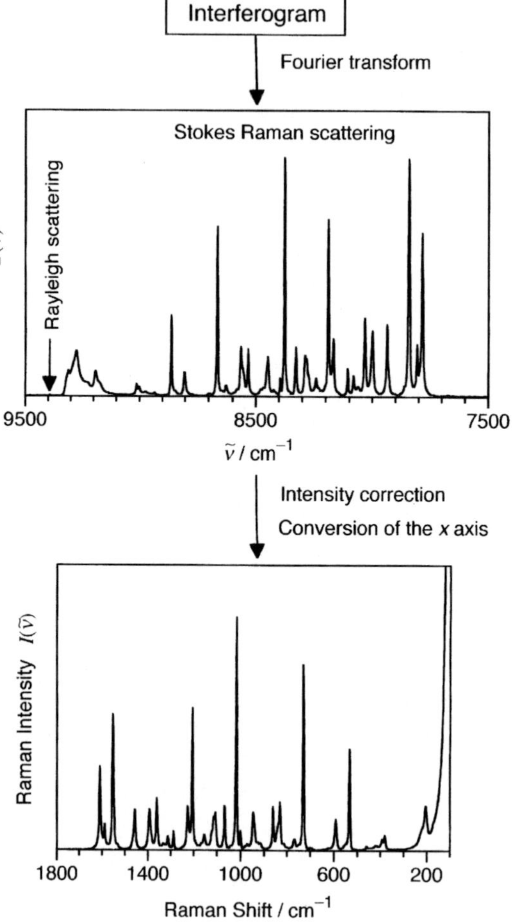

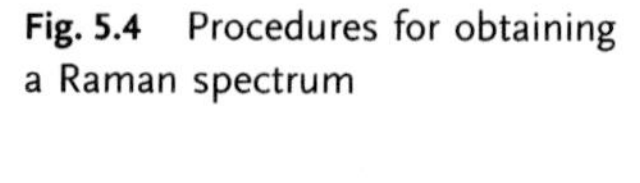
Fig. 5.4 Procedures for obtaining a Raman spectrum

$$x = 2vt \tag{5.7}$$

Thus, the interferogram is a function of time. On the other hand, any cosine wave of frequency f can be represented as follows:

$$A(t) = A_0 \cos 2\pi ft \tag{5.8}$$

where A is a general physical quantity. Thus, the relation between the audio frequency f in the interferogram and the wavenumber $\tilde{\nu}$ of light is given by

$$f = 2v\tilde{\nu} \tag{5.9}$$

For example, light of 9400 cm^{-1} is converted to an electric signal of 1880 Hz by the interferometer, provided that v is equal to 0.1 cm s^{-1}.

5.2.3
Apodisation Function and Line Shape

In Eq. (5.5), the limits of the integration are from $x=-\infty$ to $x=+\infty$. This is physically impossible. In actual measurements, the movable mirror is moved from a position $-L$ to L, where L is the maximum retardation. By analogy with Eq. (5.5), the spectrum in this case is given by the following equation:

$$B(\tilde{\nu}) = \int_{-L}^{L} F(x) \cos 2\pi\tilde{\nu}x dx = \int_{-\infty}^{\infty} U(x)F(x) \cos 2\pi\tilde{\nu}x dx \qquad (5.10)$$

where $U(x)$ is a boxcar truncation function expressed as

$$U(x) = \begin{array}{ll} 1 & x \leq |L| \\ 0 & x > |L| \end{array} \qquad (5.11)$$

It can be shown that the FT of the product of two functions is the *convolution* of the FT of each function [17, 18]. Thus, the effect of multiplying $F(x)$ by the boxcar function $U(x)$ is to yield the spectrum that is the convolution of the FT of $F(x)$ and the FT of $U(x)$. The FT of $F(x)$ is the spectrum $B(\tilde{\nu})$, while the FT of $U(x)$ is given by

$$\int_{-\infty}^{\infty} U(x) \cos 2\pi\tilde{\nu}x dx = \frac{2L \sin 2\pi\tilde{\nu}L}{2\pi\tilde{\nu}L} \equiv 2L\,\mathrm{sinc}(2\pi\tilde{\nu}L) \qquad (5.12)$$

The sinc function versus wavenumber is shown in Fig. 5.5(a2). There are side lobes (or feet) around the peak centred at 0. The sinc function intersects the wavenumber axis at wavenumbers of $n/2L$ ($n=1,2,3,\ldots$) at either side of the peak. The first intersection occurs at a wavenumber of $1/2L$. The full width at half height (FWHH) is $0.60/L$. Even when the width of a true spectral line is infinitely narrow, the line therefore shows the shape of the sinc function. The side lobes in the sinc function originate from the sudden decreases of the boxcar function at $x=\pm L$ (Fig. 5.5(a1)). We can reduce the amplitudes of the side lobes by multiplying the interferogram by a weighting function which goes gradually to zero. The suppression of the magnitude of these side lobes is known as *apodisation*. The instrument line-shape functions obtained from the FTs of the triangular and the Happ-Genzel apodisation functions are shown in Fig. 5.5(b2) and 5.5(c2), respectively. The amplitudes of the side lobes are reduced in these line-shape functions. On the other hand, the FWHHs of these line shapes become broader. In most Raman spectra containing both intense and weak bands, especially when their widths are of the same order as the instrumental resolution, strong apodisation can be applied. Parker et al. [20] have studied the effect of apodisation and finite resolution on FT-Raman spectra for the triangular and the Norton-Beer apodisation functions.

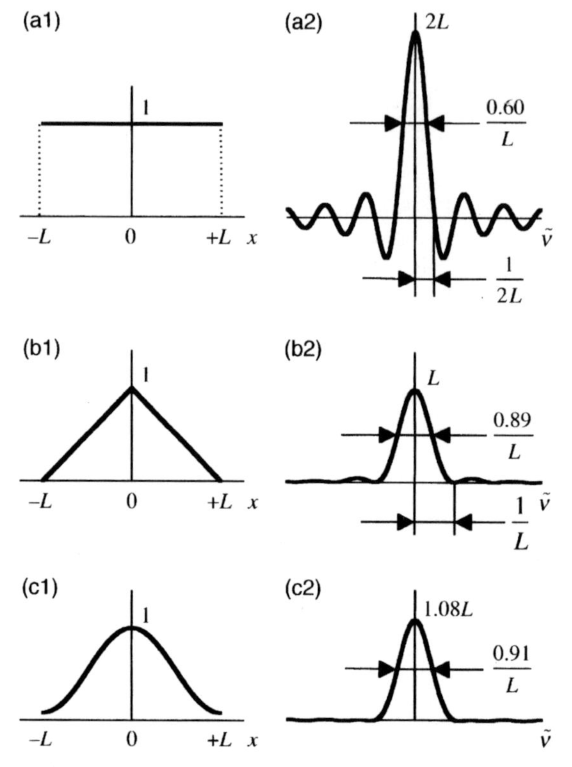

Fig. 5.5 Apodisation functions, (a1)–(c1), and line shape functions, (a2)–(c2): (a1) boxcar function; (b1) triangular function; (c1) Happ-Genzel function

5.2.4
Resolution

When the triangular function is used for apodisation, the instrument line shape is expressed by the following sinc^2 function, as shown in Fig. 5.5(b2).

$$f(\tilde{\nu}) = L\,\text{sinc}^2(\pi\tilde{\nu}L) \tag{5.13}$$

This is also the instrument line-shape function of a diffraction-limited grating monochromator. The sinc^2 function first becomes zero at $1/L$ on either side of the peak; the FWHH of the line shape function is $0.89/L$. Let us consider the case of a spectrum consisting of two lines that have equal intensity and are separated by $0.89/L$ (FWHH), $1/L$ and $2/L$, as shown in Fig. 5.6(a), 5.6(b) and 5.6(c), respectively. When the two lines are separated by $0.89/L$, these lines are just resolved, as shown in Fig. 5.6(a). We may define the *resolution* $\Delta\tilde{\nu}$ of a spectrometer by using the FWHH value, which is referred to as the *Taylor criterion*. When the two lines are separated by $1/L$, in other words, the centre of one line is at the same wave-

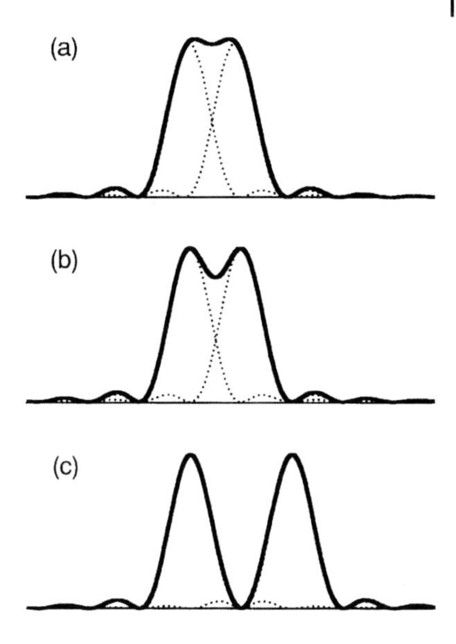

Fig. 5.6 Line shapes of a doublet separated at (a) 0.89/L, (b) 1/L and (c) 2/L

number as the first zero value of the line-shape function of the other, a 20% dip is found in this case, as shown in Fig. 5.6(b), and these two lines are considered to be resolved. We may alternatively define the resolution by this condition, which is referred to as the *Rayleigh criterion*. When we use the triangular apodisation function, under the Rayleigh criterion, the resolution is given by the following equation:

$$\Delta\tilde{\nu} = \frac{1}{L} \tag{5.14}$$

When L is 0.25 cm, the resolution is 4 cm^{-1}. The resolution becomes smaller with increasing L value.

5.2.5 Sampling Frequency

In order to compute a complete spectrum from the FT of an interferogram, it must be digitised. In dealing with discrete data, it is important to realise that we no longer know what a waveform is, except at the instants of measurement. It can be shown that a discrete FT still yields the correct spectrum, provided that we have enough data points per unit time. Any waveform that is a sinusoidal function of time or distance can be sampled unambiguously by using a sampling frequency (f_s) greater than or equal to twice the bandwidth 0 to f_{max} of the system [18]; this is known as the *Nyquist criterion*:

$$f_s \geq 2f_{max} \tag{5.15}$$

The signal may then be effectively recorded without any loss of information. The frequency $f_s/2$ is called the *Nyquist frequency*. Consider the case of the spectrum in which the highest wavenumber is $\tilde{\nu}_{max}$ cm^{-1}. The frequency of the cosine wave in the interferogram corresponding to $\tilde{\nu}_{max}$ is $f_{max} = 2v\tilde{\nu}_{max}$ Hz from Eq. (5.9). According to the Nyquist criterion, the interferogram must be digitised at a frequency greater than or equal to $4v\tilde{\nu}_{max}$ Hz or once every $(4v\tilde{\nu}_{max})^{-1}$s. This is equivalent to digitising the signal at retardation intervals of $(2\tilde{\nu}_{max})^{-1}$cm.

When we sample an analogue interferogram at regular, discrete retardations, we effectively multiply the interferogram by a repetitive impulse function. The repetitive impulse function is an infinite series of Dirac delta functions spaced at an interval Δx, as shown in Fig. 5.7(a). That is:

$$\text{Ш}(x) = \sum_{n=-\infty}^{\infty} \delta(x - n\Delta x) \tag{5.16}$$

where n is integer, δ is the Dirac delta function and Ш is the "shah" function, also known as the "comb" function from its appearance. The FT of the shah function is another shah function of period $1/\Delta x$:

$$\text{Ш}(\tilde{\nu}) = \frac{1}{\Delta x} \sum_{n=-\infty}^{\infty} \delta(\tilde{\nu} - n\Delta\tilde{\nu}) \tag{5.17}$$

n is integer, in which:

$$\Delta\tilde{\nu} = \frac{1}{\Delta x} \tag{5.18}$$

This shah function is shown in Fig. 5.7(b). The FT of the digitised interferogram, the product of the analogue interferogram $F(x)$ and the shah function Ш(Δx), is the convolution of the spectrum $B(\tilde{\nu})$ and the other shah function Ш$(\tilde{\nu})$. The effect of this convolution is to repeat the spectrum obtained from an analogue inter-

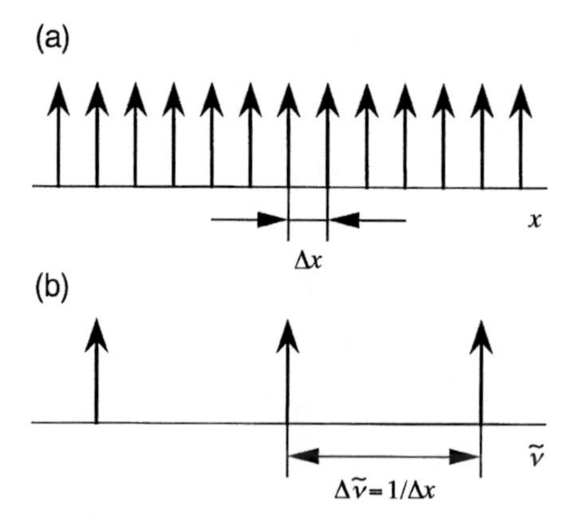

Fig. 5.7 (a) Shah function of period Δx and (b) its Fourier transform, another shah function of period $1/\Delta x$

ferogram (Fig. 5.8(a)) ad infinitum, as shown in Fig. 5.8(b). If the spectrum ranges from 0 to $\tilde{\nu}_{max}$, the transformed shah function must have a period of at least $2\tilde{\nu}_{max}$. In other words:

$$\Delta x \leq \frac{1}{2\tilde{\nu}_{max}} \tag{5.19}$$

If the sampling intervals were increased so that the signal was sampled at an interval larger than $1/2\tilde{\nu}_{max}$, the transformed spectra would overlap and spectral features would appear at incorrect wavenumbers. This phenomenon is known as *aliasing* or *folding*.

Most commercial FT spectrometers use a He-Ne laser operating at about 633 nm (ca. 15800 cm^{-1}) to trigger data collection. The laser light passing through the interferometer generates a cosine wave. In each cycle, the signal passes through zero twice. By using these zero crossing points, we can collect equidistant data points. In spectral measurements of the mid-infrared region between 4000 and 400 cm^{-1}, the interferogram is sampled at every other zero crossing (λ sampling) and the retardation interval is 633 nm, which corresponds to the maximum wavenumber 7900 cm^{-1}. In spectral measurements of the NIR region between 10000 and 4000 cm^{-1}, the interferogram is sampled at every zero crossing ($\lambda/2$ sampling). In this case, the maximum wavenumber is 15800 cm^{-1}. When we observe Stokes Raman lines with the excitation wavelength of 1064 μm, the highest wavenumber is 9395 cm^{-1}. Thus, we must choose $\lambda/2$ sampling.

5.2.6
Intensity Calibration

It is necessary to calibrate the intensity of an observed FT-Raman spectrum because the observed Raman spectrum $B(\tilde{\nu})$ is affected by the spectral response of the spectrophotometer. We can calibrate the intensity of the observed Raman spectrum by the use of Eq. (5.6). The instrument response function $C(\tilde{\nu})$ in Eq. (5.6) can be obtained from the spectrum of a precisely known light distribution, for ex-

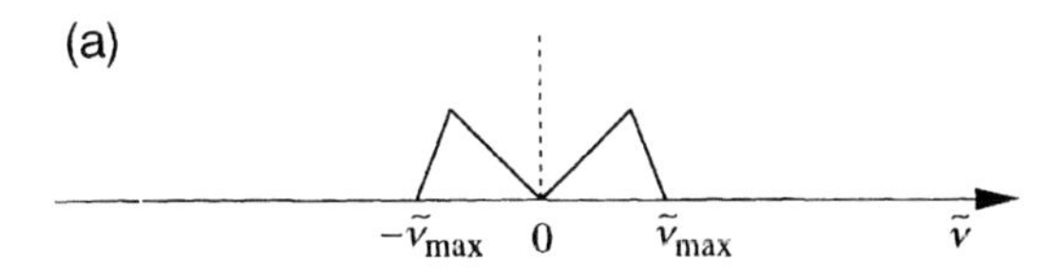

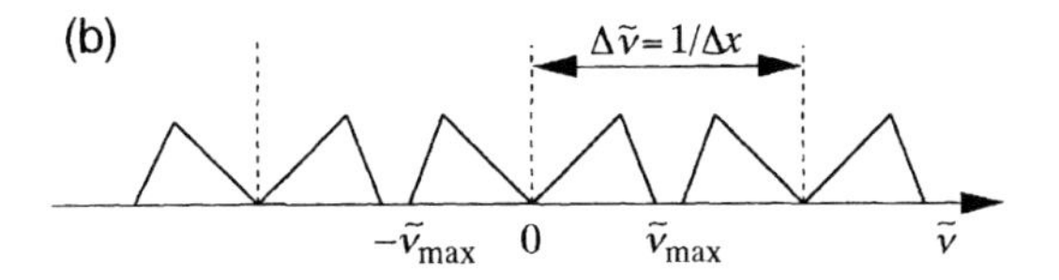

Fig. 5.8 Schematic illustrations of (a) a spectrum obtained from an analogue interferogram and (b) spectra obtained from a discrete interferogram

ample blackbody emitters, standard lamps, fluorescence spectra or rotational Raman spectra. The method described here uses a blackbody emitter [21]. The brightness of a blackbody emitter as a function of photon number is given by:

$$dL_{\mathrm{p}}(\tilde{\nu}, T) = L_{\mathrm{p}}(\tilde{\nu}, T)d\tilde{\nu} = 2c\tilde{\nu}^2 \left[\exp\left(\frac{hc\tilde{\nu}}{kT}\right) - 1\right]^{-1} d\tilde{\nu} \tag{5.20}$$

where c is the velocity of light in vacuum; h, Planck's constant; k, Boltzmann's constant; T, radiation temperature and $hc/k = 1.4388$ K cm. The instrument response function $C(\tilde{\nu})$ can be determined from the observed spectrum of the blackbody emitter $R(\tilde{\nu})$ by using the following equation:

$$C(\tilde{\nu}) = K\frac{R(\tilde{\nu})}{L_{\mathrm{p}}(\tilde{\nu}, T)} \tag{5.21}$$

where K is an arbitrary constant. Therefore, the Raman spectrum $I(\tilde{\nu})$ can be obtained from

$$I(\tilde{\nu}) = \frac{B(\tilde{\nu})}{C(\tilde{\nu})} = \frac{B(\tilde{\nu})L_{\mathrm{p}}(\tilde{\nu}, T)}{KR(\tilde{\nu})} \tag{5.22}$$

where an appropriate value of K is chosen according to the relative intensity of the observed Raman scattering.

5.3 Instrumentation

An FT-Raman spectrophotometer is shown schematically in Fig. 5.9. A sample is irradiated with the 1.064 μm laser line provided from a continuous-wave (CW) Nd:YAG laser. This laser is available in both a lump-pumped and a diode-laser-pumped configuration. (The diode-laser-pumped Nd:YAG laser has the advantages of air cooling, lower noise amplitude and smaller physical size.) Even with 1.064 μm excitation, certain materials emit luminescence so we cannot observe their Raman spectra. Longer wavelength excitation may reduce the luminescence background. The 1.32 μm line [22, 23] and the 1.339 μm line [24] which are provided by a Nd:YAG laser have also been used as excitation source.

Rayleigh- and Raman-scattering light is collected with a 90° off-axis parabolic mirror in the back-scattering configuration. Of course, we can construct the optical system for collecting Raman scattering with lenses instead of mirrors. We can also adopt the forward-scattering and the 90°-scattering configuration. Collected light is passed through optical filters (F_2 in Fig. 5.9) to eliminate only the Rayleigh scattering. Two or three holographic notch filters are used for Rayleigh line filtering, because their cut-off is sharp and the absorbance at the incident wavenumber is extremely high. The transmission spectrum of a holographic Super-

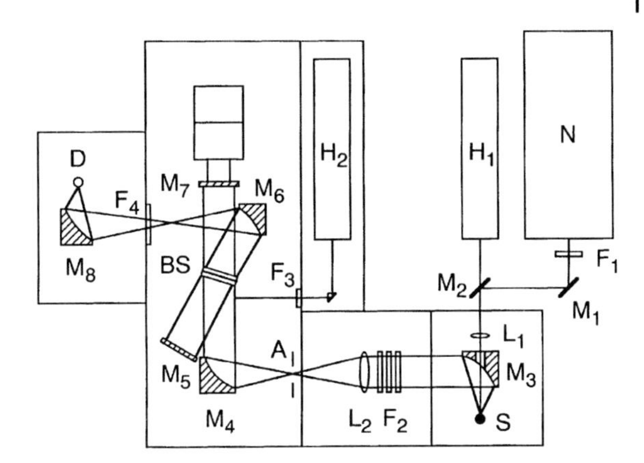

Fig. 5.9 Schematic illustration of an FT-Raman spectrophotometer. N, Nd:YAG laser; H_1 and H_2, He-Ne lasers; F_1–F_4, optical filters; M_1–M_8, mirrors; BS, beam splitter; L_1 and L_2, lenses; A, aperture; D, detector; S, sample

notch® filter (Kaiser Optical Systems) is shown in Fig. 5.10. The detectable limit of the low wavenumber in Raman spectra depends on the optical filters. The development of the holographic notch filtering technique lowers the limit to about 50 cm^{-1}, although the limit also depends on the state of the sample.

After the light has been passed through an aperture, it is introduced into a Michelson interferometer. The output light from the interferometer is collected onto an indium-gallium-arsenide (InGaAs) detector. In most cases the detector is operated at liquid-nitrogen temperature to reduce noise and to maximise sensitivity. The spectral response curves of an InGaAs detector are shown in Fig. 5.11(a). It should be noted that at 77 K, at which point sensitivity is maximised, the Raman spectral region is limited to lower than about 3000 cm^{-1}. When the detector is operated at room temperature, the upper limit goes up slightly and the Raman spectral region is, in this case, lower than 3600 cm^{-1}. The noise-equivalent power of the InGaAs detector is less than 10^{-14} W $Hz^{-1/2}$. The extremely high feedback resistors employed in the first stage preamplification, coupled with any stray capacitance, serve to reduce the frequency response of the detector, which is shown in

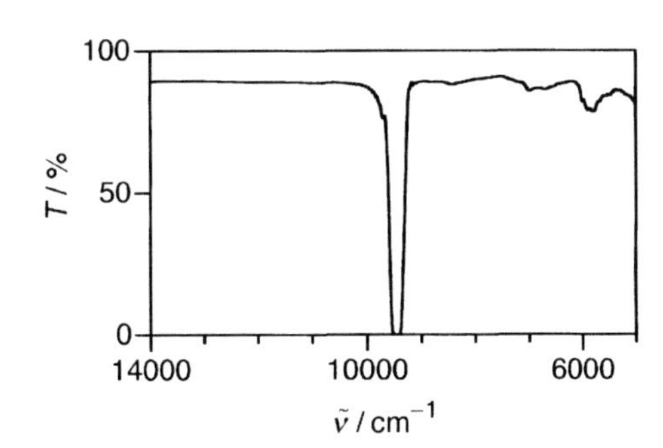

Fig. 5.10 Transmission spectrum of a holographic Supernotch® filter from Kaiser Optical Systems

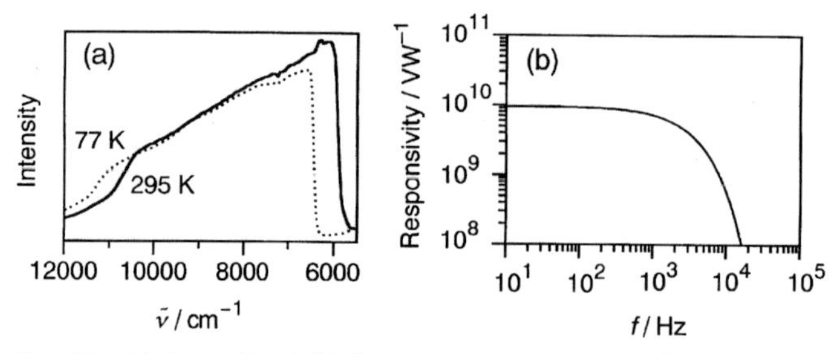

Fig. 5.11 (a) Spectral and (b) frequency response curves of an InGaAs detector

Fig. 5.11(b). Mirror velocities of 0.10 cm s^{-1} or less are required in order to keep the Fourier modulation frequencies within the detector's response range. Instead of an InGaAs detector, a germanium (Ge) detector is also available. The noise-equivalent power of the Ge detector is about 10^{-15} W Hz$^{-1/2}$, and the spectral range of the detector extends out to 3400 cm^{-1} Stokes shift at 77 K.

5.4
Applications

In order to demonstrate the high wavenumber accuracy of FT-Raman spectrometry, the NIR FT-Raman spectrum [25] of liquid indene, which is a wavenumber standard, is shown in Fig. 5.12. The observed wavenumbers are listed in Table 5.1, with those in the literature [26–28]. The Raman shift of each band is calculated by subtracting the wavenumber of the band from that of the Rayleigh line. The resolution is 2 cm^{-1}, data points are added by the zero-filling method and exist every 0.06 cm^{-1}. Thus, errors arising from readings are less than 0.12 cm^{-1}.

Spectral subtractions, which have become quite commonplace in FT-IR spectrometry, are useful in qualitative or quantitative analysis beyond our expectation. Walder

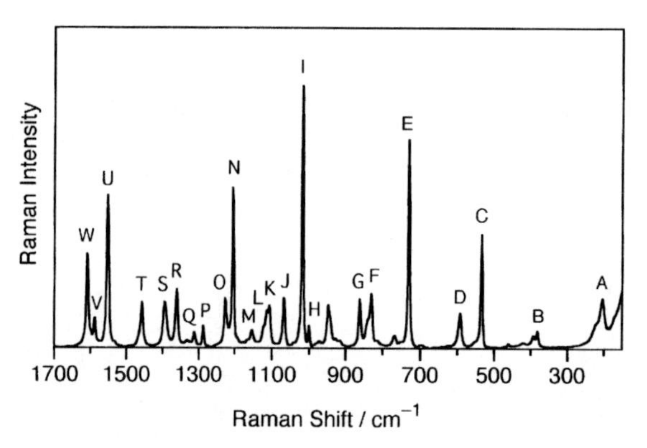

Fig. 5.12 FT-Raman spectrum of liquid indene: resolution, 2 cm^{-1}; data point, 0.06 cm^{-1}

Tab. 5.1 Observed wavenumbers of the Raman spectrum of liquid indene

Band	Observed wavenumber [cm^{-1}]			
	Raman			Infrared
	Ref. 25	Ref. 26	Ref. 27	Ref. 28
A	203.88	203.5±1	203.8±0.3	–
B	381.42	381.3±0.5	381.8±0.3	–
C	533.95	533.8±0.5	534.4±0.4	–
D	591.75	591.9±1	592.3±0.4	592.1
E	730.30	730.4±0.3	730.7±0.3	–
F	830.58	830.4±0.5	831.4±0.3	830.5
G	861.31	861.3±0.5	861.7±0.3	861.3
H	1000.71	1001.0±0.5	1001.5±0.2	–
I	1018.67	1018.6±0.5	1019.0±0.2	1018.5
J	1067.60	1067.6±0.5	1068.1±0.3	1067.7
K	1106.59	1106.7±1	1107.5±0.2	–
L	1113.28	1112.9±1	1112.9±0.3	–
M	1153.54	1153.5±1	1153.9±0.3	–
N	1204.89	1205.2±0.3	1205.3±0.3	1205.1
O	1226.10	1226.4±0.5	1226.6±0.4	1226.2
P	1287.69	1287.5±0.5	1288.1±0.2	1288.0
Q	1312.34	1312.5±1	1312.8±0.2	1312.4
R	1361.03	1361.0±0.5	1361.5±0.4	1361.1
S	1393.70	1393.7±1	1393.2±0.4	1393.5
T	1457.04	1457.1±1	1457.5±0.4	1457.3
U	1552.92	1553.2±1	1552.2±0.5	1553.2
V	1587.87	1587.8±1	1588.4±0.4	1587.5
W	1609.63	1609.7±0.5	1609.7±0.3	1609.8

and Smith [29] have evaluated the reproducibility of FT-Raman spectra and the ability to subtract spectra in the case of a mixture of *p*-xylene, *o*-xylene and *m*-xylene with overlapping bands. Detection limits for subtractions are at the 1% level. The difference FT-Raman technique is one of the advantages of the FT-Raman spectrometry.

5.4.1
Various Materials

It has been demonstrated in a large number of articles [3–10] that the nonresonant NIR FT-Raman method gives high-quality spectra of many materials such as synthetic polymers, inorganic compounds, biological materials, catalysts, etc. In this section, NIR FT-Raman measurements with preresonant or rigorous resonant conditions are presented.

The NIR FT-Raman technique has proved advantageous for several problems in ordinary synthetic polymers: polyethylene, polystyrene, polyimide, etc. Amongst organic polymers, a group of conjugated polymers, shown in Fig. 5.13, shows high conductivities when doped with an electron acceptor or an electron donor. These polymers are called *conducting polymers* and have attracted much attention from many researchers in various fields [30]. They are also expected to be materials for light emitting diodes. The properties of these polymers have been explained in terms of electronic excitations, accompanied by structural changes (such as solitons, polarons and bipolarons) [30, 31]. Vibrational spectroscopy is useful in studying the structures of conjugated polymers. In particular, NIR FT-Raman spectroscopy gives information about the electronic excitation in doped states [32–34], because 1.064 and 1.3 μm laser lines are to be found within the electronic absorptions of doped conjugated polymers, which are concomitant with the appearance of high conductivity.

The NIR FT-Raman spectra of neutral and BF_4^--doped polythiophenes [35] are shown in Fig. 5.14(a) and 5.14(b), respectively. Neutral polythiophene films contain all *s-trans* sequences of thiophene rings linked at the α- and α'-positions with a distribution of sequential conjugation lengths [36]. The Raman bands of doped polythiophene have been analysed on the basis of the data of model compounds; the radical cations and the dications of α-oligothiophenes are models of a positive polaron and a positive bipolaron, respectively. As a result, the observed Raman bands of doped polythiophene have been attributed to positive polarons [35]. It is considered that a positive polaron can move on a polymer chain with structural changes and positive charge; in other words, a positive polaron is a charge carrier. The NIR FT-Raman spectra of doped states of poly(*p*-phenylenevinylene) [37–40] and poly(*p*-phenylene) [41, 42] have been explained in terms of polarons and bipolarons.

Polyacetylene has a structure of alternating C=C and C–C bonds. The electrical properties of doped polyacetylene have been discussed in terms of solitons and polarons. The 1.32 μm excited FT-Raman spectra of *trans*-polyacetylene doped with Na at various concentrations are shown in Fig. 5.15 [23]. At low Na concentrations (Fig. 5.15(b)–(e)) small but significant spectral changes are observed, whereas these spectral changes are not observed with visible laser lines. This is because the 1.32 μm line is located within the doping-induced electronic absorption, and the Raman bands arising from electronic excitations generated by doping appear by the reso-

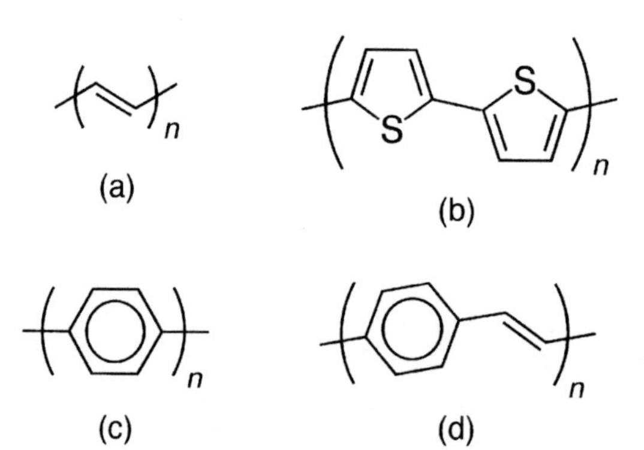

Fig. 5.13 Structures of conjugated polymers: (a) *trans*-polyacetylene; (b) polythiophene; (c) poly(*p*-phenylene); (d) poly(*p*-phenylenevinylene)

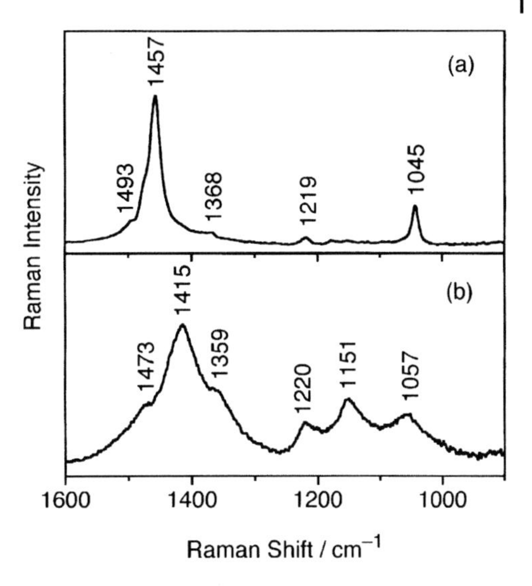

Fig. 5.14 FT-Raman spectra of (a) a neutral polythiophene film and (b) a $BF4^-$-doped polythiophene film [35]. Excitation wavelength is 1.064 μm

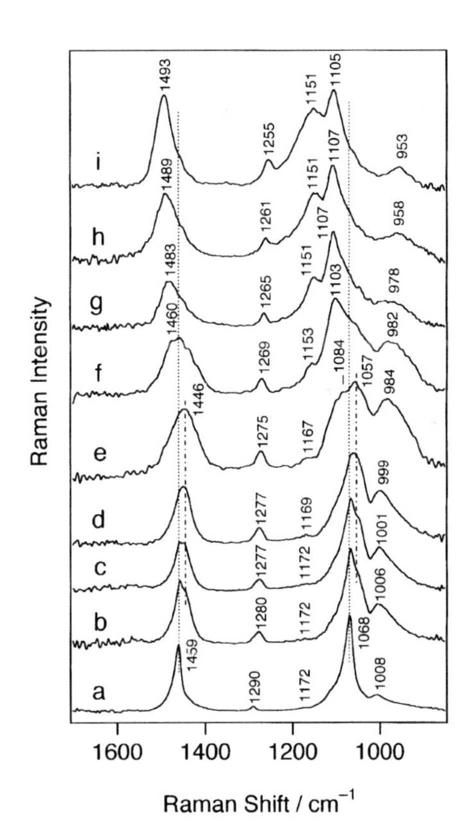

Fig. 5.15 FT-Raman spectra of (a) a neutral *trans*-polyacetylene film and (b–i) Na-doped *trans*-polyacetylene films at various dopant concentrations. The Na concentration increases from b to i. (Reproduced with permission from ref. [23])

nance Raman effect. At high Na concentrations (Fig. 5.15(f)–(i)), large spectral changes are observed. The observed wavenumbers of the C=C and C–C stretching vibrations of heavily doped *trans*-polyacetylene are higher than those of the corresponding bands of neutral *trans*-polyacetylene, whereas the opposite is the case for lightly doped *trans*-polyacetylene. These differences may be associated with the semiconductor-metal transition at a Na concentration of about 6 mol% per CH unit.

It has been considered difficult to obtain the NIR FT-Raman spectra of doped polymers. This is because the apparent intensities of the Raman bands of doped polymers are weaker than those of neutral polymers. Since a doped polymer has an intense absorption band in the NIR region (resonance Raman case), absorption of the excitation light and reabsorption of the Raman scattering reduce the intensity of the Raman scattering. Since doped polymers have electronic absorptions in the NIR region, they are heated by the irradiation of the 1.064 and 1.32 μm laser light. Laser-induced heating causes sample degradation or thermal background radiation in Raman spectra. Thus, the laser power at the sample stage must be reduced. However, high-quality Raman spectra have been obtained by accumulation even with low excitation laser power.

High-quality NIR FT-Raman spectra of biological materials such as proteins, nucleic acids and lipids have been obtained in the solid state. On the other hand, the NIR FT-Raman intensities from aqueous solutions of biological materials are weak, because the concentrations of the samples are not high. Thus, various types of multipass cells [43] are used for obtaining high-quality Raman spectra of these materials. NIR FT-Raman spectrometry has potential in biomedical applications [44] because fluorescence rejection is important in this field.

Bacteriochlorophyll-*a* (BChl) (Fig. 5.16), which is contained in most photosynthetic bacteria, functions as an antenna pigment in light-harvesting systems and plays an important role in charge separation and following the formation of membrane potential in a photosynthetic reaction centre. Although most photosynthetic bacteria contain the pigments of carotenoids and bacteriochlorophylls, a blue-green mutant of *Rhodobacter sphaeroides* does not contain carotenoids and only has one type of light-harvesting pigment-protein complex (B870 complex). The NIR FT-Raman spectrum of chromatophores from a blue-green mutant of *R. sphaeroides* is shown in Fig. 5.17 [45]. In this spectrum, only the Raman bands due to BChl are observed, although proteins are the major species in the cell. The photosynthetic bacteria show the Soret, Q_x and Q_y electronic absorptions due to BChl. The Soret band is observed in the visible region, and the Q_x and Q_y bands in the NIR region. The spectral features of the 1.064 μm excited Raman spectrum are quite different from those of the Soret and Q_x resonance-Raman spectra; several bands in the wavenumber region below 1200 cm^{-1} are particularly enhanced. Thus, it is considered that the NIR FT-Raman spectrum is pre-resonant with the Q_y absorption. The Q_y resonance-Raman spectra have been measured by using excitation wavelengths in the region from 750 to 920 nm [46, 47]. Since these wavelengths are almost rigorously resonant with the Q_y band, the quality of the spectra obtained is not high due to interference from strong luminescence. The Raman bands due to C=O stretching vibrations of the 9-keto and 2-acetyl groups appear at 1665 and

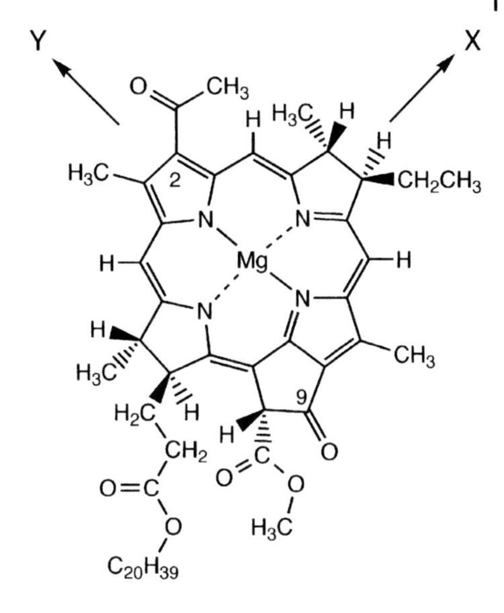

Fig. 5.16 Structure of bacteriochlorophyll-*a* (BChl)

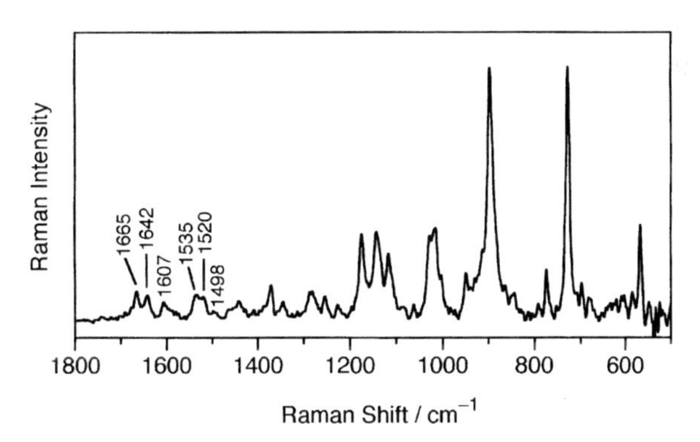

Fig. 5.17 FT-Raman spectrum of chromatophores from a blue-green mutant of *Rhodobacter sphaeroides* [45]

1642 cm^{-1}, respectively, in Fig. 5.17. Both the observed wavenumbers are 10–20 cm^{-1} lower than the values expected for the corresponding free C=O stretches. These downshifts indicate that each C=O group is involved in hydrogen bonding. The correlation between band positions and the coordination number of the central magnesium can be deduced from the C=C stretching bands observed in the region between 1620 and 1490 cm^{-1}. The band positions at 1607, 1520 and 1498 cm^{-1} in Fig. 5.17 lead to the conclusion that the BChl in the B870 light-harvesting complex of this mutant is 5-coordinate. The 1535 cm^{-1} band is not sensitive to the coordination number. Information about the structure of the B870 complex has been obtained from analysis of the FT-Raman spectrum. NIR FT-Raman spectra have been reported for living cells of photosynthetic bacteria [48].

5.4.2
Double Modulation Measurements

Even with NIR excitation, a broad background is superimposed on Raman bands in some cases. Raman band shapes are distorted and errors in the band position must be considered. This broad background arises from sample heating induced by laser irradiation. When the sample absorbs at the wavenumber of the excitation laser line (1.064 μm), it emits laser-induced thermal radiation. Similar backgrounds are also observed from samples at high temperature. Fuchs et al. [49] and Bennett [50] have demonstrated that the thermal background radiation can be removed by the double modulation method.

The excitation laser intensity $L(t)$ is sinusoidally modulated as follows:

$$L(t) = L_0\{1 + \cos(2\pi f_m t + \phi)\} \tag{5.23}$$

where L_0 is the average laser power. The modulated interferogram $F_m(x)$ can be written by

$$F_m(x) = \int_{-\infty}^{\infty} B(\tilde{\nu}) \cos 2\pi\tilde{\nu}x d\tilde{\nu} + \frac{1}{2}\int_{-\infty}^{\infty} B(\tilde{\nu}) \cos\{2\pi(\tilde{\nu}_m + \tilde{\nu})x + \phi\} d\tilde{\nu} + \frac{1}{2}\int_{-\infty}^{\infty} B(\tilde{\nu}) \cos\{2\pi(\tilde{\nu}_m - \tilde{\nu})x + \phi\} d\tilde{\nu} \tag{5.24}$$

where $B(\tilde{\nu})$ is the spectrum obtained with the average laser power, and

$$\tilde{\nu}_m = \frac{f_m}{2v} \tag{5.25}$$

The Raman spectra obtained from the modulated interferogram from a continuous-scan interferometer are schematically shown in Fig. 5.18, in comparison with the spectrum obtained from the unmodulated interferogram. The Raman spectrum obtained with average laser power is located around the incident wavenumber $\tilde{\nu}_0$ (9395 cm^{-1}) that corresponds to the wavenumber of the Rayleigh scattering, whereas the modulated interferogram gives the Raman bands shifted by $\pm\tilde{\nu}_m$. When the Raman spectral region ranges from 0 to $\Delta\tilde{\nu}_{max}$, in order to avoid the overlap of the spectrum around $\tilde{\nu}_0$ and the modulated spectrum, the following equation must be satisfied.

$$\tilde{\nu}_m \geq 2\Delta\tilde{\nu}_{max} \tag{5.26}$$

Then, from Eq. (5.25) the following equation can be derived.

$$f_m \geq 4v\Delta\tilde{\nu}_{max} \tag{5.27}$$

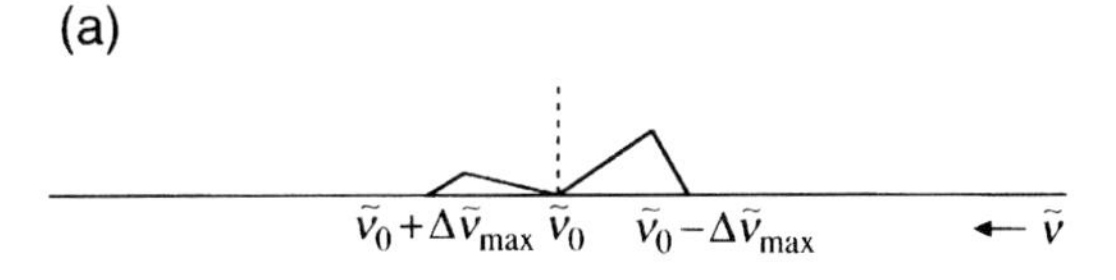

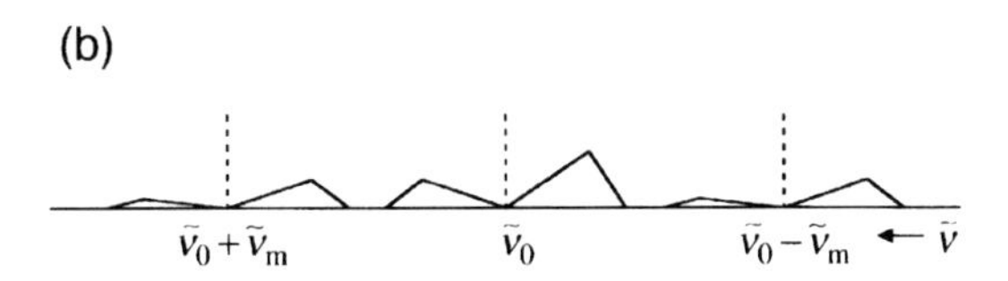

Fig. 5.18 Graphical illustration of the Raman spectra obtained from a modulated interferogram (b) and the same spectrum obtained from an unmodulated interferogram (a)

When the anti-Stokes bands are completely blocked by optical filters, the modulation frequency can be reduced to half of the value determined by Eq. (5.27).

The modulated Raman spectrum can be observed by the following three techniques. Firstly, we can obtain the spectrum by taking the FT of the modulated interferogram directly (side-band detection). Secondly, the modulated interferogram can be demodulated first by using a lock-in amplifier and then transformed (phase-sensitive detection). Thirdly, the phase of modulation ϕ can be kept constant relative to the interferogram by using a He-Ne laser that is fed through the interferometer (phase-locked excitation technique) [49]. The basic idea of the third method is to cause a phase-sensitive excitation instead of phase-sensitive detection.

A modulated laser will generate modulated Raman scattering because Raman scattering occurs instantaneously. On the other hand, a thermal process with a response time significantly slower than the period of the modulation will not depend on the modulation. It follows that the modulation technique can separate the Raman spectrum and the thermal background radiation. Fuchs et al. [49] have observed weak photoluminescence of ZnS:Ni by the phase-locked excitation technique. Bennett [50] has obtained a high-quality Raman spectrum of polystyrene by the side-band detection and the phase-sensitive detection methods.

The double modulation technique can be applied to dynamic studies of various materials under an external stimulus such as stretching, photoexcitation, voltage, etc. as well as to the removal of thermal background radiation. Two-dimensional Raman spectroscopy is feasible as a complementary method of two-dimensional infrared spectroscopy.

5.4.3 Pulsed Excitation – Synchronous Sampling

FT-Raman measurements with a pulsed excitation source have recently been developed. Cutler et al. [51], Cutler and Petty [52] and Bennett [53] have demonstrated that the use of pulsed lasers in NIR FT-Raman spectrometry can give a significant increase in the signal-to-noise ratio (SNR) over CW measurements made by using the same average laser power.

The principle of a Raman measurement with pulsed excitation [51] is illustrated in Fig. 5.19. In the Q-switched pulsed mode (pulse width, ca. 200 ns), the excitation laser is triggered immediately prior to the analogue-to-digital converter (ADC) that is digitising the interferogram while the instrument is scanning. A detector signal consists of a set of narrow pulses broadened by the response time of the detector-preamplifier system. The response of the Ge-detector-preamplifier system to a pulse of about 200 ns duration has a width of about 1 μs. By adjusting the ADC sampling point to coincide with the signal maximum, a digitised interferogram is produced whose envelope is identical to that recorded with CW excitation. In this measurement, the pulsed excitation or the timing of Raman measurement is synchronous with the ADC sampling. Thus, this measurement is one of *synchronous* sampling techniques.

An improvement in SNR is expected for pulsed excitation when the same average laser powers are used for the pulsed and CW experiments. In the pulsed mode, the amplitude of the interferogram arising from Raman scattering is inversely proportional to the time constant of the detector-preamplifier system and can be significantly enhanced. Let us assume that the ADC sampling and the repetition frequency of the pulsed laser is 3 kHz, and that the Raman signal from the detector-preamplifier system is 1 μs wide. Then, if in the pulsed or CW mode the sample receives the same number of photons per integration period, the peak

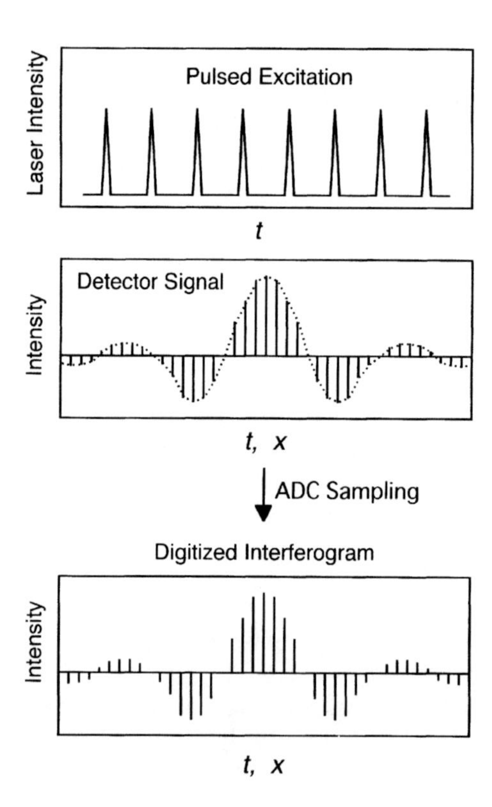

Fig. 5.19 Schematic diagram showing the principle of a synchronous FT-Raman measurement with pulsed excitation

level of the Raman signal in the pulsed mode will be, potentially, about 333 times larger than that in the CW mode. On the other hand, the detector-preamplifier system must have an increased bandwidth, which is unfortunately followed by an increase in noise, in order to cope with the faster signals arising from the pulsed operation. Experimentally, a significant SNR improvement (typically two to four times) has been obtained for pulsed experiments [51–53].

The use of pulsed excitation is effective in removing thermal background radiation [51–53]. When the sample is excited with a pulsed laser Raman scattering occurs instantaneously while thermal radiation is emitted later because of the finite time of thermal conduction. When we take the signal in the course of the pulse duration, there will be a relative suppression of the thermal background radiation in favour of the Raman scattering. Fig. 5.20(A) shows the FT-Raman spectrum of $CaSO_4$ at 140 °C. A thermal background is observed. When the sample is excited by a pulsed laser, the thermal background has been significantly reduced, as shown in Fig. 5.20(B). It seems that the major problem when operating with pulsed lasers is the pulse-to-pulse amplitude stability, which is normally poor. Cutler and Petty [52] have demonstrated the feasibility of ratioing to correct for laser pulse-to-pulse energy fluctuations.

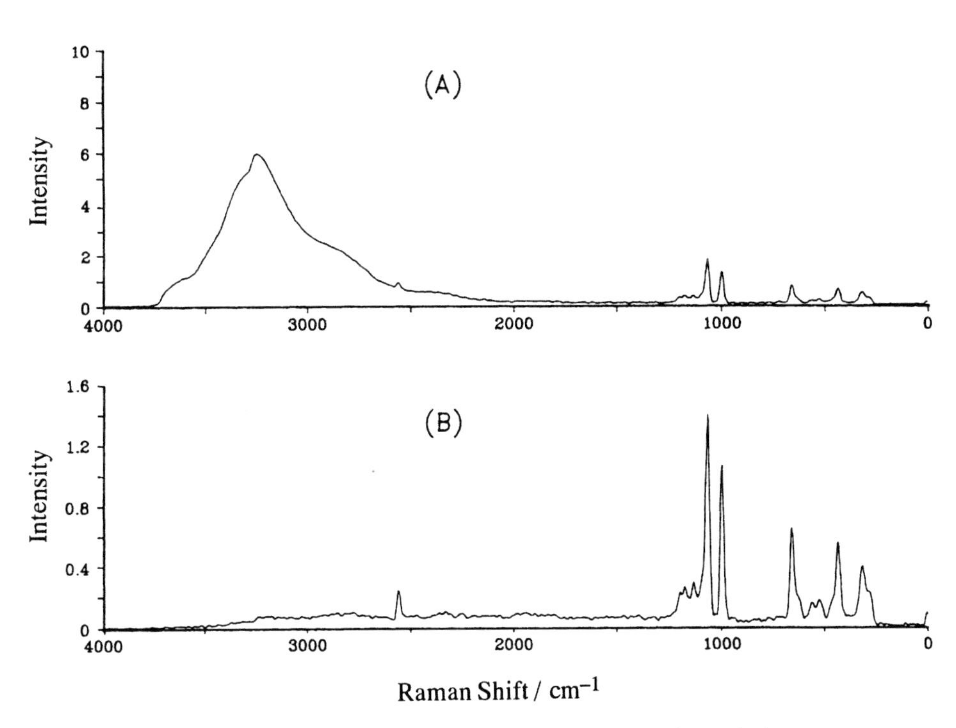

Fig. 5.20 FT-Raman spectra of $CaSO_4$ at 140 °C with (A) CW and (B) pulsed excitation. The average powers of (A) and (B) are 22.5 mW. The repetition of the pulsed excitation is 3 kHz. (Reproduced with permission from Ref. [51])

In infrared spectrometry, a step-scan interferometer has recently been developed and is widely used for dynamic measurements. Jas et al. [54] have reported FT-Raman measurements with a step-scan interferometer and a pulsed laser. In the step-scan mode, the retardation is kept fixed while the Raman signal is recorded. The movable mirror is fixed at one sampling position while data are collected by the ADC; after the data at this position have been collected, the movable mirror is translated to the next sampling position and held at that position for data collection and so on. In these measurements, the timing of Raman measurements is synchronous with the ADC sampling. Thus the method is one of synchronous sampling techniques. In the measurements on a step-scan interferometer, the repetition rate of the pulsed laser is decoupled from the scanning rate of the interferometer, that is, the velocity of the moving mirror, in contrast to an asynchronous method described in the next section.

5.4.4
Pulsed Excitation – Asynchronous Sampling

Masutani et al. [55] have proposed an *asynchronous* sampling technique for time-resolved FT-IR spectrometry, while Sakamoto et al. [56] have demonstrated the feasibility of asynchronous FT-Raman spectrometry and the suppression of thermal background radiation by using a Q-switched laser. The principle of the asynchronous Raman measurement with pulsed excitation is illustrated in Fig. 5.21. A sample is excited with a pulsed laser whose repetition frequency is f_s Hz; this means that the sampling interval is $1/f_s$ s. A detector signal (conceptually discrete interferogram) is fed into a boxcar integrator (or a sample and hold circuit), and then the output of the boxcar integrator is passed through a low-pass filter. The function of the low-pass filter is to remove higher frequencies (potential-aliased signals) which originate from the use of pulsed excitation (sampling). In other words, the discrete interferogram is converted into the analogue one by convoluting it with the impulse response function obtained from the FT of the frequency-response function of the low-pass filter. In order to recover the analogue interferogram without any loss of information, the Nyquist criterion requires that the frequency f_s is at least twice the bandwidth 0 to f_{max} of the interferogram (Eq. (5.15)). The relationship between the audio frequency in the interferogram and the wavenumber of light is given by Eq. (5.9). Thus, the following equation can be derived.

$$f_s \geq 4v\tilde{\nu}_{max} \tag{5.28}$$

where

$$\tilde{\nu}_{max} = \frac{f_{max}}{2v} \tag{5.29}$$

When $\tilde{\nu}_{max}$ is 9400 cm^{-1} and v is 0.1 cm s^{-1}, f_s should be set at greater than 3.76 kHz. The recovered analogue interferogram is digitised by the ADC in the

Fig. 5.21 Schematic diagram showing the principle of the asynchronous FT-Raman measurement with pulsed excitation

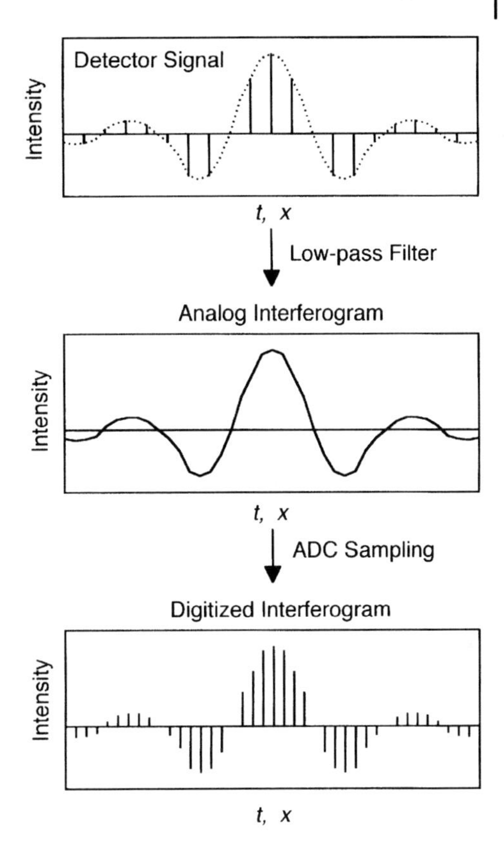

normal way. The spectrum can be obtained by the FT of the digitised interferogram. In this measurement, the pulsed excitation is not synchronised with the ADC sampling. Asselin and Chase [57] have examined experimental conditions (repetition rate, pulse width, laser power, low-pass filter setting, etc.) of asynchronous FT-Raman measurements. They have observed undefined noise sources which limit signal averaging and the achievable SNR. Probably, those noises come from the incomplete removal of higher frequencies by the low-pass filter.

5.4.5 Time-Resolved Measurements

Jas et al. [57–59] have demonstrated the feasibility of picosecond time-resolved FT-Raman measurements based on a step-scan spectrometer with a pulsed laser. The schematic diagram of their system is shown in Fig. 5.22. Raman spectra are observed with pulses of 100 ps duration at 1.064 μm provided by a mode-locked and Q-switched (at 2 kHz) Nd:YAG laser. Laser pulses of 355 nm, obtained from the third harmonic generation of the 1.064 μm pulses, are used for the generation of electronically excited states. The time delay of 1.064 μm probe pulses with respect

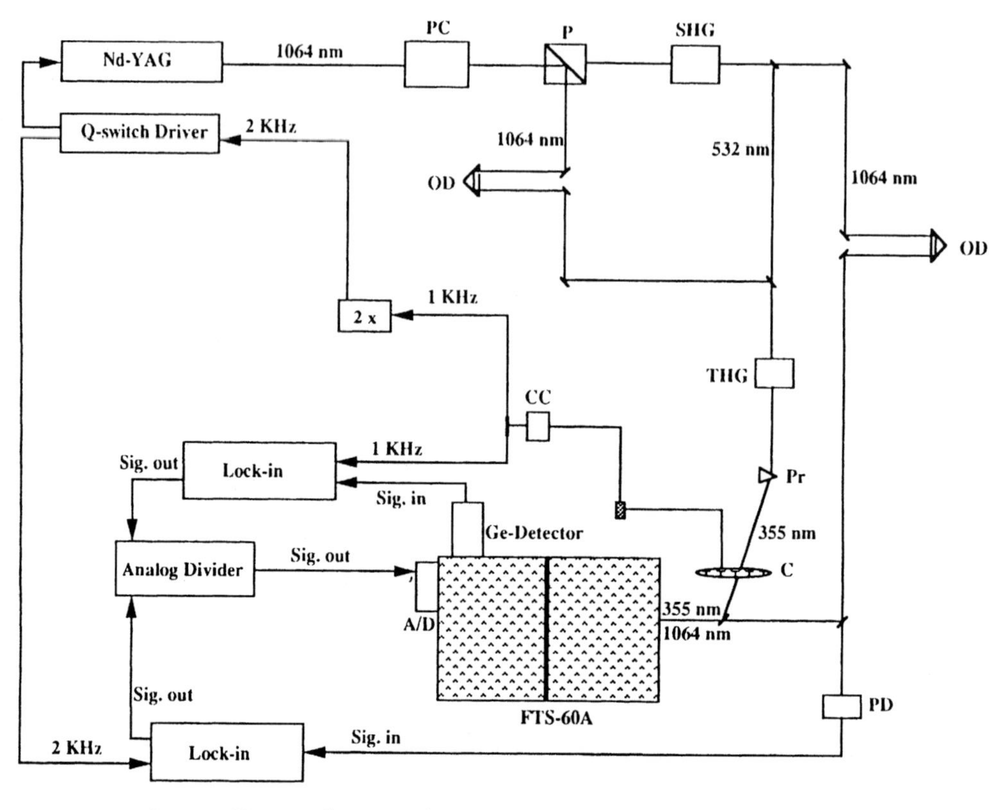

Fig. 5.22 Diagram of picosecond time-resolved FT-Raman spectrophotometer based on a step-scan interferometer. PC, Pockels cell; P, polarizer; SHG, second harmonic generator; OD, optical delay; THG, third harmonic generator; Pr, prism; C, chopper; CC, chopper controller; PD, photodiode. (reproduced with permission from Ref. [54])

to the 355 nm pump pulses is controlled by using optical delay lines. The spectrometer is operated in the step-scan mode with a step rate of 100 Hz. The pump-pulse repetition rate is reduced to half that of the 2 kHz probe pulses with a chopper. The Q-switch of the Nd:YAG laser and the chopper are synchronised by triggering the Q-switch externally at twice the chopper frequency. The signal from a detector is processed by a lock-in amplifier referenced at 1 kHz in order to amplify changes induced in the signal by the pump pulses. The Raman signal output is divided by a reference signal in order to cancel out the pulse-to-pulse energy fluctuation.

In these time-resolved FT-Raman measurements with the pump-probe technique, the time resolution is determined by the probe pulse width of about 100 ps because Raman scattering is instantaneous. The response time of the detector plays no role in the time resolution of this method. There is no fundamental limit on the time scale achievable, except the Fourier-transform limit imposed on the resolution. The extent of overlap of the pulsed light from the two arms of a

Michelson interferometer is given by the autocorrelation of the electric field of the pulsed light. Let us assume that the electric field profile of the pulse is Gaussian and that the FWHH of the intensity profile of the pulse is Δt ps. The autocorrelation of the Gaussian pulse is also Gaussian, and its FWHH is $2\Delta t$ ps. Thus, the effect of using pulsed excitation is to weight the interferogram with a Gaussian function, in effect imposing Gaussian apodisation. The FT of a Gaussian apodisation function with a width of $2\Delta t$ gives the Gaussian line-shape function with a width of $2\ \ln 2/\pi\Delta t$. When the resolution of an observed spectrum mainly depends on the pulse width, the resolution under the Taylor criterion $\Delta\tilde{\nu}$ cm^{-1} is given by

$$\Delta\tilde{\nu} = \frac{2\ln 2}{10^{-12}\pi c\Delta t} \approx \frac{14.7}{\Delta t} \tag{5.30}$$

When Δt is 100 ps, $\Delta\tilde{\nu}$ is 0.15 cm^{-1}.

Jas et al. [57–59] have reported the time-resolved FT-Raman spectra of electronically excited states (S_1) of 9,10-diphenylanthracene and anthracene. Time-resolved FT-Raman spectra of the S_1 state of anthracene in cyclohexane at several time delays [59] are shown in Fig. 5.23. The 1.064 μm probe is resonant with the $S_3 \leftarrow S_1$ absorption. Positive and negative Raman bands appear in the spectra, obtained at time delays of –100 and –50 ps. At longer time delays, these bands disappear with the 4.6 ns life time. Most of the positive bands have been attributed to vibrations of the S_1 state of anthracene. The negative bands at 800, 1027, 1265 and 1443 cm^{-1} are due to the solvent. Their intensity decreases are a result of the S_3–S_1 absorption of the 1.064 μm light. An apparently anomalous feature appears in the solvent bands at 2851, 2897 and 2934 cm^{-1}. Unlike the solvent bands in the lower wavenumber region, these bands show increases in intensity following exci-

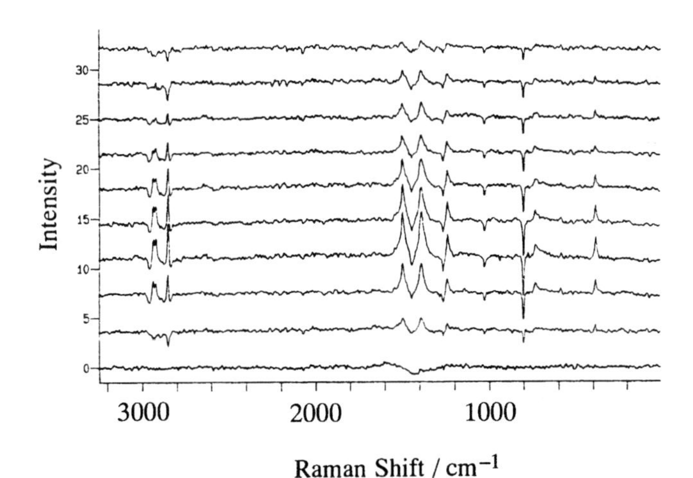

Fig. 5.23 Time-resolved FT-Raman spectra of the S_1 state of anthracene in cyclohexane at time delays (from bottom) of –200, –100, 150, 0, 1000, 2000, 3000, 3500, 4000 and 5500 ps. A total of 170 scans is summed at each time delay (reproduced with permission from Ref. [59])

tation. This may be due to the interaction of the excited state and the solvent. The total collection time for a spectrum at each time delay is 6.5 h, although the SNR obtained is high. Improvement in the SNR is needed for the further development of time-resolved FT-Raman spectrometry. Improved time resolution can be achieved by the use of lasers with shorter pulse widths, such as Ti:sapphire lasers. Sakamoto et al. [60] have reported picosecond time-resolved measurements by using the pump-probe method based on the asynchronous sampling technique.

Time-resolved FT-Raman spectroscopy will also be useful for studies of dynamics-induced temperature jump, pH jump, voltage jump, etc. of fluorescence materials.

5.5 Conclusion

NIR FT-Raman spectrometry is a powerful tool for fundamental research and application in industry because a combination of near-infrared excitation and FT spectrophotometry gives high-quality spectra without luminescence backgrounds, and high wavenumber accuracy gives various advantages: no wavenumber calibration, difference-spectrum technique, accumulation, spectral search, etc. It should be noted that the sensitivity of the NIR FT-Raman spectrophotometer is low in comparison with that of the dispersive spectrometer equipped with a charge-coupled device detector. Increases in sensitivity are expected as the detector and the preamplifier are improved. The double modulation method and NIR FT-Raman measurements with pulsed excitation will open up new applications of FT-Raman spectrometry.

5.6 References

[1] G. W. Chantry, H. A. Gebbie, C. Helsum, *Nature*, **1964**, *203*, 1052.

[2] T. Hirschfeld, B. Chase, *Appl. Spectrosc.* **1986**, *40*, 133.

[3] P. J. Hendra, Ed.: Special issue, *Spectrochim. Acta* **1990**, *46A*(2).

[4] P. J. Hendra, Ed.: Special issue, *Spectrochim. Acta* **1991**, *47A*(9/10).

[5] P. J. Hendra, Ed.: Special issue, *Spectrochim. Acta* **1993**, *49A*(5/6).

[6] P. J. Hendra, Ed.: Special issue, *Spectrochim. Acta* **1994**, *50A*(11).

[7] P. J. Hendra, Ed.: Special issue, *Spectrochim. Acta* **1995**, *51A*(12).

[8] P. J. Hendra, Ed.: Special issue, *Spectrochim. Acta* **1997**, *53A*(1).

[9] P. J. Hendra, C. Jones, G. Warnes, *FT-Raman Spectroscopy*, Ellis Horwood, Chichester, **1991**.

[10] *Fourier Transform Raman Spectroscopy from Concept to Experiment* (Eds. D. B. Chase, J. F. Rabolt), Academic Press, San Diego, **1994**.

[11] P. R. Griffiths, J. A. de Haseth, *Fourier Transform Infrared Spectrometry*, Wiley, New York, **1986**.

[12] D. A. Long, *Raman Spectroscopy*, McGraw-Hill, New York, **1977**.

[13] J. Tang, A. C. Albrecht, *Raman Spectroscopy, Vol. 2* (Ed. H. Szymanski), Plenum, New York, **1970**, pp. 33–68.

[14] B. B. Johnson, W. L. Peticolas, *Ann. Rev. Phys. Chem.* **1976**, *27*, 465.

[15] H. HAMAGUCHI, *Advances in Infrared, Raman Spectroscopy, Vol. 12* (Eds. R. J. H. CLARK, R. E. HESTER), Wiley-Heyden, London, **1985**, pp. 273–310.
[16] R. J. H. CLARK, T. J. DINES, *Angew. Chem. Int. Ed. Engl.* **1986**, *25*, 131.
[17] R. BRACEWELL, *The Fourier Transform and Its Applications*, McGraw-Hill, New York, **1986**.
[18] A. G. MARSHALL, F. R. VERDUN, *Fourier Transforms in NMR, Optical, and Mass Spectrometry*, Elsevier, Amsterdam, **1990**.
[19] Presentation of Raman Spectra in Data Collections, *Pure Appl. Chem.* **1981**, *53*, 1879.
[20] S. F. PARKER, V. PATEL, P. B. TOOKE, K. P. J. WILLIAMS, *Spectrochim. Acta* **1991**, *47A*, 1171.
[21] C. J. PETTY, G. M. WARNES, P. J. HENDRA, M. JUDKINS, *Spectrochim. Acta* **1991**, *47A*, 1179.
[22] V. PETROV, J. T. HUPP, C. MOTTLEY, L. C. MANN, *J. Am. Chem. Soc.* **1994**, *116*, 2171.
[23] J.-Y. KIM, S. ANDO, A. SAKAMOTO, Y. FURUKAWA, M. TASUMI, *Synth. Met.* **1997**, *89*, 149.
[24] K. ASSELIN, B. CHASE, *Appl. Spectrosc.* **1994**, *48*, 699.
[25] Y. FURUKAWA, UNPUBLISHED DATA.
[26] I. HARADA, J. HIRAISHI, *Laser Raman Spectroscopy and Its Application* (Eds. T. SHIMANOUCHI, M. TASUMI, I. HARADA), Nanko-do, Tokyo, **1976**, p. 18.
[27] H. HAMAGUCHI, *Appl. Spectrosc. Rev.* **1988**, *24*, 137.
[28] IUPAC Commission on Molecular Structure and Spectroscopy, Tables of Wavenumbers for the Calibration of Infrared Spectrometers, Pergamon, New York, **1977**.
[29] F. T. WALDER, M. J. SMITH, *Spectrochim. Acta* **1991**, *47A*, 1201.
[30] *Conjugated Conducting Polymers* (Ed. H. KIESS), Springer, Berlin, **1992**.
[31] *Primary Photoexcitations in Conjugated Polymers: Molecular Exciton versus Semiconductor Band Model*, (Ed. N. S. SARICIFTCI), World Scientific, Singapore, **1997**.
[32] Y. FURUKAWA, H. OHTA, A. SAKAMOTO, M. TASUMI, *Spectrochim. Acta* **1991**, *47A*, 1367.
[33] Y. FURUKAWA, *J. Phys. Chem.* **1996**, *100*, 15 644.
[34] Y. FURUKAWA, M. TASUMI, *Modern Polymer Spectroscopy*, (Ed. G. ZERBI), Wiley-VCH, **1999**, pp. 207–237.
[35] N. YOKONUMA, Y. FURUKAWA, M. TASUMI, M. KURODA, J. NAKAYAMA, *Chem. Phys. Lett.* **1996**, *255*, 431.
[36] Y. FURUKAWA, M. AKIMOTO, I. HARADA, *Synth. Met.* **1987**, *18*, 151.
[37] A. SAKAMOTO, Y. FURUKAWA, M. TASUMI, *J. Phys. Chem.* **1992**, *96*, 3870.
[38] A. SAKAMOTO, Y. FURUKAWA, M. TASUMI, *J. Phys. Chem.* **1994**, *98*, 4635.
[39] Y. FURUKAWA, A. SAKAMOTO, M. TASUMI, *Macromol. Symp.* **1996**, *101*, 95.
[40] A. SAKAMOTO, Y. FURUKAWA, M. TASUMI, *J. Phys. Chem. B* **1997**, *101*, 1726.
[41] Y. FURUKAWA, H. OHTSUKA, M. TASUMI, *Synth. Met.* **1993**, *55*, 516.
[42] Y. FURUKAWA, H. OHTSUKA, M. TASUMI, I. WATARU, T. KAMBARA, T. YAMAMOTO, *J. Raman Spectrosc.* **1993**, *24*, 551.
[43] B. SCHRADER, A. HOFFMANN, *Vibrational Spectrosc.* **1991**, *1*, 239.
[44] Y. OZAKI, A. MIZUNO, H. SATO, K. KAWAUCHI, S. MURAISHI, *Appl. Spectrosc.* **1992**, *46*, 533.
[45] T. NOGUCHI, Y. FURUKAWA, M. TASUMI, *Spectrochim. Acta* **1991**, *47A*, 1431.
[46] D. F. BOCIAN, N. J. BOLDT, B. W. CHADWICK, H. A. FRANK, *FEBS Lett.* **1987**, *214*, 92.
[47] R. J. DONOHOE, R. B. DYER, B. I. SWANSON, C. A. VIOLETTE, H. A. FRANK, S. F. BOCIAN, *J. Am. Chem. Soc.* **1990**, *112*, 6716.
[48] K. OKADA, E. NISHIZAWA, Y. FUJIMOTO, Y. KOYAMA, S. MURAISHI, Y. OZAKI, *Appl. Spectrosc.* **1992**, *46*, 518.
[49] F. FUCHS, A. LUSSON, J. WAGNER, P. KOIDL, *Fourier Transform Spectroscopy*, SPIE Vol. *12*, **1989**, *145* 323.
[50] R. BENNETT, *Spectrochim. Acta* **1994**, *50A*, 1813.
[51] D. J. CUTLER, H. M. MOULD, B. BENNET, A. J. TURNER, *J. Raman Spectrosc.* **1991**, *22*, 367.
[52] D. J. CUTLER, C. J. PETTY, *Spectrochim. Acta* **1991**, *47A*, 1159.
[53] R. BENNETT, *Spectrochim. Acta* **1995**, *51A*, 2001.

[54] C. S. Jas, C. Wan, C. K. Johnson, *Spectrochim. Acta* **1994**, *50A*, 1825.

[55] K. Masutani, H. Sugisawa, A. Yokota, Y. Furukawa, M. Tasumi, *Appl. Spectrosc.* **1992**, *46*, 560.

[56] A. Sakamoto, Y. Furukawa, M. Tasumi, K. Masutani, *Appl. Spectrosc.* **1993**, *47*, 1457.

[57] K. Asselin, B. Chase, *J. Mol. Struct.* **1995**, *347*, 207.

[58] C. S. Jas, C. Wan, C. K. Johnson, *Appl. Spectrosc.* **1995**, *49*, 645.

[59] C. S. Jas, C. Wan, K. Kuczera, C. K. Johnson, *J. Phys. Chem.* **1996**, *100*, 11857.

[60] A. Sakamoto, H. Okamoto, M. Tasumi, *Appl. Spectrosc.* **1998**, *52*, 76.

6
Sampling and Sample Presentation

Sumio Kawano

6.1
Sampling

For quantitative and qualitative analyses, NIR spectroscopy needs a calibration equation. The calibration procedure involves collecting a number of samples, obtaining both reference and NIR data on each sample and deriving a calibration equation from these data by using chemometrics (described in Chapter 7). The calibration equation derived may be used to predict the reference result for future samples.

Sampling is the most important factor in making a good calibration equation. The calibration sample set which is used to make a calibration equation should be representative of the population of future samples that will be predicted by NIR spectroscopy. In the case of agricultural products, the sample set should be sufficiently variable in respect of variety, producing area, producing year and maturity stage to meet the conditions mentioned above. The distribution of the constituent being calibrated for over the calibration samples is also important. The range of the variability should be as large as that expected in any future sample, and it is usually better to keep a more uniform spread of values over the whole range.

6.2
Sample Preparation

Sample preparation for NIR spectroscopy is not as important as it is for other methods such as IR and NMR spectroscopy. However, a little sample preparation is needed, depending on the case.

6.2.1
Grinding

In the early stages of NIR development, ground samples were commonly used. The mean and the distribution of the particle size depend on the grinder used. Even if the same one is used, these properties will be affected by grinding conditions such as warming-up time of the grinder and sample loading. According to the Grain Inspection Handbook – Book V, the Udy-Cyclone Mill with a 1 mm round-hole screen and equipped with an automatic feeder is the only grinder approved by FGIS for official wheat protein testing [1]. The procedure of grinding is described in detail in that book. Grinding is also used for the preparation of meat samples.

6.2.2
Slicing or Cutting

In a preliminary study, sliced or cut samples are sometimes used to examine the possibility of NIR analysis. In their work on NIR analysis of soluble solids in cantaloupe, Dull et al. used intact and sliced samples to evaluate the effect of sample presentation on NIR analysis [2]. A 3.8 cm wide transverse section was removed from each fruit by slicing in planes perpendicular to the stem-blossom axis at 1.9 cm to each side of the axis midpoint. A core 2.0 cm in diameter taken from the transverse section was used for NIR transmittance measurements. NIR spectra of the sliced samples were much better than those of intact samples in this experiment.

6.2.3
Shredding and Juicing

In the case of a non-uniform solid sample, such as sugar-cane stalk, shredding and juicing may be effective for measuring the mean spectrum of the sample. In the work on NIR spectroscopic quality inspection of sugar cane by Sekiguchi et al., shredding and juicing were adopted as sample preparations for NIR analysis [3]. Figure 6.1 illustrates the procedure. A sample for quality inspection is taken up with a core sampler from the sugar cane stalks transported into a sugar-making factory. The sample is cut into small pieces with a shredder after removing trashes, and the shredded sample is pressed with an oil hydraulic press (force: 260 kg cm^{-2}, time: 1 minute) to give juice. The juice may be presented to the NIR instrument after its filtration with a 200-mesh stainless sieve (75 μm) and the removal of bubbles in it with an ultrasonic cleaner. This system has been used in Japan since 1994 as described later in subsection 12.6.

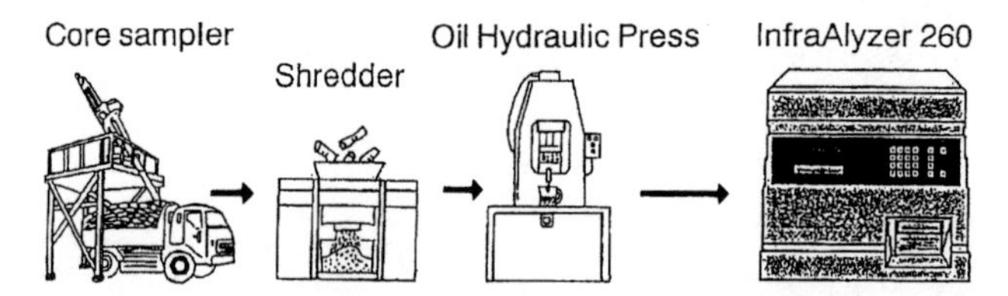

Fig. 6.1 Sample preparation of sugar cane for NIR quality inspection [3]

6.2.4 Homogenising

In the case of a non-uniform liquid sample, such as untreated milk, homogenisation may be needed prior to NIR measurements, because untreated milk has different sizes of fat globules which give fluctuations in NIR spectra. In the work on NIR analysis of milk constituents [4], a Milko-tester Mk. II homogeniser (Foss Electric, Denmark) was used to make fat globules smaller and more uniform. NIR spectra of homogenised, untreated milk were successfully taken in the longer wavelength region from 1100 nm to 2500 nm by using a sample cell with a sample thickness of 0.25 mm. If the shorter wavelength region from 700 nm to 1100 nm is used for NIR measurement, homogenisation is not necessarily needed. Chen et al. measured good NIR spectra of unhomogenised untreated milk in the shorter wavelength region using a normal test tube (12 mm inner diameter) as a sample cell. In this case, the sample in the test tube was warmed to 40 °C and then only mixed slowly by turning the top and bottom of the test tube over several times [5].

6.2.5 Temperature Control

It is well known that NIR spectra are easily affected by sample temperature. When samples that have different temperatures from that of the calibration sample set are predicted, a bias necessarily occurs. Therefore it may be better to control the sample temperature before making NIR measurements. In the case of liquids, a water bath is commonly used for that purpose. In the automatic chemical composition analyser for soy sauce described later in subsection 12.6, a water bath is used as a temperature controller [6]. As the sample goes through the glass coil in the water, the sample temperature is maintained at a constant temperature of, say, 20 °C. During NIR measurement, the sample temperature is also controlled by using water from the water bath.

When the temperature control of samples is not easy, the technique for sample temperature compensation is very useful. Kawano et al. have developed a calibration equation which can compensate for a variation of sample temperature. In this work, calibration samples having a temperature range from 21 °C to 31 °C were used. The calibration equation with temperature compensation showed a high accuracy in prediction, even if the sample temperature changed in that range [7].

6.2.6 Moisture Control

In the case of quantitative analysis, a variation of moisture content may be accepted if the variation is within the moisture range of the calibration samples. However, it should be checked that the moisture content of the sample submitted to NIR analysis is in that range because the moisture content of samples changes easily when the sample is not kept in a closed container. Sometimes the sample loses moisture and the moisture content goes below the moisture range. When grains are ground under high moisture conditions, such as on a rainy day, the ground sample absorbs moisture and the moisture content becomes higher than the moisture range. It is important to keep the moisture content of samples within the range. If not, a bias may occur.

In the case of qualitative analysis, the moisture content of the samples is the most important factor that influences the NIR spectra. If there are differences in moisture content among the groups of samples, each group can be classified by chance if NIR spectra that are influenced by moisture content are used. In order to avoid this possibility, it is better to perform a moisture control of the samples. In the study on the classification of normal and aged soybean seeds by NIR discriminant analysis, the moisture content of each single kernel was controlled in a dessicator with silica gel at room temperature until each one had a constant moisture content of 14.6%. Thanks to the moisture control, the discriminant analysis of NIR spectra went well [8].

6.3 Sample Presentation

6.3.1 Relative Absorbance

In an NIR instrument, relative rather than absolute absorbance is usually measured by using a reference material such as a ceramic plate. The intensity of light transmitted through a ceramic plate is generally measured, and then the same NIR measurement is performed by using samples in place of the ceramic plate. Relative absorbance can be obtained by the following equation,

$$As(\lambda) = \log(Ir(\lambda)/Is(\lambda)) \tag{1}$$

in which $As(\lambda)$ is the relative absorbance of samples at λ nm, $Ir(\lambda)$ is the intensity of light transmitted through the reference at λ nm and $Is(\lambda)$ is the intensity of light transmitted through the samples at λ nm. Equation (2) can then be derived as follows:

$$As(\lambda) = As^*(\lambda) - Ar^*(\lambda) \tag{2}$$

Here $As^*(\lambda)$ is the absolute absorbance of the samples at λ nm and $Ar^*(\lambda)$ is the absolute absorbance of the reference at λ nm. To keep the NIR measurements consistent, $Ar^*(\lambda)$ should be constant, indicating that you have used the identical reference during a series of your work. The above description can also be adapted in the reflection mode.

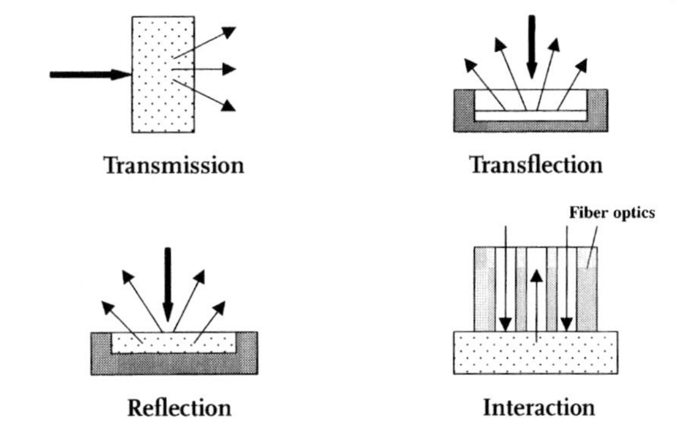

Fig. 6.2 Sample presentations of transmission, reflection, transflection and interaction

6.3.2
Transmission, Reflection, Transflection and Interaction

"Sample presentation" or how to present or set a sample to a NIR instrument is one of the important factors affecting NIR measurements. Fig. 6.2 illustrates some sample presentations known as "transmission", "reflection", "transflection" and "interaction". In the case of transmission, incident light illuminates one side of the sample and the transmitted light may be detected from the other side. This presentation is widely used for liquids, say, by using a cuvette as described below. In the case of reflection, incident light illuminates the surface of the sample and the diffusely reflected light from the surface, or from the portion near the surface, may be detected. In this case, the sample should be opaque, say a powdered sample, and have more than 1 cm depth. Transflection was originally developed by Technicon for the InfraAlyzer and combines transmission and reflection. Incident light is transmitted through the sample and then scattered back from a reflector, which is made of ceramic or aluminium to be compatible with the diffuse reflection characteristics of the instrument. In the case of interaction, an interaction probe having a concentric outer ring of illuminator and an inner portion of receptor is usually used. The end of the probe should be in contact with the surface of the sample. Therefore only the light transmitted through the sample can be detected. The interaction mode was used to determine the Brix value of intact peaches [9].

6.3.3
Sample Cell or Sample Holder

To measure good NIR spectra, various kinds of sample cells or sample holders, mentioned below, have been developed by the manufactures and the users of NIR instruments. Many types of fibre-optic probes have also been developed for online measurement and other purposes.

Fig. 6.3 Sample cells for whole grains

6.3.3.1 **Sample Cell for Whole Grains**

Sample cells for whole grains are shown in Fig. 6.3. The sample cell (right), developed by NIRSystems for the NIRS6500 instrument, has a sample loading portion 37 mm (w) × 15 mm (d) × 200 mm (h) that can be used for NIR measurements of whole grains and seeds. The cell, mounted vertically with the NIR instrument, moves up and down slowly during NIR measurement to compensate for the heterogeneity of the samples. To measure a NIR spectrum, 30–50 scans are usually performed. Another type of the cell designed for, say, the InfraAlyzer 500 is also commercially available. The cell, 85 mm in diameter and 12 mm in depth, is rotated quickly during NIR measurement. NIR radiation is consequently illuminating a ring portion of the cell.

An important point in measuring good and stable spectra with the above-mentioned cell is to keep consistency in the sample loading. Sample settling is one of important practices in this case.

6.3.3.2 **Sample Cell for a Powdered Sample**

Each sample cell for powdered samples has a similar construction. The cell has a specially designed rubber pad to ensure constant and reproducible packing density; because packing density affects scattering conditions. The scattering conditions of each sample should be identical for a series of NIR experiments. A typical one, designed for the InfraAlyzer 500, is shown in Fig. 6.4. A circular quartz window 3.5 cm in diameter is mounted in a black plastic vessel 1 cm deep. The sample is scooped into the cell and levelled by using a special spoon known as a "scoopla". The filled cell is then placed in the drawer of the instrument.

Fig. 6.4 Sample cell for powdered sample

Fig. 6.5 Sample cell for pastes; left is the so-called "Open cups" and right the so-called "High fat/high moisture cell"

6.3.3.3 Sample Cell for Pastes

A sample cell for pastes is usually used for samples such as dough, ground meat and fermented soybean paste (Miso). Fig. 6.5 illustrates the sample cells; those for reflection (left, so-called "Open cup") and the other for transmission (right, so-called "High fat/high moisture cell"). The former ones, which are designed specially for the InfraAlyzer 500, may be positioned without a glass cover into the drawer of the instrument. The surface of the sample in the open cup should be smoothed. The latter one is the sample cell designed specially for the NIRS 6500. Samples are packed into polyethylene bags and placed into a hinged, elongated sample cell. Closing the cell forces the sample to a constant thickness of 1 cm. While the spectrum is continuously scanned, the sample is transported across the NIR beam.

Fig. 6.6 Sample cells for liquids; cuvettes and aluminum cell

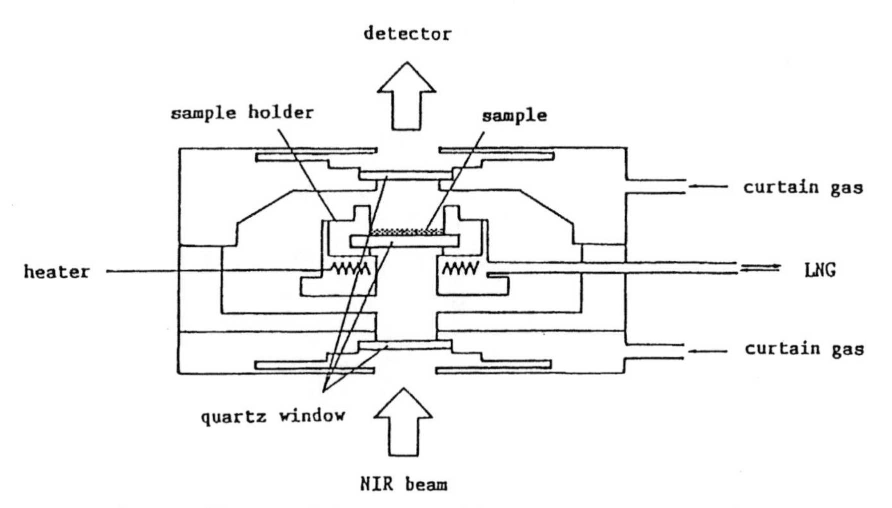

Fig. 6.7 Schematic of the cryo-cell that can control the sample temperature in the range from −150°C to +80°C [10]

6.3.3.4 **Sample Cell for Liquids**

A cuvette cell made of quartz for the transmission mode (Fig. 6.6) is popular as a sample cell in NIR spectroscopy for use with liquids such as water, liquors and juices. The width of the adopted cell depends on the wavelength region and samples used. In the case of water, the following are recommended: cells 2–3 cm in width at 970 nm, cells 2–3 mm in width at 1450 nm and cells less than 1 mm in width at 1930 nm. For measuring good NIR spectra, a suitable width should be selected.

In a reflection instrument, the aluminum cell specially designed for the Infra-Alyzer 500 and shown in Fig. 6.6 is widely used. A small amount of sample is taken and dropped onto the central portion. The cell is covered with a glass plate and may now be placed in the sample drawer of the instrument. The usual path length of the cell is 0.1 mm.

To perform basic research on the state of water, a cryo-cell has been specially designed which can control the sample temperature in the range from –150 °C to +80 °C with an accuracy of 0.1 °C (Fig. 6.7). The sample holder is cooled with liquid nitrogen gas (LNG) and heated with electric heaters. The outer surface of each quartz window is blown with drying nitrogen gas to prevent water-condensation. Using such a cell, Iwamoto et al. performed basic research on hydrogen bonding related to water in foods [10].

6.3.3.5 **Fruit Holder**

As for NIR applications to big samples, such as fruits, a specially designed fruit holder has been developed. Fig. 6.8 illustrates the sample placement for NIR interaction measurements with fibre optics. A commercially available "Interaction Probe", having a concentric outer ring of illuminator and an inner portion of receptor, is used as the fibre optics. A cushion made of urethane foam is pasted onto the end of the probe to hold a sample. For NIR measurement it is only necessary to place a sample on the end of the probe. Using such a sample holder, Kawano et al. worked on NIR determination of Brix value of intact peaches [9].

Other types of fruit holders for transmission [11] and reflection [12] have also been developed.

6.3.3.6 **Sample Holder for Single Kernels**

For NIR application to small samples such as single kernels of grains, a specially designed sample holder has been developed. Fig. 6.9 illustrates the single kernel clamp developed by Delwiche [13]. Each kernel is positioned in the clamp. Monochromatic light is focused on the kernel constrained within the clamp, and the light transmitted through the sample is detected by a silicon detector.

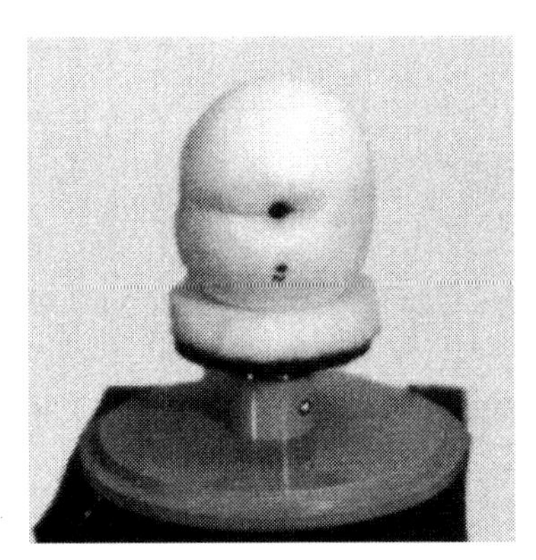

Fig. 6.8 Sample placement for NIR interaction measurements using fibre optics [9]

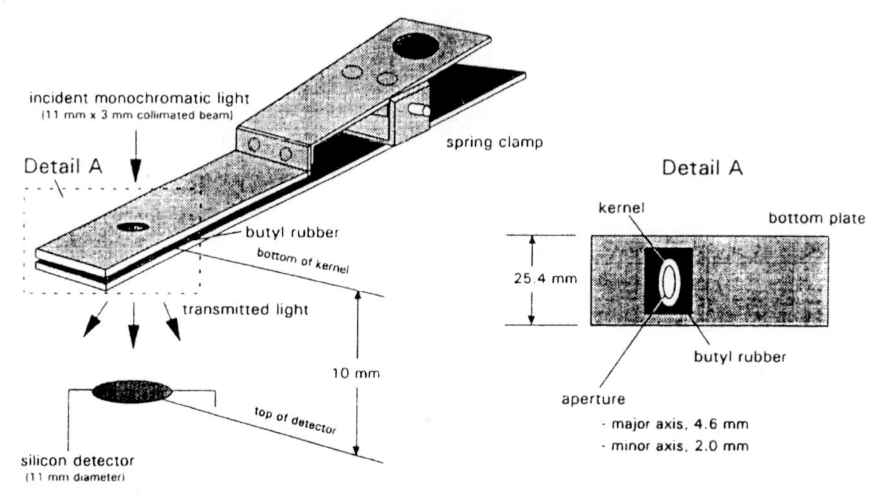

Fig. 6.9 Single kernel clamp [13]

6.3.3.7 **Fibre-Optic Probes**

Various kinds of fibre-optic probes for NIR measurements are commercially available which can be used to perform on-line monitoring in food industries.

6.4 References

[1] *USDA Grain Inspection Handbook – Book V: Wheat protein*, USDA, FGIS, USA **1987**.

[2] G. G. Dull, G. S. Birth, D. A. Smittle, R. G. Leffler, *J. Food Sci.* **1989**, *54*, 393–395.

[3] R. Sekiguchi, K. Fuchigami, S. Hara, C. Tutumi, *Near Infrared Spectroscopy. The Future Waves* (Eds. A. M. C. Davies, P. Williams), NIR Publication, Chichester, UK, **1996**, pp. 632–637.

[4] T. Sato, M. Yoshino, S. Furukawa, Y. Someya, N. Yano, J. Uozumi, M. Iwamoto, *Jpn. J. Zooteh. Sci.* **1987**, *58*, 698-706.

[5] J. Y. Chen, C. Iyo, S. Kawano, F. Terada, *J. Near Infrared Spectrosc.* **1999**, *7*, 265–273.

[6] K. Kobayashi, K. Iizuka, T. Okada, H. Hashimoto, *Proc. 2nd International NIRS Conference* (Ed. M. Iwamoto, S. Kawano), Korin, Tokyo, Japan, **1990**, p. 178–189.

[7] S. Kawano, H. Abe, M. Iwamoto, *J. Near Infrared Spectrosc.* **1995**, *3*, 211–218.

[8] T. Kusama, H. Abe, S. Kawano, M. Iwamoto, *Nippon Shokuhin Kogyo Gakkaishi (J. Japanese Society for Food Science and Technology)*, **1997**, *44*, 569–578.

[9] S. Kawano, H. Watanabe, M. Iwamoto, *J. Japan. Soc. Hort. Sci.* **1992**, *61*, 445–451.

[10] M. Iwamoto, S. Kawano, H. Abe, *NIR News* **1995**, *6(3)*, 10–12.

[11] S. Kawano, T. Fujiwara, M. Iwamoto, *J. Japan. Soc. Hort. Sci.* **1993**, *62*, 465-470.

[12] S. Kawano, H. Watanabe, M. Iwamoto, *Proc. 2nd International NIRS Conference* (Ed.: M. Iwamoto, S. Kawano), Korin, Tokyo, Japan, **1990**, pp. 343–352.

[13] S. R. Delwiche, *Trans. ASAE* **1993**, *36*, 1431–1437.

7
Chemometrics in Near-Infrared Spectroscopy

H. M. HEISE *and* R. WINZEN

7.1
Introduction

Analytical spectroscopy can provide the necessary information regarding composition and properties of materials or manufactured products, such as polymers, drugs, agricultural products or beverages, to be used for quality control and optimisation. Spectroscopy in process monitoring is another area that is nowadays receiving much attention. Research and development processes can be guided by new spectroscopic measurement techniques. Cost effectiveness is another catchword and leads to faster and less expensive measurements being planned than previously done. There is a tendency to perform fewer experiments, but still record more and more data during each of them. Referring to a different field, clinical analyses and medical diagnostics nowadays rely on spectroscopic multivariate measurements, e.g., for classifying different diseases or the physiological state of a patient's body tissue. The rapid and ongoing evolution of near-infrared (NIR) spectroscopic applications in research and daily routine would have been impossible without the parallel development of chemometric evaluation methods. This chapter is meant to introduce the most important techniques currently in use and also shows recent trends in multivariate data analysis.

Svante Wold coined the term chemometrics in the year 1972, after which a rapid development started [1]. The field of chemometrics is tremendously wide, and mathematical and statistical tools are often related to those in biometrics or econometrics [2, 3]. In general, this field provides methods that can aid chemists to move more efficiently on the path from measurements to information to knowledge. Recent reviews that illustrate the enormous activities can be found in several journals [4–7]. A main part of chemometrics is multivariate data analysis, which is essential for qualitative and quantitative assays based on NIR spectroscopy. The term "multivariate" is synonymously used for "multidimensional" or "multichannel", the latter is often found in spectroscopy. Multivariate techniques for quantitative work have been covered intensively by several books [8–11] or recent book chapters [12, 13].

An interesting classification of multivariate methods, which deal with the quantitative analysis of multicomponent systems, based just on the different amount of information available with respect to component spectra and concentrations has been presented by Liang et al. [14]. They classified white multicomponent systems as systems for which, e.g., the spectra of the chemical compounds present in the sample are known and the training set would provide all concentrations of the analytes to be investigated. Grey multicomponent systems are characterised by incomplete knowledge of the component spectra and concentrations, whereas systems are called black when complete a priori information with respect to chemical composition is missing. A strategy applied in such a case is self-modelling curve resolution, which aims at resolving spectral data, obtained, for example, during reaction monitoring, into concentration profiles and pure constituent absorbance spectra without a priori knowledge about the system under observation [15]. Such interesting chemometric developments, however, are outside the scope of this chapter.

During the last decade, the interest in multivariate quantitative analysis of spectroscopic data has increased significantly. This trend is due, in part, to improvements in the manifold spectroscopic techniques used for the plethora of analytical applications, as well as in computer hard- and software. Multivariate calibration has become a valuable tool for tackling multicomponent assays. The classical and inverse forms of Beer's law, known as K- and P-matrix approaches, have been used to form linear calibration matrix equations. A short introduction to quantitative methods will be given, to help understand the fundamental ideas behind the use of multivariate algorithms. Nowadays, most NIR-spectroscopic applications are carried out by using so-called bilinear factor methods, in particular partial least squares (PLS) regression, which will also be covered. Non-linear calibrations are touched on briefly. Still a topic of hot debate is the set-up of robust calibrations, and within this subject the selection of important spectral variables is discussed. Another important aspect is calibration transfer and stability, for example, when starting with laboratory spectrometers and preparing finally for in-line process monitoring.

After a discussion of qualitative methods, a section on signal processing will follow. It is not the purpose of this section to explain all the mathematical details behind the individual techniques for data pretreatment, but to list various existing methods with respect to the effects they have on the data. In addition, special signal processing tools are discussed in more detail, as well as questioning whether any preprocessing of data is necessary at all. The last section is intended to provide some insights into methods that have been discussed among researchers for many years, but have not yet found their way into commercial NIR spectroscopic applications on a broad basis.

7.2 Quantitative Analysis

Near-infrared spectra of solid and liquid samples are characterised by numerous bands that are usually broad compared with the range of this spectral region. The bands have their origin in overtones and combinations of fundamental vibrations, and even spectra of chemically simple substances are rich in strongly overlapping bands (see Fig. 7.1). Their identification and, even more so, their assignment to vi-

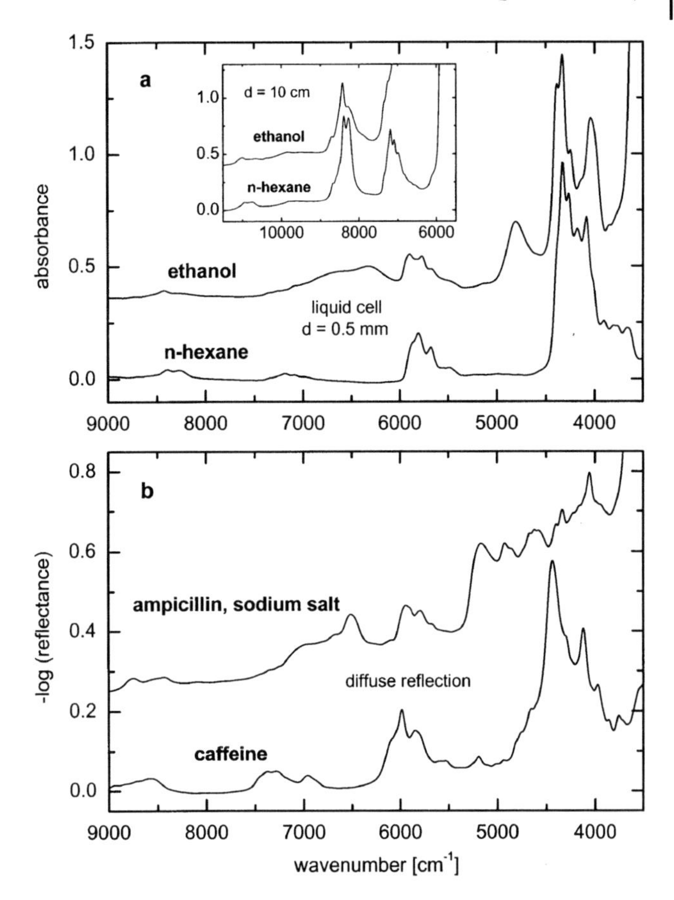

Fig. 7.1 Near-infrared spectra of pure compounds: a) spectra of two liquids measured in transmission cells of different pathlength; b) spectra of two powdered substances measured in diffuse reflection mode

brations of special molecular moieties with their unique chemical bonds is, contrary to the situation in the mid-infrared (MIR) region, often very difficult.

7.2.1
Beer's Law: a Simple Physical Model

Quantitative analysis of NIR spectroscopic data is based on Beer's law, which states that a linear relationship exists between the molar concentration y of a substance and the absorbance $x = -\log(I/I_0)$ of this substance at a given wavelength λ (for the sake of consistency with other chemometric textbooks, the absorbance is designated by the symbol x instead of A, and the concentration with the symbol y instead of c). The term I_0 denotes the spectral intensity of the radiation emitted by the spectrometer and I the intensity remaining after interaction (transmission/reflection) with the sample. The ratio I/I_0 is called transmittance (T) when the radiation transmits the sample or reflectance (R) when it is reflected by it. Thus the absorbance x can be written as $\log(1/T)$ or $\log(1/R)$. Although this quantity is by definition without any units, the abbreviation AU for absorbance units is widely used. Beer's law, see Eq. (7.1), connects concentration and absorbance with the molar absorptivity ε and the optical pathlength d, so that

$$x = \varepsilon \cdot d \cdot y \tag{7.1}$$

With a given sample pathlength d and a known molar absorptivity ε for the substance under investigation, the concentration y can be determined directly from the absorbance x by using the inverted formula Eq. (7.2) under the premises that the compound to be quantified is the only absorbing one in the mixture.

$$y = x/(\varepsilon \cdot d) \tag{7.2}$$

Modern near-infrared spectroscopic analysis is far from this classical procedure. Today we use this technique to analyse complex composite materials of various morphologies. The resulting spectra are often, at first sight, rather featureless, and the identification of bands and their direct use for quantitative analytical evaluation is nearly impossible. Furthermore, the signals may be perturbed by additional effects that arise from the interaction of radiation with the physical structure of the sample, thus complicating the analysis. It should be noted, however, that the existence of such effects enables the determination of certain physical properties of the sample, for example, the particle size of powders [16].

By combining the physical parameters absorptivity and pathlength, Beer's law can be rewritten in an inverse form to give a linear calibration model as shown in Eq. (7.3).

$$y = x \cdot b_1 + b_0 \tag{7.3}$$

The parameter b_0 compensates for additional mixture components found that have a constant concentration. The parameters b_0 and b_1 are often not accessible by direct observation, but have to be determined by calibration. The calibration model is given by Eq. (7.3), and at least two samples with known values for x (absorbance) and y (concentration) are required to solve the equation for the two unknown parameters.

Whereas within the classical approach variations in concentrations lead to changes in the measured absorbances, Eq. (7.3) states that the concentration is a function of the absorbance (so-called inverse model). When a simple calibration model like this can be applied, the "classical" form $x = (y - b_0)/b_1$ is equivalent to Eq. (7.3). However, the univariate regressions with either x or y as the independent variable assumed to be free of error show a different statistical behaviour in the classical and the inverse approach [8]. On the other hand, so-called bivariate least squares can be used for regressing data containing errors in both axes [17], but also other names such as "total least squares", "orthogonal regression" or "errors-in-variables regression" can also be found in the literature [18].

Most multidimensional or multivariate calibration models with more than one x variable are set up as inverse models, owing to the fact that the classical least squares (CLS) algorithm, which is an example of a full-spectrum method, faces limitations due to incomplete information about the multicomponent system under investigation. The multivariate total least squares approach, on the other hand, is practically unknown to the spectroscopic community, since it has rarely been applied to near-infrared spectroscopy [19].

7.2.2
A Full-Spectrum Method: CLS

A simple approach to quantitative full-spectrum evaluation can be directly derived from Beer's law under the assumption that absorbances are additive. The classical approach starts with modelling of the measured spectra by individual component spectra (in the following, uppercase bold letters denote matrices and lowercase bold letters denote column vectors, $\boldsymbol{X}'$ is a transposed matrix and $(\boldsymbol{X}'\boldsymbol{X})^{-1}$ an inverse matrix). If the spectrum of a mixture equals the sum of the spectra of the pure components of the mixture, weighted by their respective concentrations, the model can be written according to Eq. (7.4).

$$\boldsymbol{X} = \boldsymbol{Y}\boldsymbol{R}' + \boldsymbol{E} \tag{7.4}$$

Here, the $m \times n$ matrix $\boldsymbol{X}$ contains the m calibration spectra which were recorded at n wavelengths or wavenumbers, the $m \times l$ matrix $\boldsymbol{Y}$ the corresponding concentrations of all l mixture components for each of the m mixtures, $\boldsymbol{R}'$ the $l \times n$ matrix (transposed) of unit response pure component spectra (or sensitivity matrix) and $\boldsymbol{E}$ is the $m \times n$ matrix of spectral residuals due to noise or other non-systematic variations. Since mean centring is generally applied as preliminary treatment, background terms are omitted.

When the pure component spectra are available, no calibration is needed, and the model Eq. (7.4) can be solved directly by forming the prediction Eq. (7.5).

$$\hat{\boldsymbol{y}}^{u\prime} = \boldsymbol{x}^{u\prime} \cdot \boldsymbol{R} \cdot (\boldsymbol{R}' \cdot \boldsymbol{R})^{-1} \tag{7.5}$$

This solution is written for the prediction of concentrations by using the spectrum of one unknown sample. If the pure component spectra $\boldsymbol{R}$ are not accessible, which is the case when component interactions leading to spectral changes have to be taken into account, they can be estimated from the calibration data by least squares according to Eq. (7.6) and $\hat{\boldsymbol{R}}$ can be used in Eq. (7.5) instead of $\boldsymbol{R}$.

$$\hat{\boldsymbol{R}}' = (\boldsymbol{Y}^{\mathrm{cal}\prime} \cdot \boldsymbol{Y}^{\mathrm{cal}})^{-1} \cdot \boldsymbol{Y}^{\mathrm{cal}\prime} \cdot \boldsymbol{X}^{\mathrm{cal}} \tag{7.6}$$

The classical least squares (CLS) method is rather limited because the concentrations of *all* mixture components have to be known in the calibration step with this approach; this is rather inconvenient and leads to expensive calibrations. By using the strategy shown below, however, the background built from interfering spectra can be modelled. Furthermore, the classical approach assumes that only variations in $\mathbf{Y}$, usually concentration changes, affect the spectra in $\mathbf{X}$. This immediately causes problems with effects that influence the spectra, but cannot be expressed as values in $\mathbf{Y}$. Such effects are not rare in non-destructive near-infrared spectroscopy of complex composite samples. Examples are the interaction between components through hydrogen bonds, the influence of temperature and various scatter effects observed with diffuse reflection measurements.

In the case in which there are unknown component spectra for the calibration modelling, the equation for a single mixture spectrum can be written as

$$\mathbf{x} = \mathbf{R}_k \mathbf{y}_k + \mathbf{R}_{un} \mathbf{y}_{un} + \mathbf{e} \tag{7.7}$$

where the indices k and un mark the components with known and unknown spectra, respectively. If the false model $\mathbf{x} = \mathbf{R}_k \mathbf{y}_k + \mathbf{e}$ is used and solved, the result is

$$\hat{\mathbf{y}}^f = (\mathbf{R}_k' \cdot \mathbf{R}_k)^{-1} \cdot \mathbf{R}_k' \cdot \mathbf{x} \tag{7.8}$$

although the expectation value, calculated by using the complete model will read

$$\hat{\mathbf{y}}^{\exp} = \mathbf{y}_k + (\mathbf{R}_k' \cdot \mathbf{R}_k)^{-1} \cdot \mathbf{R}_k' \cdot \mathbf{R}_{un} \cdot \mathbf{y}_{un} \tag{7.9}$$

The second term in Eq. (7.9) describes a bias in the estimate of $\mathbf{c}_k$, because spectra of unknown and known components usually show some overlap ($\mathbf{R}_k' \cdot \mathbf{R}_{un} \neq \mathbf{0}$). The unlikely case is that all known component spectra are orthogonal to the unknown component spectra, which were not taken into account for calibration modelling. Furthermore, when the calibration step has to be included, the CLS method requires two matrix inversions that may lead to propagation and accumulation of errors, in particular when near singular matrices have to be handled. Therefore, a variant of the CLS method was developed. Other approaches for reducing the systematic errors from false modelling will be discussed later.

A group of chemometricians called the algorithm, which is based on a corollary of Eq. (7.4), the parallel calibration method [20]. The authors also provided a technical justification by indicating that the estimator due to spectral error components is similar to the structure of a ridge estimator, which has been used by statisticians in ill-conditioned problems. In addition, a comparison of the performance of this algorithm to that of the standard CLS and PLS methods, respectively, was given and results were always superior with the parallel method. The basic features had been published earlier already by McClure et al. [21]. An application of this Q-matrix method, as it was named by the latter authors, to the quantification of blood-plasma proteins based on this approach was published later [22]. The algebraic basis of this approach, which has not been used frequently by analytical chemists – although powerful, can be given with two equations ($\mathbf{x}_u$ and $\mathbf{y}_u$ were chosen as column vectors)

$$x_u = X'q \quad \text{and} \quad y_u = Y'q \tag{7.10}$$

with q an m component column vector. For solving the first equation it is required that the rank of the matrix X, which is set up from m mixture spectra, be m with $n>m>l$. The vector q can be estimated by least squares, $\hat{q} = (XX')^{-1}Xx_u$, and the result used for the second equation, requiring just a matrix-vector multiplication for the concentration estimate. Usually, a so-called QR factorisation of the matrix X is calculated for matrix inversion [23].

The normal limitations of CLS methods in the case of an unknown component were overcome by exploiting information from the spectral or concentration residuals from the CLS calibration or prediction, respectively. This was used for the calculation of a so-called augmented classical least squares method model to allow the modelling of interferents [24].

There are several figures of merit that are useful in characterising the calibration model [25]; besides the expected prediction error, sensitivity (i.e. the slope of the calibration curve for univariate data), selectivity (accounting for the degree of interference), and the limits of detection. For the classical approach, an expression of the standard error in the concentration estimates, under the assumption of heteroscedasticity (i.e. unequal variance among wavelengths and sample concentrations), was derived by Bauer et al. [26]. Their experimental example was from optical emission spectroscopy, but the theoretical aspects are applicable as well for multivariate near-infrared spectroscopy. (Actually their concept was similar to an errors-in-variables approach.) The result derived by error propagation was rather complex, but simpler expressions assuming homoscedasticity (i.e. constant variance) were derived by Faber and Kowalski [27]

$$\sigma_{\hat{y}k} = \{(R' \cdot R)^{-1}_{kk}[\sigma_x^2 + y' \cdot (Y'Y)^{-1} y \cdot \sigma_X^2] + y' \cdot (Y'Y)^{-1} y \cdot \sigma_Y^2\}^{1/2} \tag{7.11}$$

The formula shows that errors associated with the unknown sample, that is, in x, are propagated by the kth diagonal element of the matrix $(R'R)^{-1}$, which is known as the variance factor. Large overlap between the columns of R leads to large off-diagonal elements in $R'R$ and consequently to a large variance factor (orthogonal columns on the other hand produce diagonal elements). The uncertainties in the calibration data (i.e. in X and Y) are multiplied by the factor $y'(Y'Y)^{-1}y$ showing that the concentration design within the calibration stage is also important. Further discussion with respect to detection limits can be found in Ref. [27].

An approach by considering the "net analyte signal" (NAS), which allows an estimation of figures of merit such as sensitivity or selectivity for multivariate data, was originally developed by Lorber [28]. The net analyte signal was defined as the part of the signal that is orthogonal to the spectra of the other components. Its computation is straightforward with the classical model, for which the knowledge of the pure spectra is available. Recently, this strategy was also extended to inverse calibration models [29]; for more details, see next chapter. A recent discussion on various definitions of multivariate sensitivity was given by Faber [30], which supports the definition given previously by Lorber.

7.2.3
Inverse Multivariate Calibrations

A different approach for predicting the chemical composition from the measured spectrum can be formulated, as seen already for the simple univariate case, by using the inverse formulation of Beer's law, which is also called the P-matrix approach and by which the shortcomings of the classical model with restricted knowledge concerning the composition of the system during the calibration stage can be avoided. The so-called statistical model is obtained by

$$\begin{aligned} y_1 &= b_0 + x_{1,1} \cdot b_1 + \ldots + x_{1,n} \cdot b_n + e_1 \\ \cdot & \qquad \cdots\cdots \\ y_m &= b_0 + x_{m,1} \cdot b_1 + \ldots + x_{m,n} \cdot b_n + e_m \end{aligned} \tag{7.12}$$

where the concentrations y_i are provided for the single compound of interest and e_i are random errors, representing measurement and modelling uncertainties. Eq. (7.13) gives an equivalent form in matrix notation. The constant term b_0 can be realised by a column vector of 1 s in the spectral data matrix $\boldsymbol{X}$. However, it is usual practice to use mean-centred matrices and vectors (centred around zero by subtracting their respective mean values) before model-parameter estimation is started, so the inverse model reads

$$y = \boldsymbol{Xb} + \boldsymbol{e} \tag{7.13}$$

With the equation above, an n-dimensional parameter vector $\boldsymbol{b}$ is sought that maps the m sample spectra into estimates of their appropriate concentrations. For solving the equation by least squares, it is required that the minimum number of calibration standards is greater than or equal to the number of components or wavelengths in the analysis. The least squares estimate of $\boldsymbol{b}$ is obtained by

$$\hat{\boldsymbol{b}} = (\boldsymbol{X}' \cdot \boldsymbol{X})^{-1} \cdot \boldsymbol{X}' \cdot y \tag{7.14}$$

A problem arises from possible collinearity among the independent variables, since the variance of the least squares solution depends decisively on the invertibility of the matrix $\boldsymbol{X}'\boldsymbol{X}$. The high redundancy within the spectra should intuitively improve the result of the regression estimate, but leads to numerical complications for the least squares procedure, because of the ill-conditioned nature of matrix $\boldsymbol{X}'\boldsymbol{X}$. Under the usual assumptions of linear regression, the least squares estimator is the best linear unbiased one. Despite the absence of bias, it may possess a large variance under the condition of highly correlated regressors. Hence, there may be biased and/or non-linear estimators whose prediction performance is better than that of the least squares estimator. There are several strategies for avoiding the collinearity problem in the spectral data, for example by ridge regression [31] or regressor selection, in which there has been enormous interest, in particular in the field of near-infrared spectroscopic agricultural product analysis. "Individual wavelength" methods, also called Multiple Linear Regression (MLR), were the first to be developed; beside classical approaches for optimum wavelength selection such as considering all possible combinations, step-up and step-down searches, recent strategies include, for example, genetic algorithms or simplex optimisation [13].

Other approaches for searching for an optimum regression vector $\boldsymbol{b}$ use an orthogonal decomposition of the measurement matrix $\boldsymbol{X}$

$$X = \boldsymbol{Q}_1 \boldsymbol{Z} \boldsymbol{Q}_2' \quad \text{with} \quad \boldsymbol{Q}_1' \boldsymbol{Q}_1 = \boldsymbol{I} \quad \text{and} \quad \boldsymbol{Q}_2' \boldsymbol{Q}_2 = \boldsymbol{I} \tag{7.13}$$

Here, the matrix $\boldsymbol{Z}$ is regular, that is invertible, and the column vectors of the matrices $\boldsymbol{Q}_1$ and $\boldsymbol{Q}_2$ are orthonormal to each other. With this decomposition an inverse matrix $\boldsymbol{X}^i = \boldsymbol{Q}_2 \boldsymbol{Z}^{-1} \boldsymbol{Q}_1'$, the so-called pseudo inverse or generalised inverse, can be calculated to give the regression vector $\hat{\boldsymbol{b}} = \boldsymbol{X}^i \boldsymbol{y}$. An advantage is that rank-reduced inversions can be formulated by reducing all three matrices involved, actually with the effect of data compression and filtering. This is the idea behind the advantageous methods like principal component regression (PCR) or PLS regression. Also with orthogonal decomposition, two matrices can be multiplied to yield so-called bilinear "latent variables" that can particularly facilitate the graphical presentation of spectral data.

For PCR, the so-called Singular Value Decomposition (SVD) of the matrix $\boldsymbol{X} = \boldsymbol{U}\boldsymbol{S}\boldsymbol{V}'$ that corresponds to the eigenvalue decomposition of the matrix $\boldsymbol{X}'\boldsymbol{X} = \boldsymbol{V}\boldsymbol{S}^2\boldsymbol{V}'$, where $\boldsymbol{S}^2$ is a diagonal matrix with the eigenvalues of $\boldsymbol{X}'\boldsymbol{X}$. In SVD, the individual orthonormal $\boldsymbol{u}$- and $\boldsymbol{v}$-vectors of the matrices $\boldsymbol{U}$ and $\boldsymbol{V}$ are called left and right singular vectors, respectively, and the scalars s are named singular values. In Principal Component Analysis (PCA), the equivalent formulation found is $\boldsymbol{X} = \boldsymbol{T}\boldsymbol{P}'$, where the matrix $\boldsymbol{T} = \boldsymbol{U}\boldsymbol{S}$ consists of orthogonal, but not orthonormal vectors and $\boldsymbol{P} = \boldsymbol{V}$. The individual $\boldsymbol{t}$ vectors are called "scores", whereas for the $\boldsymbol{p}$ vectors the term "loadings" or "factor" spectra is generally used.

For PLS different algorithms have been described, and the first definitions were formulated by iterative algorithms only. However, matrix notation can also be employed to throw light onto the theoretical foundation. In this context, the most powerful PLS1 variant, as shown by Manne [32], is equivalent to the bi-diagonal orthogonal decomposition first proposed by Golub and Kahan in 1965 [33]. These investigations were useful for a better classification of the PLS algorithm into the different families of generalised inverses. An interesting recent overview on this subject has been provided by de Jong and Phatak [34].

The PLS1-decomposition of the matrix $\boldsymbol{X}$ is given by

$$\boldsymbol{X} = \boldsymbol{O}\boldsymbol{B}\boldsymbol{W}' \tag{7.14}$$

in which the first normalised vector is defined by $\boldsymbol{w}_1 = \boldsymbol{X}'\boldsymbol{y}/\|\boldsymbol{X}'\boldsymbol{y}\|$, which defines the iteration process for the orthogonal decomposition uniquely, with the Euclidean length of the vector used for normalisation. As a result, the bi-diagonal matrix $\boldsymbol{B}$ is obtained, which defines the interdependence of the orthogonal iteration vectors

$\boldsymbol{o}_i$ and $\boldsymbol{w}_i$ that build the matrices $\boldsymbol{O}$ and $\boldsymbol{W}$. Therefore, for matrix inversion the first R consecutive factors are used for a PLS-inverse matrix (of rank R).

The so-called PLS2 algorithm can handle multiple y-variables, that are component concentrations. One feature of the orthogonal decomposition used in such a case is that the corresponding matrix $\boldsymbol{B}$ is triangular [32]. PLS2 regression was stated to be useful for a preliminary overview in explorative data analysis. Another situation that was claimed to be advantageous for this algorithm was if the y variables are known to be strongly intercorrelated to each other [8].

One implication of the strategy of rank-reduced matrix inversions is that biased estimators, in contrast to those as obtained by least squares, have to be selected; this will provide superior results. However, the standard deviation of the rank-dependent fits that are calculated from the calibration residuals with m the number of standards

$$\hat{\sigma}(R) = \left\{ \sum_i [y_i - \hat{y}_i(R)]^2 / (m - R - 1) \right\}^{1/2} \tag{7.15}$$

is not appropriate for estimating the prediction performance of ill-conditioned calibration models. For optimisation, the mean squared error (MSE) is used, for which the systematic deviations (bias) and variance of the prediction error are taken into account. Practical means for MSE approximation are realised by different validation strategies.

The validation process for testing the prediction performance of calibration models of different rank has often been discussed. For testing, the data is usually split into two subsets, one for calibration modelling and the second for prediction testing with independent data. This approach has the advantage that it does not rely on any statistical assumptions derived from solving the calibration equation system. Another strategy is to have a subset of calibration samples (m–γ) and a validation subset of size γ, but replace the data after validation for repeat calibrations and select another subset of samples, not previously considered during the validation process, until all samples have been used. The root mean squared error of prediction (RMSEP) or standard error of prediction (SEP) is calculated from the prediction residual sum of squares (PRESS). When the validation subset size is one, this is called cross-validation by using the "leave-one-out" strategy, which is a reasonable approach when only a limited number of samples is available for model training [35], so that this statistics parameter reads:

$$\mathrm{SEP}(R) = \left\{ \sum_i [y_i - \hat{y}_{i,\mathrm{pred}}(R)]^2 / m \right\}^{1/2} \tag{7.16}$$

The standard error of prediction is a suitable statistic for optimum calibration-model selection, which is also called "standard error of cross-validation" (SECV) by several authors. One aspect that must be mentioned is that by using groups of several standards for independent prediction – thus reducing the size of the calibration population – calibration data quality can be tested.

Several criteria have been proposed for selecting the optimum subset of factors needed for the pseudo-inverse matrix calculation. One is the selection of the SEP minimum, but other statistical criteria, such as those derived from F testing, are used as well [36]. When dealing with PLS calibrations, the question put forward is how many factors should be implemented; some examples from one of the authors' work will be given in Section 13.2.1. However, this is not so trivial for the situation faced with PCR calibrations. The same strategy with top-down selection is certainly also applicable that has often been used in statistical calibration work [37, 38].

An interesting problem discussed by us earlier [39] was that PLS calibrations are always superior to PCR calibrations when the same number of factors "from the top" have been implemented for calibration modelling. A mathematical proof of this hypothesis could be given by de Jong [40]. Other factor-selection strategies for PCR, including several different stepwise procedures or those based on "correlation ranking" can be found in the literature, which are listed in a recent paper by Depczynski et al. [41]. The latter authors introduced the technique of parallel genetic algorithms (PGA) for the PCR-optimization task (for a discussion of genetic algorithms, see also Section 7.5.2).

The computations needed for cross-validation in estimating a realistic prediction error are rather comprehensive, especially with the "leave-one-out" strategy, so that mathematicians also searched for faster numerical methods. For full-rank matrix inversions, it can be shown that exact algebraic expressions exist for estimating the concentration prediction for a sample taken from the original calibration set, which stems from the updating of the inversion matrix when a row is deleted from the original calibration matrix. For rank-reduced systems, however, approximations can be obtained. To calculate the estimate from the calibration equation system, by using all standards, the prediction residuals $y_i - \hat{y}_{i,\text{pred}}(R)$ are replaced by the so-called leverage-corrected values $\{y_i - \hat{y}_i(R)\}/\{1 - h_{ii}(R)\}$, where $h_{ii}(R)$ are the leverage values, being the diagonal elements of the hat matrix $\boldsymbol{H}(R) = \boldsymbol{X}\boldsymbol{X}^{i}(R)$ [42]. This strategy works well for PCR calibrations with top-down factor selection when comparing the results from cross-validation and leverage correction [43]; on the other hand, further scaling factors must be derived for PLS calibrations which can, however, be estimated from over-determined linear equation systems [39].

An attractive feature of multivariate calibration methods such as PCR and PLS is the ability to detect outliers, which are samples containing interferences that are absent in the calibration sample population or samples that require model extrapolation. The property prediction for such a sample is usually degraded. Interferences can be sample contamination, measurement artifacts or temperature effects. Such samples found within the calibration samples need to be removed during calibration if they have a negative influence on the calibration modelling.

The problem of outlier detection in prediction has been studied for many years. The parameters commonly used for outlier detection are based on leverage and the magnitude of the spectral $\boldsymbol{X}$ residuals. The leverage for an observation refers to its position in relation to the others; this means that the outlier sample is extreme within the model. The latter residuals reflect the lack of fit between experimental data $\boldsymbol{X}$ and the prediction $\hat{\boldsymbol{X}}$ from the model owing to noise, unmodelled interferences or nonlinearities and others. Hottelling's T^2 test, which is based on the Mahalanobis distance [44], has been proposed. Q-residual plots on the other side are based on the sum of squares of the residuals in $\boldsymbol{X}$. In the classical approach, the critical value above which a standard is considered to be an outlier is based on well-established criteria (for example, the threshold leverage value was defined as twice the mean value of the calibration leverage, that is, $h_{\text{thres},R} = 2R/m$ [45]).

In addition to the leverage values, there are two other well-known statistical parameters, the studentised residual $t_{i,R}$, which is based on a scaled calibration residual $t_{i,R} = (y_i - \hat{y}_i)/\{\hat{\sigma}(1 - h_{ii,R})^{1/2}\}$ with $\hat{\sigma}^2$ the usual estimate of the sample variance. The residuals are scaled to unit variance, so that the $t_{i,R}$ values allow a comparison against each other, and the significance level may be tested using t statistics. Cook's distance D is another useful statistic that provides an estimate of both the outlier degree and the influence on the regression $D_{i,R} = \{t_{i,R}^2/R\} \cdot\{h_{ii,R}/(1 - h_{ii,R})\}$; the values of $D_{i,R}$ can be compared by F statistics (see also comments found in Refs. [39, 46]). These established statistics of $h_{ii,R}$, $t_{i,R}$ and $D_{i,R}$ provide enough information to estimate the influence of a *single* outlier.

Finally, some further developments in calibration-algorithm optimization will be presented. There is a fundamental paper by Marbach [47] that provides a closed form for the statistical multivariate-calibration model in terms of the pure-component spectral signal, the spectral noise and the signal and noise of the reference method. The statistical calibration models rely as much on the pure-component spectra as any physical hard modelling, that is the classical approach. The calculation of the optimum regression vector ("Wiener filter" solution) is shown to be possible by using the a priori known pure analyte spectrum and the covariance matrix of the spectral noise. Spurious correlations including the effect of reference noise can be completely avoided, and specificity proven. The concept of the application-specific signal-to-noise ratio (SNR_{as}) was introduced, which combines the univariate SNR of the reference with the multivariate SNR of the spectral data. Important statistics such as the correlation coefficient, prediction error and slope deficiency in scatter plots are described as functions of the SNR_{as}. The concept will give the basis for future work regarding calibration optimisation.

The net analytical signal (NAS) has already been mentioned above, and its use for simplifying calibration problems can be found in several papers [48, 49]. The NAS concept points into the right direction, since it judges the amount of information from the pure component spectrum of interest that is useful for calibration. However, as pointed out by Marbach [47], the NAS is identical to the classical model calibration and suffers as such from a poor estimate of the covariance of the spectral noise. The amplitudes of the interfering component spectra, which certainly influence the severity of the spectral overlap to the analyte spectrum, are not considered. The approach presented by Xu and Schechter [48] composes a spectral subspace that excludes contributions from the analyte of interest. The NAS for this compound is then calculated through an orthogonal projection to an orthogonal space. In this way, a calibration method free from an optimum-factor number-selection scheme was set up. Two experimental examples, prediction of the ethanol concentrations and octane number of gasoline samples by using near-infrared spectra, were presented. The results obtained with spectral pretreatment are similar to those based on conventional PCR-calibration modelling.

One approach recently published is based on the same NAS concept, and uses the a priori knowledge of the analyte spectrum. On this information basis, the so-called hybrid-linear-analysis algorithm has been developed [49]. The method prepares for a spectral background matrix in the calibration stage by eliminating the analyte contributions. The background matrix is used to produce an orthogonal spectral vector basis by PCA, and the projection of the analyte spectrum onto this subspace is used to subtract the result from the analyte spectrum. The residual analyte spectrum is then, by definition, orthogonal to the background basis vectors, and a normalised vector is used for prediction.

A new classical least-squares/partial least-squares hybrid algorithm has been presented by Haaland and Melgaard [50]. By adding the previously reported prediction-augmented classical least-squares approach (see previous Section 7.2.2) to the CLS/PLS-hybrid algorithm, the additional benefit was gained that known or empirically derived spectral-shape information could be inserted into the hybrid method to correct for the presence of unmodelled sources of spectral variation.

Some other aspects of multivariate calibration, such as local centring, have been treated in recent publications. This has been proposed for improving the performance of statistical calibrations in cases with dominating model errors or extrapolations. For this method, a subset of calibration samples is chosen that are closest to the concentration of the sample to be analysed [51]. It is different from locally weighted regression, since the whole calibration sample population is used for calibration modelling. However, the latter approach can be applied to non-linear calibration problems (see, for example [52]). A comparison of the results from a locally weighted regression with those from an inherent non-linear method (neural networks) was recently provided by the same group [53].

One method for non-linear modelling with latent variables was proposed in the early paper of Taavitsainen and Korhonen [54]. The most common approach to model non-linearities is to use a Taylor expansion on the set of variables. Such an approach is not satisfactory in the sense that most occurring non-linearities are not modelled by a simple polynomial. A generalised PCR and PLS method for non-linear cases has been introduced. For further details, the reader is referred to the original reference. A different approach, in particular for modelling non-linear absorption spectra or measurements with a soft saturation dependency has been presented by Robertsson [55], for which the author used an extension of the PLS-regressor variables.

7.2.4 Wavelength Selection for Multivariate Calibrations

Wavelength selection has already been mentioned for multiple linear regression (MLR) to make calibration models based on a least squares solution possible. Special selection strategies have been followed for such methods, which generally rely on a reduced number of spectral variables. The number of wavelengths optimally required for spectroscopic calibration has been discussed for many years. In fact, once one uses as many variables as there are independent spectral constituents, the addition of further wavelengths should serve to reduce effects from noise. On the other hand, as more wavelengths are used, the probability of encountering additional spectral interferences increases. This progression eventually leads to a situation where the use of more variables may start to degrade the accuracy of the result. Frequently, such observations were made and reported in the literature, that MLR-calibration models based on a few selected spectral variables would outperform full spectrum methods such as PLS (see, for example [56, 57]).

For the classical approach, various criteria have been developed, among which the determinant and the condition number of the calibration matrix served as the most accepted criteria for choosing the best wavelength combination (for a listing of previous work, see Refs. in [58]). The major disadvantage is that all components are studied simultaneously with the same matrix, whilst in most cases the optimal conditions for quantitative analysis of each compound are different and should be optimised individually. The authors of the latter paper developed a theory that models analytical uncertainty in multicomponent analysis. An error indicator function was provided to predict the analytical performance by using certain spectral variables. Most informative spectral ranges can thus be identified.

Recently, considerable effort has been placed on evaluating procedures that identify spectral variables carrying useful information for the setup of robust calibration models. A selection of papers dealing with these aspects, considering search strategies such as genetic algorithms, simulated annealing or artificial neural network approaches are cited in an interesting paper by McShane et al. [59]. One of their examples for the application of the so-called peak-hopping stepwise wavelength-selection algorithm involved the choice of glucose calibration models based on near-IR spectra (see also [60]). A rapid and reliable variable-selection algorithm for statistical calibrations based on PLS-regression vector choices has been developed, which has been tested for various calibration scenarios [61–63]. As the optimum regression vector obtained from statistical calibrations contains the weights for the spectral variables needed for concentration prediction, spectral variables were chosen pairwise to provide the minima and maxima of the regression vector, but in a ranking order decided on by their coefficient weight size. The optimisation of the analytical performance of the calibration models, based on a consecutively increasing number of spectral variable pairs and optimum PLS rank, was guided by cross-validation.

Further strategies for variable selection have recently been published. One graphically oriented approach is called interval partial least squares regression and is based on dividing the spectrum into intervals of suitable size and selecting the interval with the best prediction performance. It has been compared with variable selection methods such as principal variables, forward stepwise selection and recursively weighted regression [64]. A different variable selection approach has been presented by Westad and Martens [65]. For significance testing, the estimated uncertainty variance of the PCR and PLS regression coefficients were obtained by jack-knifing [66]. After eliminating useless variables, a best combination search based on two variables was followed and provided satisfying results.

7.3 Qualitative Analysis

Unlike concentrations in quantitative analysis, qualitative sample properties that have to be related to spectral variations cannot be expressed as continuous variables, but instead have discrete values, which, for example, can represent a product identity or a good/bad product quality. In the latter situation, the suitability of the material being measured may be simply gauged. The most widely used qualitative application of near-infrared spectroscopy is certainly the verification of the assumed identity of samples, where identity is abstractly defined as the belonging to a group with known properties. With respect to this definition, NIR identification methods can be much more powerful than traditional classifications in the mid-infrared, which focus primarily on the chemical structure of a substance.

Qualitative methods known from mid-infrared spectroscopy often involve the visual interpretation of spectral features, such as the number, positions and relative intensities of characteristic peaks. Usually, this is not possible with near-infrared spectra, which consist of many broad and overlapping bands. Therefore, the same selectivity and interference problems have to be solved as described above for quantitative methods.

Chemometric methods used for grouping and classification of spectral objects from samples, compounds or materials are subdivided into supervised and non-supervised learning algorithms. The latter approach does not require any knowledge about the objects to be grouped, but instead produces the grouping (or clustering, as it is often called) itself. In the above-mentioned verification of identity, the group structure of the training set is known, so supervised learning algorithms can be used. An extensive comparison of different approaches for classification was presented by Walczak and Massart [67].

Most qualitative methods can operate either in wavelength space (some with whole spectra, whereas others rely on selected wavelengths only) or in a dimension-reduced factor space. In the latter case, principal component analysis (PCA) is the customary method for data compression by using an orthogonal matrix decomposition. In general, factorial methods are aimed at projecting the original data set from a high-dimensional space to a few transformed coordinates only. So-called score plots can be used for a graphical presentation of sample groupings.

For most qualitative algorithms an appropriate data library is required. Libraries of spectra that are representative of the natural variation in each product to be identified later can be built up during the calibration process. For automated and computer-aided verification of the identity of an unknown sample, the conventional routine method uses library searching based on spectral mapping algorithms. Before this step, some spectral preprocessing such as elimination of baseline effects and noise standardisation, etc. is usually performed on the sample spectrum. Comparison of the processed spectrum with a candidate library spectrum can be carried out based on different procedures such as correlation of spectra, similarity and distance measures or logical operations. The spectral correlation is a rather robust measure because it is not influenced by spectral offsets and globalintensity variations. Therefore, it is not suited to distinguishing between spectra that differ mainly in such effects, which may arise from a different particle size of the same product, for example. A mixture identification based on near-infrared spectra of liquids by an automated library-searching method was presented by Lo and Brown [68]. It has also been applied to textile dyes [69], and recently the identification of mixture components in gases was presented [70].

A simple distance measure is the Euclidean distance between two samples; this can be calculated both in wavelength or in a dimension-reduced factor space. Extending this measure to describe the distance between a single spectrum and a group of spectra or between two groups of spectra is done by using the group means as reference points. However, the natural "size" of the groups, which can differ from product to product, is not taken into account. In Fig. 7.2 this problem is illustrated with two unequally sized groups ("suns" and "moons") of objects in a two-dimensional space. The "unknown" sample would obviously be assigned to the wrong group when using the criterion of the smallest Euclidean distance ($d_1 < d_2$) between this object to each of the two group centres.

Such misclassifications can be avoided when not simply the nearest group is considered as the "correct" one, but when an upper distance limit is introduced, above which an unknown sample is not assigned to the group. When, furthermore, this limit is based on the standard deviation of the respective group, the effect of different cluster sizes is compensated for. By the *Mahalanobis* distance (MD) [71] this concept is incorporated into the distance measure itself by using the quotient of the Euclidean distance to a group mean and the standard deviation of this group (for a one dimensional problem). For the multidimensional for-

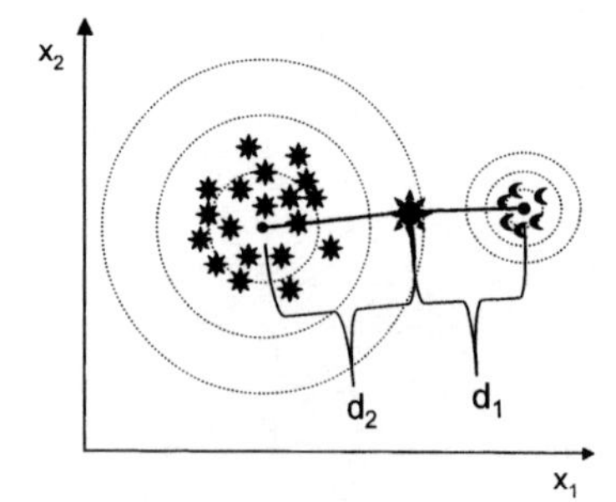

Fig. 7.2 Misclassification based upon Euclidean distance

mulation, the inverse of the variance-covariance matrix C_x is used for scaling (where $C_x = X_c' X_c$ with X_c the column-centred spectral matrix X containing n spectra in the rows measured for p variables). With this measure, the same distance limits for assignment or rejection can be used for all groups, regardless of the size of the cluster scatter. Chemometric techniques based on MD and applied in different fields such as process control and pattern recognition were presented in a recent tutorial by De Maesschalck et al. [44].

An often-used rule of thumb is to set distance limits of three times the standard deviation of the respective group (for the Euclidean distance) or, equivalently, a limit of three *Mahalanobis* distances. This works well in many cases, but especially for groups with only a few objects, it is preferable to use confidence intervals based on Fisher's t statistics, which allow the probability of a sample's belonging to a group or not to be quantified. Updated control limits were published by Whitefield et al. [72] based on the number of wavelengths and training samples.

A very simple univariate, yet full-spectrum approach is the so-called *maximum distance* method [73]. First, the univariate variance-scaled distances between the spectrum to be classified and the group mean are calculated for each data point. The resulting maximum distance of the spectrum is then defined as the maximum of all these individual distances. To avoid artificially high distances in spectral regions with a low standard variation, a minimum variance parameter is introduced as a threshold. If the standard deviation at some data points is below the minimum variance, this parameter's value is used instead of the standard deviation in the calculation of the distance. The maximum distance method is able to distinguish between very similar product groups and detects even very small deviations in product matrices but, on the other hand, reacts very sensitively to noise. Of course this approach can be used not only in wavelength space, but as well in a reduced factor space.

Distances or similarity measures are used for cluster analysis, where objects are aggregated stepwise based on the similarity of their spectral features. This technique belongs to the unsupervised pattern recognition tools (for more details, see, for example [46]). Informative dendrograms can be obtained from a hierarchical cluster analysis, in which the distances between the clusters are depicted graphically.

Within the supervised methods the linear discriminant analysis (LDA) was developed for classification, by which a way to finding an optimum decision boundary between different classes is described. The boundary or hyperplane is calculated in such a way that the variance between the classes is maximised and that within the individual classes minimised. A recent application of this technique was given by De Groot and co-workers [74] for classification of demolition waste based on near-infrared reflectance spectra. By simulated annealing [75], a number of most discriminating wavelength regions were selected, and further data preprocessing techniques, such as standard normal variate transformation and others (see also Section 7.4.2.5), were applied before classification.

Another multivariate method, PLS regression, that was originally developed for quantitative evaluation, can also be used in qualitative analysis. Its ability to extract spectral information that is correlated to the dependent variables can be utilised with discrete dependent variables as follows. The idea is to assign numbers to each group and to use these numbers as semicontinuous variables. The simplest approach, to use one y variable with values $1, 2, \ldots g$ for g groups or products, will not work properly because such an ordered sequence is very unlikely to be present (directly or in terms of linear combinations) in the spectral data. Therefore the PLS2 method, modelling several dependent variables simultaneously, is used with a setup of the $\mathbf{Y}$ matrix as pictured in Fig. 7.3. The $\mathbf{Y}$ matrix is centered and normalised before calibration, and suitable limits for identification and rejection have to be defined. One example that uses such a strategy was published by Cleve et al. [76]. The authors applied the qualitative-analysis technique to the identification of textiles, textile coatings and moisture measurements. Apart from quality control, the tools could be used for on-line textile recycling processes.

The soft independent modelling of class analogies (SIMCA) method [77] is based on a library of principal component analysis (PCA) decompositions for the spectral sets of each product. Spectra of new samples to be classified are projected into the space of each model in the library. The distance to each individual model, expressed as the residual spectral variance, and the leverage of the new sample's spectrum, compared to the mean leverage of each class, are used as criteria to determine whether the new sample belongs to one class, to several classes or to no class at all.

An obvious advantage of the SIMCA method is that the PCA models can be constructed with different numbers of principal components for each product, thus the complexity of the individual products can be modelled precisely. A drawback is that meaningful limits for residuals and leverages can only be obtained with "clean" calibration data sets, which are free of outliers. In practice it is often difficult to establish "satisfactory" PCA models for heterogeneous products. An application of this approach for the identification of pharmaceutical excipients that

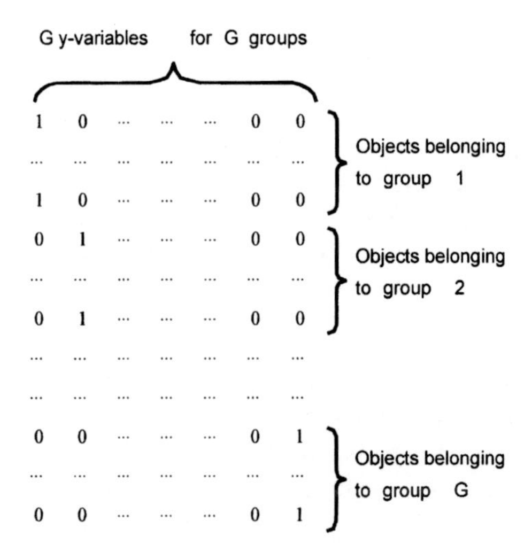

Fig. 7.3 $\mathbf{Y}$ matrix for a PLS2-based, qualitative calibration

uses near-infrared reflectance spectra was presented by Krämer and Ebel [78]. Classification was improved by spectral pretreatments of multiplicative scatter correction (see also Section 7.4.2.5), derivation and wavelength selection.

Another option for classification is the use of artificial neural networks (ANN) (see also Section 7.5.1), which have been used intensively for pattern recognition purposes [79, 80]. An advantage of ANN is that non-linear dependencies can be implemented. A quality control task was posed as a decision problem by Sánchez et al. [81], who used a so-called Genetic Inside Neural Network for the control of water content in an penicillin derivative employing near-infrared diffuse reflectance spectra.

7.4 Signal Processing

Within the context of the development of analytical methods in near-infrared spectroscopy, signal processing is used to transform (spectral) data *prior* to calibration. Therefore signal processing methods are often referred to as data pretreatment methods. Generally, if the analytical conditions have been optimised for maximum signal intensity, then any increase in signal-to-noise ratio can be further achieved by reducing the noise level arising from random noise, baseline effects or spectral interferences.

In this section, following an introductory discussion about the necessity and usefulness of data pretreatment in combination with multivariate calibration methods, the most commonly used techniques are described with respect to the effects they are able to correct. Finally, a recently presented approach for Orthogonal Signal Correction is introduced and briefly discussed.

7.4.1 Why Data Pretreatment?

There have been intensive discussions on the subject of data pretreatment, although statistical calibration methods are popular for one striking reason, namely to provide prediction models in the case of unidentified compounds, as discussed in detail above. Even of greater importance is, when spectra are influenced by effects that cannot be easily quantified by reference analysis. In practice, the value of data pretreatment must be seen in the improvements that the calibration problem is better posed after processing.

In modern NIR spectroscopy of complex composite samples, such effects are quite common. Frequent sources of spectral variations are:

- the interaction of compounds, for example, through intermolecular hydrogen bonds
- light scattering from solid samples or cloudy liquids
- poor reproducibility in the measurement process itself, for example, through pathlength variations

- spectral distortions caused by the spectrometer hardware, such as:
  - baseline drift
  - wavelength shifts
  - effects from detector non-linearity or stray light
  - noise from the detector, amplifier and analogue-digital converter.

Such interference can violate the assumptions upon which the model equations are based. For example, the simple linear relationship as stated by Beer's law does not hold completely true, and the additivity of individual spectral responses is not guaranteed. With diffuse-reflectance spectra, suitable signal transformations can yield an improved linearity on the ordinate scale. In this context, the use of –log (R) or Kubelka-Munk (K-M) transformation has been discussed [82]. As pointed out by Griffiths [83], there are more severe distortions in a K-M transformed spectrum due to baseline shifts. In general, the aim of signal preprocessing is to reduce, eliminate or standardise the impact of the above-mentioned effects on the spectral data without influencing the spectroscopic information needed for prediction.

At first glance, this may appear to be a minor problem when using bilinear modelling techniques such as Partial Least Squares (PLS) regression. It is well known that these methods are able to approximate such aberrations, at least to a certain extent, by utilising more factors (latent variables) than the degrees of freedom (chemical rank) as calculated from the number of substances in the product sample matrix would suggest.

However, the approach of including all types of spectral variations into the training set, regardless of whether they are meaningful for prediction or not, and taking account of them with additional latent variables suffers from some severe drawbacks:

- The representative training set should include all variations to be expected in the spectra of future samples. This applies to spectral effects caused by changing concentrations of components, as well as to variations caused by the above-mentioned interferences. All types of variations should be as uncorrelated as possible within the training set to avoid spurious correlations (an example for this is given in Section 13.3.3). For such reasons, the experimental design for calibration and validation tends to become more complex and the number of required calibration samples rises with the increasing number of sources of variation.
- The development of "statistical" calibrations can involve several visualisation steps. The graphical presentation of objects or variables in the factor space is an especially valuable and frequently used tool also for outlier diagnostics. Unfortunately, human visual-recognition ability is limited to a maximum of three simultaneously displayed dimensions. Therefore, these graphical tools tend to loose their practicability as the number of factors increases.

In conclusion, carefully designed data pretreatment algorithms can help to reduce the model complexity so that more easily interpretable methods are achieved. Often these methods are more robust against unexpected perturbations in future spectra than models based upon non-pretreated spectra. On the other hand, redirecting this task to latent variables in a multivariate calibration requires all "non-predictive" variation to be represented in the calibration data set. However, and this will be discussed in more detail in the algorithm descriptions below, most pretreatment methods also bear the potential danger of influencing a useful part of spectral information. Thus, the spectroscopist is still responsible for the choice of a suitable technique, including the proper optimisation of pretreatment parameters.

7.4.2 Techniques and Algorithms

7.4.2.1 Local Filters

For special spectral data pretreatment techniques, the calculation of the processed signal values $\overset{*}{x}_1, \overset{*}{x}_2, \ldots, \overset{*}{x}_K$ from the original data $x_1, x_2, \ldots, x_K$ cannot be described by a simple point-to-point mapping function f with $\overset{*}{x}_k = f(x_k)$. Instead, each $\overset{*}{x}_k$ is expressed as a function of the local neighbouring original signal values $x_{k-s}, \ldots, x_{k+s}$ within an interval of $2s + 1$ points situated symmetrically around the data point to be calculated. This interval is referred to as a spectral *window*. Consequently, s data points at each end of the spectrum cannot be transformed because the "neighbourhood" is missing, with the consequence that these points have to be omitted or extrapolated.

In its most general form, filtering can be expressed as the moving inner (dot) product of a filter coefficient vector with the spectral data vector (within the actual spectral window) normalised by a term that adjusts for the scaling introduced by the filter coefficients (also called convolution).

7.4.2.2 Smoothing

The objective of smoothing spectral data is the reduction of noise, which can be described as random high-frequency perturbations. This random behaviour allows for an effective noise suppression by averaging multiple measurements. The co-addition of interferograms in FT-NIR spectroscopy, for example, serves such a purpose. For obvious reasons (measurement time, data storage and computational speed) this solution is not applicable in many situations. All other reduction methods make use of the high-frequency band-limited nature of noise and, therefore, can be regarded as low-pass filters. The drawback is that a sub-optimal cut-off frequency will also influence useful high-frequency parts of the signal (in spectroscopy: narrow peaks), so that the observed spectral resolution decreases. These effects have to be balanced against each other by a proper adjustment of the algorithm's parameters.

Two techniques that use local smoothing filters in the frequency domain have become popular. The simple filter coefficient vector for the moving average method is $[1 \ldots 1]/N$, so that the filtered signal is the arithmetic mean of all original N signal values within the corresponding spectral window. This method achieves satisfactory smoothing results, but strongly influences the height as well as the area of spectral peaks.

The idea behind a more advanced approach, known as the Savitzky-Golay method [84], is to fit a low-degree polynomial through the data points within the local spectral window and to derive the processed signal values from the polynomial's function. This can equally be expressed as a local convolution filter. For a polynomial degree of zero, all filter coefficients are the same and the moving average, as described above, is achieved. With higher-order polynomials, the individual weights derived from the polynomial coefficients are not the same for all data points within the spectral window to give a weighted moving average. Quadratic and cubic polynomials usually yield the best results for which the window width certainly has to be adjusted according to the band halfwidth of the spectral features to be smoothed. The segment size also controls the extent of the smoothing effect. As the segment increases, both noise and spectral resolution decrease. In Fig. 7.4 the effects of smoothing methods are illustrated. Sets of Savitzky-Golay filter coefficients for several segment sizes, as well as for the derivatives discussed in the following section can be found, for instance, in Ref. [46]. The use of the Savitzky-Golay procedure is as much traditional as representing any theoretical optimum. Therefore, Kawata and Minami [85] proposed an adaptive smoothing method by taking into account the local statistics of the observed waveform and the same criterion as the optimal Wiener filter [86], as used in signal processing techniques, which minimises the mean-squared error between estimated and true waveforms.

As is obvious from the discussion above, filtering in the Fourier domain is also possible, see for example [9] and is nowadays often implemented in conventional spectrometer software. Smoothing, derivatives calculation or even spectral resolution enhancement by Fourier self-deconvolution is possible after Fourier transformation and appropriate apodisation. Examples with a special filtering application have been published recently: Raw and digitally filtered data within the spectral range between 5000 and 4000 cm^{-1} have been used as PLS calibration input for several blood substrates in blood serum [87]. For the latter spectra, Gaussian-shaped filter re-

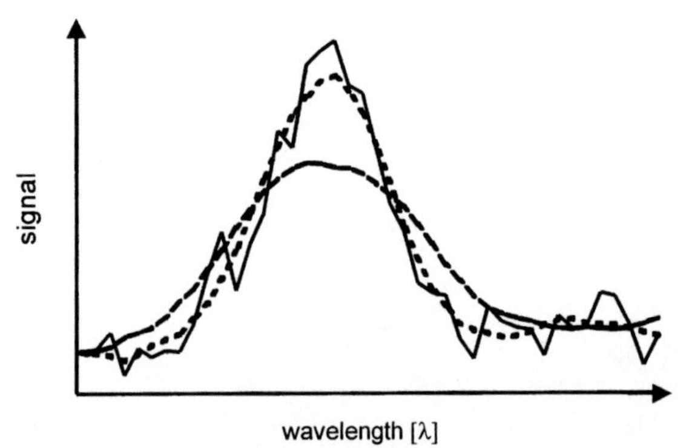

Fig. 7.4 Noisy signal of an absorption band (solid line), smoothed with a moving average filter (dashed line) and a Savitzky-Golay filter (short dashed line)

sponse functions for different digital Fourier filters were calculated to simplify the PLS calibration modelling. Thus the number of PLS factors was decreased, but it achieved the same level of performance as accessible with spectral raw data. Previous work of this group – see Refs. in [87] – was in developing temperature-insensitive calibration models, by using Fourier filtering for spectrum preprocessing.

Through Fourier transformation an option is given to compress the relevant data into a smaller number of coefficients; this had also been looked at in the past for multivariate calibration or library data compression. Recently, wavelet methods have been introduced into chemometrics, providing powerful tools for analysing, encoding, compressing, reconstructing and modelling signals. Compared with traditional Fourier techniques, signals with narrow lines and small or irregular features are often better analysed with wavelets [88]. There is also a fast wavelet transform providing an orthogonal decomposition with the wavelet basis function set. The special properties were recently demonstrated by using a numerical spectroscopy-related example [89]. Signal denoising properties and data compression were also illustrated by the group, with tests carried out on NIR spectra of wheat and different polymers [90].

An application to multivariate calibration with wavelet coefficient regression (WCR) in combination with a genetic algorithm selection scheme (see also Section 7.4.2) was presented by the same authors, comparing these results with those obtained from Fourier coefficient regression, multiple linear regression (MLR) and those from PCR and PLS [91]. Their conclusion was that WCR-calibration can be regarded as self-adaptive with the chosen strategy. Finally, an informative review on the applications of wavelet transform techniques in chemical analysis can be mentioned [92]. A recent paper also shows some tutorial character, including an application of wavelet transformations to wheat near-infrared spectra [93].

7.4.2.3 **Derivatives**

Derivatives of spectral data are used to remove or suppress constant background signals and to enhance the visual resolution. Background signals and global baseline variations are low-frequency phenomena, so derivatives can be interpreted as high-pass filters. Their corrective effects are easy to understand when one looks at the influence of (subsequent) derivatives on the graph of a polynomial. Each derivative reduces the polynomial order by one, so a constant offset is removed. It transforms the linear term into a constant one, thus removing linear tilting of the graph and so on. In spectroscopic applications, the second derivative is popular but it transfers peak maxima into minima and vice versa. In combination with the above-mentioned resolution enhancement, this is a valuable tool for identifying weak peaks that are not visible in the original spectrum. Fig. 7.5 shows the resolution enhancement with synthetically constructed spectra.

"True" derivatives of spectral data cannot be calculated because spectra are digitally stored as a sequence of (mostly equally spaced) data points and not as continuous functions. However, the local filters used for smoothing can also be used to calculate a derivative-like pretreatment of spectral data by applying modified coef-

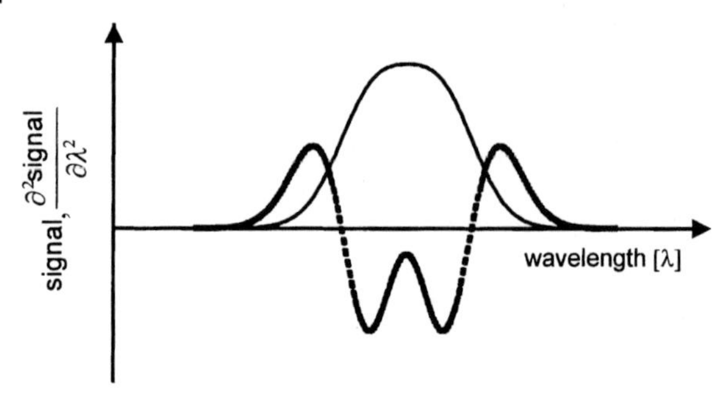

Fig. 7.5 Resolution enhancement by use of derivatives: original signal, composed of two overlapping Gaussian peaks (solid line), and scaled second derivative of the original signal (dotted line)

ficients. As a result, weighted local differences are calculated that are proportional to the gradient in the respective spectral segment. For example, the filter coefficient vector [–1 0 1] (without scaling) produces a first derivative without previous smoothing and the set [–1 –1 –1 0 1 1 1] results in the first derivative of the original spectrum, which was smoothed with a three-point moving average filter. Duplicate application of the latter coefficients is equivalent to the single use of the filter vector [1 2 3 –1 –4 –6 –4 –1 3 2 1] as can be easily verified by manual calculation.

When local filters are used for derivatives, the terminology changes slightly. Whereas the segment size equals the windows size in smoothing filters, a new parameter, the *gap* size, is introduced for derivatives. In the vector we used above for the first derivative of the smoothed original spectra, the three-point smoothing segment can be recognised as the blocks of –1 and 1, respectively. The distance between the two segments is the gap. The technique described above for calculating derivatives from spectral data was introduced by Karl Norris and is therefore often referred to as Norris differentiation.

In analogy to the smoothing techniques, the Savitzky-Golay method or the Fourier domain filtering technique can be used for the calculation of derivatives as well. An intuitive way to understand this is on the basis that the equation of the fitted polynomial is differentiated by ordinary calculus. Suitable filters can be constructed from the various polynomial coefficients. Corresponding to the smoothing techniques is the observation that Savitzky-Golay derivatives (with quadratic/cubic polynomials) yield curves that are closer to the "true" derivative than those achieved with the Norris method. In practice, these differences between the two methods do not have a significant impact on the quality of subsequent multivariate calibrations.

Side effects of derivatives on spectroscopic data are the loss of the original shape of the spectral curve and the reduction of the signal-to-noise ratio. It should be noted, however, that the latter aspect is often overvalued. The signal-to-noise ratio is certainly a useful measure for comparing, for example, the performance of spectrometers, but it is not an adequate tool for comparing the quality of spectra with or without data pretreatment. For the sake of argument, the signal of a derivative spectrum is close to zero at many points, so the signal-to-noise ratio is artificially low due to mere numerical reasons. Furthermore, the absolute signal level is rather unimportant for subsequent calibrations. Derivatives are used to remove *low-information* parts of the signal, so the decrease of a signal level runs actually parallel to a reduction to the *useful* part of the signal. Thus, the ratio of *information-bearing* signal to noise remains unchanged by derivatives, and it is this ratio that affects the quality of subsequent calibrations.

Derivative preprocessing of diffuse reflectance NIR spectra was recently compared by Brown et al. [94] with a maximum likelihood principal component analysis (MLPCA) method, which uses error covariance information obtained from sample replicate measurements. In simulation and experimental studies, the prediction performance of calibration models from derivative-preprocessed PCR and MLPCR were compared, with the result that the latter method performed as well or even better than the conventional one.

Recently, the concept of multivariate sensitivity was applied to the interpretation of the effect of spectral pretreatment methods [95]. For the examples shown, derivative methods substantially reduced the multivariate net analyte signal and, consequently, a smaller multivariate sensitivity is obtained; this results in a larger uncertainty for the regression vector estimate and consequently in larger prediction errors.

7.4.2.4 **Baseline Correction Methods**

Constant underlying spectra and other non-systematic effects that influence the global shape and the absolute level of a spectrum can easily be removed or reduced by derivative spectroscopy as mentioned above. However, often some a priori knowledge about the product sample matrix and about the nature of spectra can help in applying methods that are more simple, that preserve the main spectral shapes and suffer from less side-effects than derivative processing. For example, random offsets in spectra can be eliminated by shifting each spectrum so that the signal value at a given wavelength becomes constant for all spectra. This one-point baseline correction requires the existence and recognition of a spectral interval that is influenced only by offset shifts.

A straightforward extension towards a two-point baseline correction can eliminate linear tilting of the spectra. Here the correction for each data point is linearly interpolated between two reference points. The use of even more reference points leads to a classical method, the so-called "rubber band" baseline correction, which can be further refined by replacing the linear interpolation with the fitting of a low-degree polynomial through the reference points and subtracting its curve from the spectrum. Because the rather featureless NIR spectra from, for example, solid samples make the selection of reference points very tentative, these manually performed baseline corrections should normally be restricted to a simple offset or a linear baseline.

A practicable alternative to the rubber-band approach is to fit a low-degree polynomial through *all* data points of the spectrum and then subtract the resulting function's curve from the spectrum. This method is known as *detrending* and is often used together with the standard normal variate (SNV) method, which will be discussed in the following section.

7.4.2.5 **Multiplicative Corrections**

Frequent sources of multiplicative perturbations are changes in the optical pathlength and sometimes in the sensitivity of the detector and the amplifier. Light scattering produced from solid samples (powders, granules etc.) as well as from emulsions and dispersions results in multiplicative deviations that are dependent on the wavelength. These effects are difficult to approximate with linear factor combinations during the calibration (parameter estimation) process and none of the currently popular bilinear calibration methods is able to model them explicitly.

At this point it should be mentioned, however, that an early calibration method developed by Karl Norris, the so-called *derivative rationing regression* (or for convenience: the Norris regression) [8] is able to take into account even wavelength-dependent multiplicative perturbations. This is achieved by replacing the absorbance vectors in a multiple linear-regression equation by vectors of the ratio of the absorbances at two wavelengths. In addition, these vectors are taken from differentiated spectra, so additive perturbations are eliminated before the ratios are calculated. An approach to embed multiplicative corrections into a full spectrum factor method has been presented by Manne et al. [96].

Several ways to correct multiplicative effects within a data pretreatment step are used. The common idea is to multiply the whole spectrum that is exposed to changes from varying pathlength by a suitable value to fix a parameter to a constant value after calculating it from the raw spectral data. For example, the classical approach in mid-infrared spectroscopy is to find a peak that is influenced by pathlength changes but not by variations in the constituents, and to scale all data points so that the maximum of this peak becomes constant for all spectra. In the near-infrared spectral region, however, this is very difficult owing to overlapping peaks. Therefore the largest peak, or alternatively both the global maximum and minimum of each spectrum, is often used to calculate the scaling parameter. These approaches suffer from uncertainties caused by random errors in the spectral data. More robust methods utilise standardisation values that can be calculated from all spectral data points. Often the area between the spectral curve above the abscissa or the length of the spectral data vector is normalised.

A different approach is multiplicative scatter correction (MSC) [97]. Here each spectrum is shifted and scaled to fit a given target spectrum. The goodness of the fit is expressed as the sum of the squared differences between the data points of the transformed spectrum and the corresponding data points of the target spectrum, hence a linear regression is used to estimate the parameters for shifting and scaling. In practice, the mean spectrum of the calibration set is often chosen as the target spectrum. Generally, the target spectrum should correlate well with the spectra to be corrected. This provides a (more philosophical than technical) problem in qualitative applications in which the objective is to verify the identity of an unknown sample. Those applications are usually based on libraries with many products. When applying MSC, one would have to chose the appropriate target spectrum, in this case the mean spectrum of the "correct" product sample, which however is not known at that point in the analysis.

In such a situation (and in most other cases as well) the standard normal variate (SNV) method [98] gives nearly equivalent results without requiring additional product-dependent information. The SNV method centers each spectrum around zero by subtracting the mean and then divides each signal value by the standard deviation of the whole spectrum. If the standard deviation correlates with, for example, pathlength changes, this procedure results in a multiplicative standardisation. Finally, an interesting review of MSC-related correction methods can be found in Ref. [99].

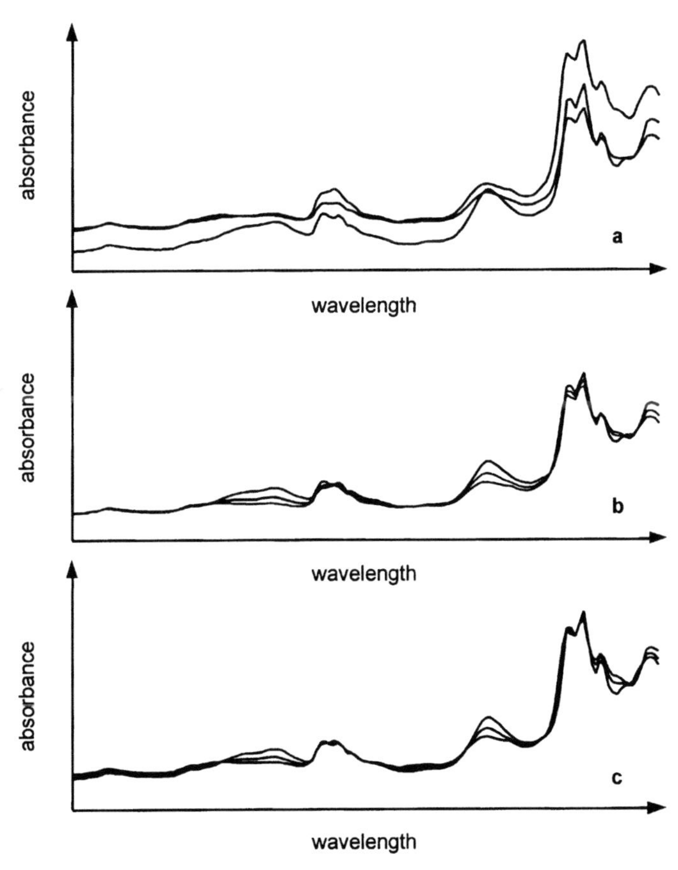

Fig. 7.6 Multiplicative correction with the MSC method: a) raw spectral data after multiplication by random scalars to simulate different optical sample pathlengths; b) original spectra; and c) corrected spectra obtained after MSC-processing

As an example of multiplicative correction methods, Fig. 7.6 demonstrates the corrective effect of MSC. The undisturbed spectra, as shown in Fig. 7.6 (b), were multiplied by random values (a) and corrected by using MSC (c).

For instrument standardisation, Brown and co-workers [100] presented an interesting modification of the MSC method. Instead of estimating the scaling parameters globally for the full spectrum, a separate scaling is calculated for each data point by restricting the regression on the target spectrum to a small spectral window around each individual data point that is to be scaled. The analogies to smoothing and differentiating filters described earlier in this chapter are obvious, and, not surprisingly, this technique can be interpreted as a finite impulse response (FIR) filter. The standardisation of instruments was achieved in this case by using the spectrum of a characteristic sample recorded on another spectrometer as the target spectrum. This localised MSC is likely to produce prominent artifacts in spectral regions that contain only small systematic variations.

A similar method for instrument standardisation, which can also be used for qualitative calibrations (see the discussion about MSC in qualitative applications above), was used by Horn and Winzen from the University of Essen [unpublished], who localised the SNV approach in an analogue way. Similar artifacts as obtained with Brown's method occur in spectral regions with a very low standard deviation. These can be avoided to a certain extent by introducing a threshold parameter. If the standard deviation in a given spectral segment drops below this predefined limit, the threshold value is used instead of the actual standard deviation.

The effect of data preprocessing methods on diffuse reflectance spectra was recently investigated for the determination of an active compound in a pharmaceutical formulation [101]. Several of the methods mentioned above were tested: normalisation, derivatives, MSC, SNV and detrending. Normalisation by dividing the spectra by the mean value of each individual spectrum was found to be least effective, whereas first and second derivatives led to best prediction PLS models, although the errors of prediction varied only little relative to calibrations based on raw data.

A recent publication on the quantitative evaluation of visible/short-wave near-infrared spectra of whole blood, which were recorded either by transmission spectroscopy or total transmission measurements with an integrating sphere, in particular for the determination of total haemoglobin concentration, also provides several clues for data preprocessing [102]. Mean centred data as input for PLS calibrations led to minimum mean-squared prediction errors, whereas other pretreatment methods yielded significantly degraded results. For conventional transmission spectra, the shortening of the broader wavelength interval of 500–800 nm to an interval of 500–560 nm with most intensive haemoglobin absorption had no effect on the average prediction error, whereas for the spectra recorded by using the integrating sphere the calibration performance was deteriorated by a factor of 1.5.

7.4.2.6 **Orthogonal Signal Correction (OSC)**

Recently, a special data pretreatment algorithm was discussed intensively for analytical near-infrared spectroscopy problems, although such projection steps have been repeatedly considered previously for removal of, for example, atmospheric water vapour absorption lines in mid-infrared spectroscopy. For the near-infrared spectroscopy of skin tissue, it is known that the first-factor spectra derived from a principal-component analysis are dominated by variance due to water absorptions (see Section 13.3.2). An orthogonal signal correction with such orthogonal spectral vectors was therefore carried out for simplifying diffuse reflectance skin spectra [103]. Regardless of these early applications, Wold's paper about orthogonal signal correction [104] started a rapid development with a series of publications and algorithms in the context of near-infrared spectroscopic calibration [105–109]. An application especially to calibration transfer was presented by Sjöblom et al. [110].

As mentioned above, NIR spectra often contain large sources of variation that are of no predictive value. Therefore, it appears intuitive to pretreat the spectral matrix by projection to remove such variance. Actually, one subtracts from the spectral matrix $\boldsymbol{X}$ such factors whose scores represent as much as possible of the variability in X, while being orthogonal to y (as in most commonly occurring cases, a one-component concentration vector is considered here). As a consequence, the first factor spectrum will be orthogonal to the covariance of $\boldsymbol{X}$ and y, which is proportional to $\boldsymbol{X}'y$ (after normalisation, by definition also the first PLS loading vector). A singular value decomposition can be used to calculate further factor spectra; for a clear description of the mathematical problem, see also [105]. OSC can be considered a variant of matrix decomposition, as achieved by PLS.

A PLS regression based on OSC-corrected spectra needs a smaller number of factors than one based on spectral raw data, and thus might be easier to interpret. As in most cases, each OSC factor will simplify a subsequent PLS calibration by one degree of freedom, but the total prediction performance (in terms of the RMSEP) does not change significantly when compared to an optimum PLS model without preceding OSC correction. However, it has been claimed that especially the so-called direct orthogonalisation presented by Westerhuis [108] saves more than one PLS-factor after considering a factor calculated for OSC.

The reduction of model complexity has been described by the paper of Ferre and Brown [109], which follows a slightly different strategy. They used additional measurements, in which for a number of representative samples the non-predictive variance from, for example, changes in temperature or particle size, was produced explicitly, in contrast to otherwise constant experimental conditions. For orthogonalisation against spectral features not related to the property of interest, the significant factors modelling the non-relevant spectral changes were used. One advantage that is claimed is that even perturbations, which are to a certain extent correlated to the property of interest, can be removed. A similar approach to that of Ferre and Brown [109] has been recently proposed by Hansen [111].

The most outstanding feature of OSC over traditional signal-processing methods is that it leaves predictive variances completely untouched. However, the simplification of a PLS calibration by a preceding OSC correction has to be seen in relation to a possibly more extensive (and expensive) setup of calibration spectra within the calibration stage. Recently, a comparison of orthogonal signal correction and net analyte signal preprocessing has been given (the example was for sample spectra recorded in the visible range). If enough factors are extracted by either or both methods, the remaining calibration problem is amenable to a classical least squares solution [112]. Orthogonal signal correction and wavelet analysis in combination with multivariate calibration was presented by Eriksson et al. [113] and demonstrated that a significant data compression down to 4% of the original matrix size is possible without loss of predictive power.

7.4.2.7 **Instrument Standardisation and Calibration Transfer**

Recalibrations that are necessary due to instruments changes with time and calibration transfer are very important topics, since industrial and research laboratories spend a lot of effort in calibrating instruments for best performance. Often a calibration method has been developed by using the laboratory research spectrometer, and calibration transfer is needed for the in-line spectrometers intended for process monitoring. An obvious example is the transfer from a Fourier-Transform spectrometer to a dispersive instrument, for example, due to different inherent instrumental line shapes and other effects [114].

In general, the responses from two spectrometers for the same sample under the same conditions will be different. Several parameters and aspects have been discussed already in Section 7.4.1; however, such differences can be traced back to major instrumental effects resulting from changes in the wavelength scale and the ordinate axis. Photometric accuracy may be affected due to stray light or detector non-linearity. Furthermore, there may be differences in the accessory optics that influence spectral intensities. For example, diffuse-reflectance spectra are influenced by parameters such as different Fresnel reflection suppression or a solid angle change in collecting backscattered radiation.

A chemometric strategy for characterising the main sources of variation affecting near-infrared spectroscopic measurements was presented by Rutan et al. [115]. For their transmittance measurements, the most significant sources of variation were found to arise from changes in cell pathlength and a variable curved background. Another comprehensive approach to instrument characterisation was presented by Workman and Coates [116], leading finally to spectrometer standardisation and calibration transferability.

If instrumental responses are very similar and remain stable enough, calibration transfer will not be an analytical performance issue. In the case of different instrument responses, one solution is the application of mathematical methods to match the instrument responses that allows the calibration transfer (so-called *instrument matching*). For example, wavelength scale calibration can be achieved by measuring an internal standard such as a polystyrene film. Systematic near-infrared calibration studies were carried out by Shenk and Westerhaus [117] for diffuse reflectance spectra of agricultural products by using simple univariate regression. In a recent publication, the optical matching of NIR reflectance monochromator instruments was presented, with either 30 sealed powdered samples of diverse types or a single ground-wheat sample and including a wavelength standardisation [118]. Yang and Kowalski treated the instrument standardisation by using multivariate regression techniques (see for example [119]). With so-called direct standardisation, response matrices on both spectrometers to be matched can be related to each other by a transformation matrix.

Several variants of these algorithms have been published in the last few years. In particular, the piece-wise direct standardisation (PDS) based on multivariate correction of the spectra should be mentioned (see, for example [120]). For the derivation of the transformation matrix, each data point on the primary instrument is reconstructed from data within a small window on the secondary instrument. The scaling coefficient vector is obtained by low-dimensional multivariate regression. This is helpful since the standardisation model is usually derived from a small subset of samples from the calibration training set. Often selection is done on the basis that most informative objects are taken into account. An informative tutorial on the principles of calibration transfer has been published by de Noord [121].

Meanwhile Fourier- and wavelet-transform approaches for the purpose of standardisation have also been investigated [122, 123]. In addition, a calibration transfer based on neural networks was previously presented [124].

A different algorithm was developed by Brown and co-workers to accomplish calibration transfer between spectrometers by application of a finite-impulse-response (FIR) filter to map the spectra from one instrument to a representative target spectrum from another instrument (often the mean of a calibration set) without the need for transfer standards [100, 125]. The FIR filter serves as a preprocessing technique, analogously to a piece-wise multiplicative scatter correction.

One approach to reducing the impact from spectrometer-to-spectrometer variations is to include these inter-instrument spectral changes to allow for a more robust calibration modelling. This strategy has been tested by Despagne et al. [126]; the authors also compared the performance of a global linear PLS method, locally weighted regression and a neural networks approach. The opposite strategy is to eliminate the sources of variation that are not intrinsically related to the analyte of interest by means of spectral preprocessing (see preceding Section). A compromise is to select spectral intervals that are least influenced by instrumental variations and experimental conditions. Finally, a comparison of ten different calibration transfer methods applied to spectral data of soluble-solids content in melons, recorded by a FT-NIR and scanning-grating-based instruments was given, in which a modified wavelet transform performed slightly better than all other chemometric algorithms [127].

7.5 New Developments

Usually, it takes several years or even decades from the first time a new method is presented until it is established and used in analytical daily routine. So the attribute "new" for the techniques introduced in this section does not contradict the fact that these methods have been topics of research for many years. However, as opposed to well-known tools such as PLS, they can only occasionally be found in commercial applications. From the overwhelming variety of new developments, the authors have selected the topics of Artificial Neural Networks (ANN) and Genetic Algorithms (GA), because each of them represents a basically new concept that is not merely a refinement or modification of existing ideas. It is certainly outside the scope of this book chapter to cover the area exhaustively, but some recent literature in the context of multivariate calibrations will be given.

7.5.1 Artificial Neural Networks

The main motivation in searching for alternatives to traditional computer architectures is the experience that computers, in spite of their superiority in speed and accuracy of numeric calculations, perform rather poorly in some tasks that are very simple for human beings. Whereas, for example, every child can distinguish between dogs and cats on seeing photos of these animals, this is a surprising difficult task for a computer.

Such seemingly simple things obviously require other techniques than the traditionally used approaches for constructing computers and for designing software. The artificial neural networks to be introduced in this section have been proven useful not only for topics of the above-mentioned type, but also for handling more classical questions, among them spectroscopic calibration problems. They are often superior to established methods when the exact nature of the observed phenomena is not completely understood and the assumed models are too far from reality. For instance, this can be the case in near-infrared spectroscopy of products with complex sample matrices.

Artificial neural networks are designed to roughly mimic the structure of biological neural networks (which form the human or animal brain and nervous system) and the way they work or are supposed to work. They consist of layers of *neurons*, which can be regarded as very simple processing units. Each neuron receives several input values and produces a single output. In Fig. 7.7 a schematic neural network with three layers is shown, which is the usual setup when a non-linear relationship between independent and dependent variables is assumed.

In Fig. 7.8 a neuron's functionality, which consists of two consecutive steps, is illustrated. First the weighted input signals are summed, then the result is transformed by a non-linear threshold function that produces the final output. Most authors regard the weights not as a property of the neuron itself, but of the receiving dendrites. For formal description, however, it is easier to use the given nota-

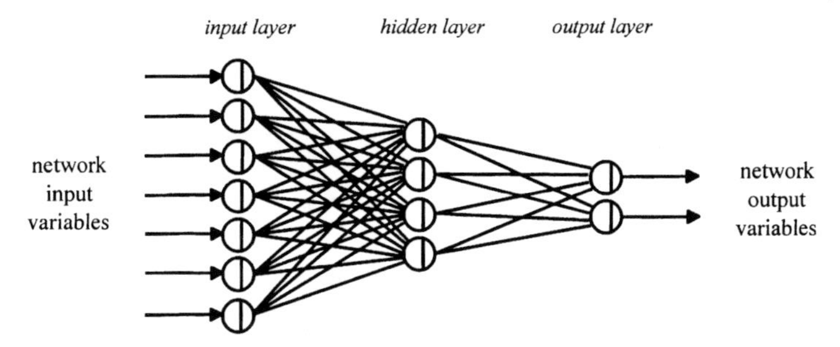

Fig. 7.7 Example of a three-layer-network topology

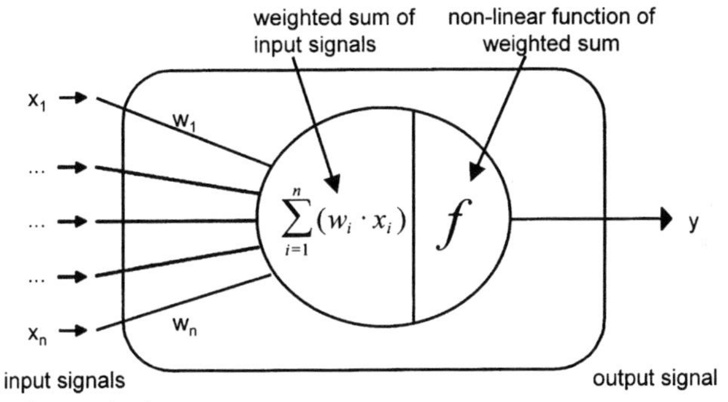

Fig. 7.8 Simplified scheme of an artificial neuron

tion. Compared with traditional computer topologies, the sum and the transfer function could be regarded as the processing units, whereas the weights of the input signals represent the memory. The transfer function itself is a threshold function that computes a "low" output below and a "high" output above a certain input value. The function types used differ mainly in their gradient.

As in other methods, calibrating means to estimate the parameters of the calibration model, based on a representative set of training data pairs x and y. In artificial neural networks, the parameters used for subsequent predictions are the weights assigned to the input signals of each neuron. The training process is iterative. Beginning with an initial set of weights, a predicted output is calculated and the error compared with the "true" y values is used to improve the value of the weights, which are again used to compute a predicted output and so on. Similarly to the numbers of independent variables of bilinear factors in other calibration techniques, the number of training cycles is a critical parameter that has to be balanced between proper modelling of systematic relations and unwanted overfitting of random relations between x and y values.

In the prediction steps, the network's input values are passed through the layers from left to right, and in the weight adjustment steps the resulting errors are passed back in the opposite direction. Therefore the most commonly used neural networks are referred to as *backpropagation networks.*

Neural networks are especially suited to modelling non-linear relationships. Basically, they can describe any functional relationship without assuming a fixed mathematical model. The drawback is that neural networks have a strong tendency to model even random, non-systematic information, that is, to overfit the calibration data. Therefore, they usually need very large training data sets. Not surprisingly, they are often used to establish global calibrations, for instance for analyser networks. An attempt to abstract from the various network topologies and learning rules has been presented by Anderson and Kaufmann [128].

The use of artificial neural networks in multivariate calibrations has been proposed by several authors. Næs et al. discussed network architecture, estimation process, as well as the relationship to linear and non-linear techniques [129]. Further applications were meanwhile published, but only two representative papers are listed here. For example, non-linear modelling was achieved with a coupled neural-network PLS-regression approach aiming at three real-life industrial data sets [130]. For process control, rugged calibrations were presented that were based on near-infrared data sets, although extreme care must be taken to avoiding any overfitting, which can especially be observed for samples clearly diagnosed as outliers [131].

Classifications and spectral library search assistance based on ANN have already been mentioned in Section 7.3 [79–81]. The use of another popular neural-network approach, the so-called *Kohonen networks,* for a qualitative problem in near-infrared spectroscopy has also been demonstrated [132]. Further applications to calibration transfer can be found in Refs. [124, 133].

7.5.2 Genetic Algorithms

Another completely different approach to copying a natural mechanism is presented by the so-called genetic algorithms. According to Darwin's theories, an evolutionary circle of reproduction, mutation and selection is the reason for the arising, developing and vanishing of species as well as individuals. The driving force behind this process is the "struggle for life" within a hostile environment. The "survival of the fittest" ensures that in the long run only those species that are best adapted to their environment have a chance. On a more abstract level, evolution can be understood as an iterative process for solving optimisation problems. If the possible solutions of a given problem are regarded as "individual beings", the parameters that could influence the solution are their individual chromosome sets.

The idea is to start with a handful of solutions with randomly selected (and therefore most likely not optimal) parameter sets and allow the individuals to evolve, that is, to reproduce, thereby combining their chromosomes. Mutations are realised by random perturbations of the chromosomes. A special function, the fitness function, is used to rate the "quality" of the individuals. Based upon this rating a decision is made as to who "survives" and who has to "die". As in nature, individuals with "good" genetic material will live longer and will have better chances to reproduce and to combine their chromosomes, so in the long run, the genetic quality of new individuals will improve.

In spectroscopic calibration problems, a chromosome set could contain the values of all the parameters one would usually vary manually to optimise the result, such as the wavelength variables used, the type of data pretreatment or the number of latent variables. The fitness function would calculate the *RMSEP* (Root Mean Square Error of Prediction) or a similar value. As this example shows, genetic algorithms are especially useful in solving combinatorial problems in which the sheer number of possible parameter combinations makes it impossible to calculate them all explicitly.

Finally, a few examples from the literature for successful GA implementation are given. A comparative study of a genetic algorithms, simulated annealing, and stepwise elimination for wavelength selection in multicomponent analysis was presented by Lucasius et al. [134]. It was found that the genetic algorithm used performed best for this optimisation. Another application - that should be mentioned from among many others – was for the selection of optimum factors needed in principal component regression, which was presented by Depczynski et al. [41]. The authors also provided an exhaustive review on recent usage of this powerful optimisation strategy. It is to be expected that such efficient chemometric algorithms will also soon be implemented in routine multivariate data analysis software packages.

7.5 References

[1] K. Esbensen, P. Geladi, *J. Chemom.* **1990**, *4*, 389.

[2] D.L. Massart, B.G.M. Vandeginste, S.N. Deming, Y. Michotte, L. Kaufmann, *Chemometrics: a Textbook*, Elsevier, Amsterdam, **1988**.

[3] D.L. Massart, B.G.M. Vandeginste, L.M.C. Buydens, S. de Jong, P.J. Lewi, J. Smeyers-Verbeke, *Handbook of Chemometrics and Qualimetrics*, Elsevier, Amsterdam, Parts A, **1997** and B, **1998**.

[4] J.J. Workman, Jr., P.R. Mobley, B.R. Kowalski, R. Bro, *Appl. Spectrosc. Rev.* **1996**, *31*, 73.

[5] P.R. Mobley, B.R. Kowalski, J.J. Workman, Jr., R. Bro, *Appl. Spectrosc. Rev.* **1996**, *31*, 347.

[6] R. Bro, J.J. Workman, Jr., P.R. Mobley, B.R. Kowalski, *Appl. Spectrosc. Rev.* **1997**, *32*, 237.

[7] B.K. Lavine, Chemometrics, *Anal. Chem.* **1998**, *70*, 209R; **2000**, *72*, 91R.

[8] H. Martens, T. Næs, *Multivariate Calibration*, Wiley, Chichester, **1989**.

[9] M.J. Adams, *Chemometrics in Analytical Spectroscopy*, Royal Soc. Chem. Cambridge, **1995**.

[10] R. Kramer, *Chemometric Techniques for Quantitative Analysis*, Marcel Dekker, New York, **1998**.

[11] K. R. Beebe, R. J. Pell, M. B. Seasholtz, *Chemometrics – A Practical Guide*, Wiley, New York, **1998**.

[12] J.H. Duckworth in *Applied Spectroscopy – A Compact Reference for Practitioners* (Eds. J. Workman, Jr., A. Springsteen), Academic Press, San Diego, Chapters 4 and 5, **1998**, pp. 93–176.

[13] H. MARK in *Encyclopedia of Analytical Chemistry* (Ed. R.A. MEYERS), Wiley, Chichester, **2000**, pp. 13587–13606.

[14] Y.-Z. LIANG, O. M. KVALHEIM, R. MANNE, *Chemom. Intell. Lab. Syst.* **1993**, *18*, 235.

[15] E. FURUSJÖ, L.-G. DANIELSSON, E. KÖNBERG, M. RENTSCH-JONAS, B. SKAGERBERG, *Anal. Chem.* **1998**, *70*, 1726.

[16] J.L. ILARI, H. MARTENS, T. ISAKSSON, *Appl. Spectrosc.* **1988**, *42*, 722.

[17] J. RIU, F.X. RIUS, *Anal. Chem.* **1996**, *68*, 1851.

[18] S. VAN HUFFEL, J. VANDEWALLE, *The Total Least Squares Problem-Computational Aspects and Analysis*, SIAM, Philadelphia, **1991**.

[19] S. VAN HUFFEL, *Recent Advances in Total Least Squares Techniques and Errors-in-Variables Modeling*, SIAM, Philadelphia, **1997**.

[20] C.H. SPIEGELMAN, J.F. BENNETT, M. VANNUCCI, M.J. MCSHANE, G.L. COTÉ, *Anal. Chem.* **2000**, *72*, 135.

[21] G.L. MCCLURE, P.B. ROUSH, J.F. WILLIAMS, C.A. LEHMANN, *Computerized Quantitative Infrared Analysis* (Ed. G.L. MCCLURE), American Society for Testing and Materials, Philadelphia, PA, **1987**, ASTM STP 934, p. 131.

[22] M.R. NYDEN, G.P. FORNEY, K. CHITTUR, *Appl. Spectrosc.* **1988**, *42*, 588.

[23] G. STRANG, *Linear Algebra and its Application*, Academic Press, New York, **1976**.

[24] D.M. HAALAND, D.K. MELGAARD, *Appl. Spectrosc.* **2000**, *54*, 1303.

[25] K. FABER, A. LORBER, B.R. KOWALSKI, *J. Chemom.* **1997**, *11*, 419.

[26] G. BAUER, W. WEGSCHEIDER, H.M. ORTNER, *Fresenius J. Anal. Chem.* **1991**, *340*, 135.

[27] K. FABER, B.R. KOWALSKI, *Fresenius J. Anal. Chem.* **1997**, *357*, 789.

[28] A. LORBER, *Anal. Chem.* **1986**, *58*, 1167.

[29] A. LORBER, K. FABER, B.R. KOWALSKI, *Anal. Chem.* **1997**, *69*, 1620.

[30] N.M. FABER, *Anal. Chim. Acta* **1999**, *381*, 103.

[31] I. FRANK, J. FRIEDMAN, *Technometrics* **1993**, *35*, 109.

[32] R. MANNE, *Chemom. Intell. Lab. Syst.* **1987**, *2*, 187.

[33] G.H. GOLUB, W. KAHAN, *SIAM J. Numerical Anal. Ser. B* **1965**, *2*, 205.

[34] S. DE JONG, A. PHATAK in *Recent Advances in Total Least Squares Techniques and Errors-in Variables Modelling* (Ed. S. VAN HUFFEL), SIAM, Philadelphia, **1997**, pp. 25–36.

[35] M. STONE, *J. Royal Statist. Soc.* **1974**, *B36*, 111.

[36] K. FABER, B.R. KOWALSKI, *Anal. Chim. Acta* **1997**, *337*, 57.

[37] T. NÆS, H. MARTENS, *J. Chemom.* **1988**, *2*, 155.

[38] A. LORBER, B.R. KOWALSKI, *Appl. Spectrosc.* **1990**, *44*, 1464.

[39] R. MARBACH, H.M. HEISE, *Chemom. Intell. Lab. Syst.* **1990**, *9*, 45.

[40] S. DE JONG, *J. Chemom.* **1993**, *7*, 551.

[41] U. DEPCZYNSKI, V.J. FROST, K. MOLT, *Anal. Chim. Acta* **2000**, *420*, 217.

[42] R.F. GUNST, R.L. MASON, *Regression Analysis and its Application*, Marcel Dekker, New York, **1980**, p. 258.

[43] R. MARBACH, H.M. HEISE, *Trends Anal. Chem.* **1992**, *11*, 270.

[44] R. DE MAESSCHALCK, D. JOUAN-RIMBAUD, D.L. MASSART, *Chemom. Intell. Lab. Syst.* **2000**, *50*, 1.

[45] D.C. HOAGLIN, R. WELCH, *American Statistician* **1978**, *32*, 17.

[46] M. OTTO, *Chemometrics – Statistics and Computer Application in Analytical Chemistry*, Wiley-VCH, Weinheim, **1999**.

[47] R. MARBACH, *J. Biomed. Optics* **2002**, 7 (1), in press.

[48] L. XU, I. SCHECHTER, *Anal. Chem.* **1997**, *69*, 3722.

[49] A.J. BERGER, T.-W. KOO, I. ITZKAN, M.S. FELD, *Anal. Chem.* **1998**, *70*, 623.

[50] D.M. HAALAND, D.K. MELGAARD, *Appl. Spectrosc.* **2001**, *55*, 1.

[51] L. LORBER, K. FABER, B.R. KOWALSKI, *J. Chemom.* **1996**, *10*, 215.

[52] V. CENTNER, D. L. MASSART, *Anal. Chem.* **1998**, *70*, 4206.

[53] F. DESPAGNE, D.L. MASSART, P. CHABOT, *Anal. Chem.* **2000**, *72*, 1657.

[54] V.-M. TAAVITSAINEN, P. KORHONEN, *Chemom. Intell. Lab. Syst.* **1992**, *14*, 185.

[55] G. ROBERTSSON, *Appl. Spectrosc.* **2001**, *55*, 98.

[56] D. JOUAN-RIMBAUD, D.L. MASSART, R. LEARDI, O.E. DE NOORD, *Anal. Chem.* **1995**, *67*, 4295.

[57] J.M. Brenchley, U. Hörchner, J.H. Kalivas, *Appl. Spectrosc.* **1997**, *51*, 689.
[58] L. Xu, I. Schechter, *Anal. Chem.* **1996**, *68*, 2392.
[59] M.J. McShane, B.D. Cameron, G.L. Coté, C.H. Spiegelman, *Proc. SPIE* **1999**, *3599*, 101.
[60] M.J. McShane, B.D. Cameron, G.L. Coté, C.H. Spiegelman, *Appl. Spectrosc.* **1999**, *53*, 1575.
[61] H.M. Heise, A. Bittner, *Fresenius J. Anal. Chem.* **1997**, *359*, 93.
[62] H.M. Heise, A. Bittner, R. Marbach, *J. Near Infrared Spectrosc.* **1998**, *6*, 349.
[63] H.M. Heise, A. Bittner, *Fresenius J. Anal. Chem.* **1998**, *362*, 141.
[64] L. Nørgaard, A. Saudland, J. Wagner, J.P. Nielsen, L. Munck, S.B. Engelsen, *Appl. Spectrosc.* **2000**, *54*, 413.
[65] F. Westad, H. Martens, *J. Near Infrared Spectrosc.* **2000**, *8*, 117.
[66] R. Wehrens, H. Putter, L.M.C. Buydens, *Chemom. Intell. Lab. Syst.* **2000**, *54*, 35.
[67] B. Walczak, D.L. Massart, *Chemom. Intell. Lab. Syst.* **1998**, *41*, 1.
[68] S.C. Lo, C.W. Brown, *Appl. Spectrosc.* **1992**, *46*, 790.
[69] C.-S. Chen, C.W. Brown, M.J. Bide, *J. Soc. Dyers Colourists* **1997**, *113*, 51.
[70] C.W. Brown, J. Zhou, *Appl. Spectrosc.* **2001**, *55*, 44.
[71] P.C. Mahalanobis, *Proc. Nat. Inst. Sci. India* **1936**, *12*, 49.
[72] R.G. Whitefield, M.E. Gerger, R.L. Sharp, *Appl. Spectrosc.* **1987**, *41*, 1204.
[73] Foss NIR Systems, *Vision-Manual*, January **1998**, Ver 1.0, Th-22.
[74] P.J. de Groot, G.J. Postma, W.J. Melssen, L.M.C. Buydens, *Appl. Spectrosc.* **2001**, *55*, 173.
[75] H. Swierenga, P.J. de Groot, A.P. de Weijer, M.W.J. Derksen, L.M.C. Buydens, *Chemom. Intell. Lab. Syst.* **1998**, *41*, 237.
[76] E. Cleve, E. Bach, E. Schollmeyer, *Anal. Chim. Acta* **2000**, *420*, 163.
[77] S. Wold, M. Sjöström, *ACS Symposium Series 52*, **1977**.
[78] K. Krämer, S. Ebel, *Anal. Chim. Acta* **2000**, *420*, 155.
[79] W. Wu, B. Walczak, D.L. Massart, S. Heuerding, F. Erni, I.R. Last, K.A. Prebble, *Chemom. Intell. Lab. Syst.* **1998**, *33*, 35.
[80] C. Klawun, C.L. Wilkins, *Anal. Chem.* **1995**, *67*, 374.
[81] M.S. Sánchez, E. Bertram, L.A. Sarabia, M.C. Ortiz, M. Blanco, J. Coello, *Chemom. Intell. Lab. Syst.* **2000**, *53*, 69.
[82] H.M. Heise, in *Encyclopedia of Pharmaceutical Technology* (Eds. J. Swarbrick, J.C. Boylan), Marcel Dekker, New York, 2nd ed., **2002**, pp. 2499–2510.
[83] P.R. Griffiths, *J. Near Infrared Spectrosc.* **1995**, *3*, 60.
[84] A. Savitzky, M.J.E. Golay, *Anal. Chem.* **1964**, *36*, 1627.
[85] S. Kawata, S. Minami, *Appl. Spectrosc.* **1984**, *38*, 49.
[86] A. Papoulis, *Probability, Random Variables, and Stochastic Processes*, McGraw-Hill, Singapore, 2nd ed., **1984**.
[87] K.H. Hazen, M.A. Arnold, G.W. Small, *Anal. Chim. Acta* **1998**, *371*, 255.
[88] M. Misiti, Y. Misiti, G. Oppenheim, J.-M. Poggi, *Wavelet Toolbox for Use with MATLAB*, The Mathwork, 1996; Web: http://www.mathworks.com
[89] U. Depczynski, K. Jetter, K. Molt, A. Niemöller, *Chemom. Intell. Lab. Syst.* **1997**, *39*, 19.
[90] U. Depczynski, K. Jetter, K. Molt, A. Niemöller, *Chemom. Intell. Lab. Syst.* **1999**, *49*, 151.
[91] U. Depczynski, K. Jetter, K. Molt, A. Niemöller, *Chemom. Intell. Lab. Syst.* **1999**, *47*, 179.
[92] A.K.-M. Leung, F.-T. Chau, J.-B. Gao, *Chemom. Intell. Lab. Syst.* **1998**, *43*, 165.
[93] K. Jetter, U. Depczynski, K. Molt, A. Niemöller, *Anal. Chim. Acta* **2000**, *420*, 169.
[94] C.D. Brown, L. Vega-Montoto, P.D. Wentzell, *Appl. Spectrosc.* **2000**, *54*, 1055.
[95] N.M. Faber, *Anal. Chem.* **1999**, *71*, 557.
[96] R. Manne, T. Karstang, *Chemom. Intell. Lab. Syst.* **1992**, *14*, 165.
[97] P. Geladi, D. MacDougall, H. Martens, *Appl. Spectrosc.* **1985**, *39*, 491–500.
[98] R.J. Barnes, M.S. Dhanoa, S.J. Lister, *Appl. Spectrosc.* **1989**, *43*, 772.
[99] I.S. Helland, T. Næs, T. Isaksson, *Chemom. Intell. Lab. Syst.* **1995**, *29*, 233.

[100] T.B. Blank, S.T. Sum, S.D. Brown, S.L. Monfre, *Anal. Chem.* **1996**, *68*, 2987.

[101] M. Blanco, J. Coello, H. Iturriaga, S. Maspoch, C. de la Pezuela, *Appl. Spectrosc.* **1997**, *51*, 240.

[102] Y.J. Kim, S. Kim, J.-W. Kim, G. Yoon, *J. Biomedical Optics* **2001**, *6*, 177.

[103] H.M. Heise, *Mikrochim. Acta [Suppl.]* **1997**, *14*, 67.

[104] S. Wold, H. Antii, F. Lindgren, J. Öhman, *Chemom. Intell. Lab. Syst.* **1998**, *44*, 175.

[105] T. Fearn, *Chemom. Intell. Lab. Syst.* **2000**, *50*, 47.

[106] C.A. Andersson, *Chemom. Intell. Lab. Syst.* **1999**, *47*, 51.

[107] B.M. Wise, N.B. Gallagher, http://www.eigenvector.com/MATLAB/OSC.html

[108] J.A. Westerhuis, S. de Jong, A.K. Smilde, *Chemom. Intell. Lab. Syst.* **2001**, *56*, 13.

[109] J. Ferre, S.D. Brown, *Appl. Spectrosc.* **2001**, *55*, 708.

[110] J. Sjöblom, O. Svensson, M. Josefson, H. Kullberg, S. Wold, *Chemom. Intell. Lab. Syst.* **1998**, *44*, 229.

[111] P.W. Hansen, *J. Chemom.* **2001**, *15*, 123.

[112] H.C. Goicoechea, A.C. Olivieri, *Chemom. Intell. Lab. Syst.* **2001**, *56*, 73.

[113] L. Eriksson, J. Trygg, E. Johansson, R. Bro, S. Wold, *Anal. Chim. Acta* **2000**, *420*, 181.

[114] J. Lin, S.-C. Lo, C.W. Brown, *Anal. Chim. Acta* **1997**, *349*, 263.

[115] S.C. Rutan, O.E. de Noord, R.R. Andréa, *Anal. Chem.* **1998**, *70*, 3198.

[116] J. Workman, J. Coates, *Spectroscopy* **1993**, *8*, 36.

[117] J.S. Shenk, M.O. Westerhaus, *Crop Sci.* **1991**, *31*, 1694.

[118] B.G. Osborne, Z. Kotwal, I.J. Wesley, L. Saunders, P. Dardenne, J.S. Shenk, *J. Near Infrared Spectrosc.* **1999**, *7*, 167.

[119] Y. Wang, B.R. Kowalski, *Anal. Chem.* **1993**, *65*, 1301.

[120] E. Bouveresse, C. Hartmann, D.L. Massart, I.R. Last, K.A. Prebble, *Anal. Chem.* **1996**, *68*, 982.

[121] O. de Noord, *Chemom. Intell. Lab. Syst.* **1994**, *25*, 85.

[122] C.S. Chen, C. Brown, C.-S. Lo, *Appl. Spectrosc.* **1997**, *51*, 744.

[123] B. Walzak, E. Bouveresse, D.L. Massart, *Chemom. Intell. Lab. Syst.* **1997**, *36*, 41.

[124] F. Despagne, B. Walczak, D.L. Massart, *Appl. Spectrosc.* **1998**, *52*, 732.

[125] S.T. Sum, S.D. Brown, *Appl. Spectrosc.* **1998**, *52*, 869.

[126] F. Despagne, D.L. Massart, P. Chabot, *Anal. Chem.* **2000**, *72*, 1657.

[127] C.V. Greenssill, P.J. Wolfs, C.H. Spiegelman, K.B. Walsh, *Appl. Spectrosc.* **2001**, *55*, 647.

[128] G.G. Andersson, P. Kaufmann, *Chemom. Intell. Lab. Syst.* **2000**, *50*, 101.

[129] T. Næs, K. Kvaal, T. Isaksson, C. Miller, *J. Near Infrared Spectrosc.* **1993**, *1*, 1.

[130] G. Anderson, P. Kaufmann, L. Renberg, *J. Chemom.* **1996**, *10*, 605.

[131] P.J. Gemperline, *Chemom. Intell. Lab. Syst.* **1997**, *39*, 29.

[132] A. Niemöller, *Wavelet-Analyse als chemometrisches Werkzeug: Analytische Anwendungen in der NIR-Spektrometrie*, Dissertation, Universität-GH Duisburg, Germany, **1999**.

[133] L. Duponchel, C. Ruckebusch, J.P. Huvenne, P. Legrand, *J. Mol. Struct.* **1999**, *480/481*, 551.

[134] C.B. Lucasius, M.L.M. Beckers, G. Kateman, *Anal. Chim. Acta* **1994**, *286*, 135.

8
Two-Dimensional Near-Infrared Correlation Spectroscopy

YUKIHIRO OZAKI

8.1
Introduction

The idea that one can simplify the visualisation of complex spectra consisting of many overlapped bands by spreading spectral peaks over the second dimension was first realised in NMR spectroscopy about twenty years ago [1–4]. Nowadays, two-dimensional correlation analysis is an essential technique in NMR. The idea of 2D correlation spectroscopy in vibrational spectroscopy was proposed by Noda in 1986 as 2D mid-infrared (MIR) correlation spectroscopy [5–7]. In this 2D MIR, a system is excited by an external perturbation that induces dynamic fluctuations of MIR signals, and a simple cross-correlation analysis is applied to sinusoidally varying dynamic MIR signals to generate a set of 2D MIR correlation spectra [5–7]. Since dynamic 2D MIR spectra are powerful in emphasising spectral features not readily observable in conventional one-dimensional spectra, 2D MIR correlation spectroscopy has been successful in the investigation of systems stimulated by a small-amplitude mechanical or electrical perturbation [8–17]. However, in this original approach, the time-dependent behaviour (i.e., waveform) of dynamic spectral-intensity variations must be a simple sinusoid in order to allow the original data analysis scheme to be employed effectively [5–7]. Therefore, in 1993 Noda [18] presented a more generally applicable, yet reasonably simple, mathematical formalism for constructing 2D correlation spectra from any transient or time-resolved spectra having an arbitrary waveform. He named the new 2D correlation spectroscopy generalised 2D correlation spectroscopy [18–20]. The newly proposed formalism can be applied to various types of spectroscopy, including near-IR (NIR) and Raman spectroscopy [21–50].

Another 2D vibrational spectroscopy was proposed by Barton et al. [51–54] in 1992. This approach to 2D spectroscopy is a statistical one and has its basis in the one-dimensional correlations that are obtained through the application of chemometrics. The method utilises cross correlation by least squares to assess changes in two regions, such as NIR and MIR, that result from changes in sample composition. Recently, 2D correlation spectroscopy has made two significant leaps forward [55–61]. One is the proposal of sample-sample 2D correlation spectroscopy [55–58, 60, 61] and the other is the birth of statistical 2D correlation spectroscopy

[59, 61]. In sample-sample correlation spectroscopy, generalised 2D correlation spectra with sample axes are generated instead of creating 2D spectra with axes from the variable (wavenumber, wavelength, etc.) [55]. In the usual generalised 2D correlation spectroscopy (variable-variable correlation spectroscopy), a correlation between bands is discussed, while in sample-sample correlation spectroscopy, one can discuss the concentration dynamics directly. Statistical 2D correlation spectroscopy differs from generalised 2D correlation spectroscopy in that the former abstracts spectral features by pretreatment and by 2D maps that are limited by the correlation coefficients in the range from 1 to –1 [59]. The idea of statistical 2D correlation spectroscopy extensively exploits the original approach by Barton et al. [51]. Relative to the 2D approach of Barton et al., several improvements have been made for the objects and targets of correlation analysis, as well as a relatively simple linear-algebra presentation that the methodology utilised [59]. In this section the principles and advantages of generalised 2D correlation spectroscopy, the 2D correlation spectroscopy of Barton et al. [51] and sample-sample correlation spectroscopy are introduced.

8.2 Generalised Two-Dimensional NIR Correlation Spectroscopy

8.2.1 Background

A conceptual scheme for obtaining generalised 2D correlation spectra is illustrated in Fig. 8.1. In this 2D correlation method, a variety of external perturbations can be applied to stimulate the system of interest [18–20]. The excitation and subsequent relaxation processes caused by the perturbations are monitored with many different types of electromagnetic probes. In order to construct generalised 2D correlation spectra from perturbation-induced dynamic fluctuations of spectroscopic signals, a set of dynamic spectra must first be calculated (see Section 8.2.2). The generalised 2D correlation method is designed to handle signals fluctuating as an arbitrary function of time, or any other physical variable such as temperature, pressure or even concentration [18–50]. Moreover, it is also possible to construct a set of 2D correlation spectra based on two independent spectroscopic analyses by using different electromagnetic probes (2D heterospectral correlation spectrum).

The advantages of generalised 2D correlation spectroscopy may be summarised as follows [18–50]:

- Enhancement of spectral resolution by spreading spectral peaks over the second dimension.
- Simplification of complex spectra consisting of many overlapping peaks.
- Band assignments through correlation analysis of bands selectively coupled by various interaction mechanisms.

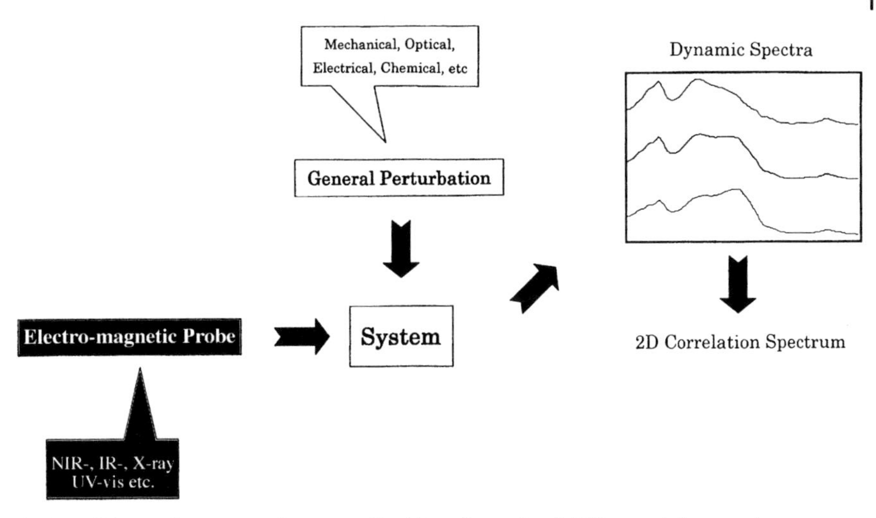

Fig. 8.1 Scheme for constructing generalised two-dimensional (2D) correlation spectra

- Investigation of various inter- and intramolecular interactions through selective correlation of peaks.
- Probing of the specific order of the spectral intensity changes.

Generalised 2D correlation spectroscopy is particularly powerful in the NIR region where spectra are often very complicated owing to the overlapping of a number of overtones and combination bands [20, 26, 27, 30–36, 39, 48–50]. Generalised 2D NIR correlation spectroscopy provides the intriguing possibility of correlating various overtone and fundamental bands to establish unambiguous assignments in the NIR region [20, 26, 27, 45, 48]. 2D NIR-MIR and 2D NIR-Raman heterospectral analyses are also useful for investigating the assignments of NIR bands [20, 35, 47].

The first example of generalised 2D NIR correlation spectroscopy was concerned with temperature-dependent spectral variations of oleyl alcohol in the pure liquid state [26]. Since that time, it has been employed for studying the temperature-dependent spectral variations of various compounds such as *N*-methylacetamide (NMA) [27] and Nylon 12 [30]. Generalised 2D NIR correlation analysis has been extensively applied to concentration-dependent spectral changes in protein solutions at various temperatures [20, 36, 49] and to composition-dependent spectral changes in polymers [20, 43, 47]. Two-dimensional NIR-MIR heterospectral correlation analyses of Nylon 12 [35] and ribonuclease A [50], and two-dimensional NIR-Raman heterospectral correlation analyses of polymer blends [20, 47] have also been reported.

8.2.2
Mathematical Treatment

Let us consider a time-dependent fluctuation of spectral intensity $y(\nu, t)$ observed for a period of time between $-T/2$ and $T/2$. The variable ν can be any appropriate physical quantity, for example, NIR wavenumber/wavelength, Raman shift, or even X-ray diffraction angle [18, 62]. In order to construct generalised 2D correlation spectra, it is first necessary to calculate the dynamic spectrum. It is defined as:

$$\tilde{y}(\nu, t) = \begin{cases} y(\nu, t) - \bar{y}(\nu) & \text{for } -T/2 \leq t \leq T/2 \\ 0 & \text{otherwise} \end{cases} \tag{8.1}$$

where $\bar{y}(\nu)$ is the reference spectrum. Although the selection of the reference spectrum is somewhat arbitrary, $\bar{y}(\nu)$ is usually set to be the static or time-averaged spectrum, which is defined as

$$\bar{y}(\nu) = \frac{1}{T} \int_{-T/2}^{T/2} y(\nu, t) dt \tag{8.2}$$

As the next step, one must Fourier transform the dynamic spectra measured in the time domain into the frequency domain [18]. The forward Fourier transform $\tilde{Y}_1(\omega)$ of the dynamic spectral intensity fluctuations $\tilde{y}(\nu_1, t)$ observed at some spectral variable ν_1 is given by:

$$\begin{aligned} \tilde{Y}_1(\omega) &= \int_{-\infty}^{\infty} \tilde{y}(\nu_1, t) \mathrm{e}^{-i\omega t} dt \\ &= \tilde{Y}_1^{\mathrm{Re}}(\omega) + i\tilde{Y}_1^{\mathrm{Im}}(\omega) \end{aligned} \tag{8.3}$$

where $\tilde{Y}_1^{\mathrm{Re}}(\omega)$ and $\tilde{Y}_1^{\mathrm{Im}}(\omega)$ are the real and imaginary components of the complex Fourier transform of $\tilde{y}(\nu_1, t)$. The Fourier frequency ω represents the individual frequency component of the time-dependent variation of $\tilde{y}(\nu_1, t)$. Similarly, $\tilde{Y}_2^*(\omega)$, the conjugate of the Fourier transform of the dynamic spectral intensity $\tilde{y}(\nu_2, t)$ at the spectral variable ν_2 is given by:

$$\begin{aligned} \tilde{Y}_2^*(\omega) &= \int_{-\infty}^{\infty} \tilde{y}(\nu_2, t) \mathrm{e}^{+i\omega t} d \\ &= \tilde{Y}_2^{\mathrm{Re}}(\omega) - i\tilde{Y}_2^{\mathrm{Im}}(\omega) \end{aligned} \tag{8.4}$$

Now one can define the complex 2D correlation intensity between $\tilde{y}(\nu_1, t)$ and $\tilde{y}(\nu_2, t)$ as follows:

$$\Phi(\nu_1, \nu_2) + i\Psi(\nu_1, \nu_2) = \frac{1}{\pi T}\int_0^{\infty} \tilde{Y}_1(\omega) \cdot \tilde{Y}_2^*(\omega) d\omega \tag{8.5}$$

The real and imaginary components of the complex 2D correlation intensities, $\Phi(\nu_1, \nu_2)$ and $\Psi(\nu_1, \nu_2)$, are referred to, respectively, as the generalised synchronous and asynchronous correlation spectra of the dynamic spectral intensity variations [18]. The synchronous spectrum represents the simultaneous or coincidental changes of spectral intensities at ν_1 and ν_2, while the asynchronous spectrum represents sequential, or unsynchronised, variations. The time used in the above 2D correlation analysis can actually be regarded as a general variable which could be replaced by any other reasonable physical variable, such as temperature, pressure or distance [18].

It is rather cumbersome to carry out the calculation of actual correlation intensities based upon the above definition for a limited number of spectral data. Therefore, a simpler algorithm that is based on the discrete Hilbert transform is often used [19, 20, 62]. It enables the 2D correlation to be calculated more efficiently and easily. In this case, the 2D correlation spectra can be calculated by the following equations:

$$\Phi(\nu_1, \nu_2) = \frac{1}{T}\int_{-T/2}^{T/2} \tilde{y}(\nu_1, t) \cdot \tilde{y}(\nu_2, t) dt$$

$$\Psi(\nu_1, \nu_2) = \frac{1}{T}\int_{-T/2}^{T/2} \tilde{y}(\nu_1, t) \cdot \tilde{z}(\nu_2, t) dt \tag{8.6}$$

where the orthogonal spectrum $\tilde{z}(\nu_2, t)$ is a time-domain Hilbert transform of $\tilde{y}(\nu_2, t)$. A more detailed description of the mathematical background of the generalized 2D correlation spectroscopy is found in reference [62].

8.2.3 Properties of Generalised Two-Dimensional Correlation Spectra

Fig. 8.2(a) and (b) shows schematic contour maps of synchronous and asynchronous correlation spectra, respectively (the terms *synchronous* and *asynchronous* are always used even when the spectral variation is measured as a function not of time but of another physical variable). A one-dimensional reference spectrum is provided above and to the left of the contour maps to show the basic features of the spectra of the system during the experiment. A synchronous spectrum is symmetric with respect to a diagonal line corresponding to the spectral coordinates $\nu_1 = \nu_2$. The intensity of peaks located at diagonal positions corresponds to the autocorrelation function of the spectral intensity variations that are observed during a period T. These peaks are therefore referred to as "autopeaks". The autopeaks represent the overall extent of the dynamic variations in spectral intensity.

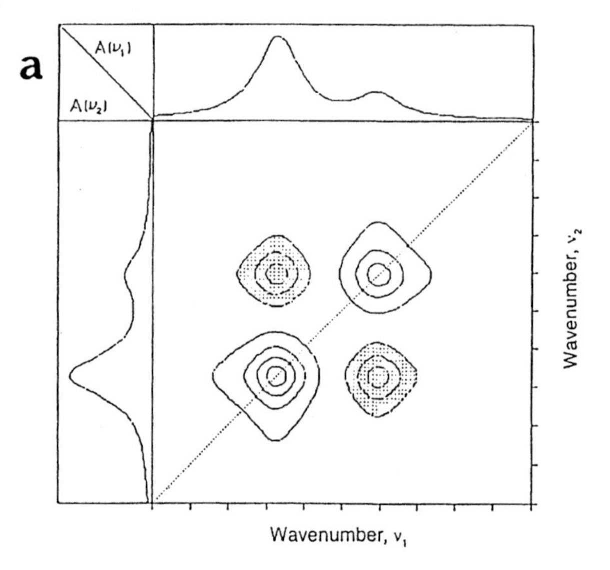

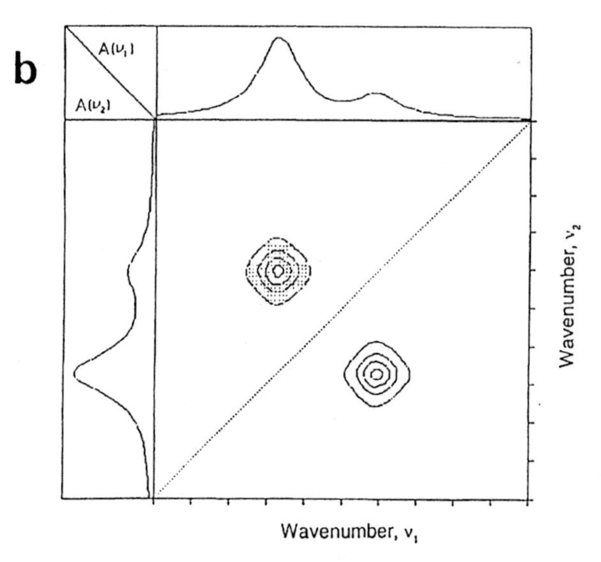

Fig. 8.2 Schematic contour maps of (**a**) synchronous and (**b**) asynchronous spectra. A one-dimensional reference spectrum is also provided above and to the left of each 2D map. (original figures were prepared by Noda)

Thus, regions of a dynamic spectrum that change intensity to a greater extent show stronger autopeaks, while those remaining constant give very weak or no autopeaks.

Synchronous cross peaks appearing at off-diagonal positions represent the simultaneous changes of spectral signals at two different wavenumbers. In other words, a synchronous change suggests the possible existence of a coupled or related origin of the spectral intensity variation. Positive cross peaks indicate that the spectral intensities at corresponding wavenumbers are either increasing or decreasing together as functions of physical variables during the observation period, while negative cross peaks (usually marked by shading) show that one of the intensities is increasing while the other is decreasing.

An asynchronous 2D correlation spectrum that consists exclusively of off-diagonal cross peaks and is antisymmetric with respect to the diagonal line provides information complementary to the synchronous spectrum. It represents sequential, or unsynchronised changes of spectral intensities measured at ν_1 and ν_2. An asynchronous cross peak develops only if the intensities of two spectral peaks vary out of phase (i.e., delayed or accelerated) with each other for some Fourier frequency components of signal fluctuation. This feature of asynchronous spectra is a powerful tool in differentiating overlapping NIR bands that arise from different spectral origins or moieties. By convention, a shaded asynchronous cross peak indicates that the event (increase or decrease of intensity) observed at wavenumber ν_1 occurs later than the event observed at wave number ν_2. The unshaded region of an asynchronous spectrum indicates the opposite.

Examples of the applications of generalised 2D correlation spectroscopy to the NIR region will be described in Chapter 9.

8.3
Two-Dimensional NIR Correlation Spectroscopy Proposed by Barton et al.

Barton et al. [51] proposed another 2D correlation spectroscopy as a means of assisting with qualitative spectral interpretation. The method employs cross correlation by least squares in order to assess spectral variations in two regions, such as the NIR and MIR regions, which arise from variations in sample composition. By using this technique, it is possible to obtain a correlation of the spectral intensities between two kinds of spectroscopy.

Fig. 8.3 shows an example of a contour map plot of the NIR versus MIR against the coefficient of determination (R^2). The map in Fig. 8.3 was obtained by Barton et al. [51] for complex agricultural samples that differed in wax (cuticle), carbohydrate, protein and lignin content. One-dimensional MIR and NIR spectra of the samples are shown above and to the right of the contour map, respectively. A high R^2 value at a particular point means that there is a high correlation of the intensities between the MIR and NIR bands that are giving that crossing point.

The correlation coefficient (r) between two sets of values $\boldsymbol{X}$ and $\boldsymbol{Y}$ is calculated by:

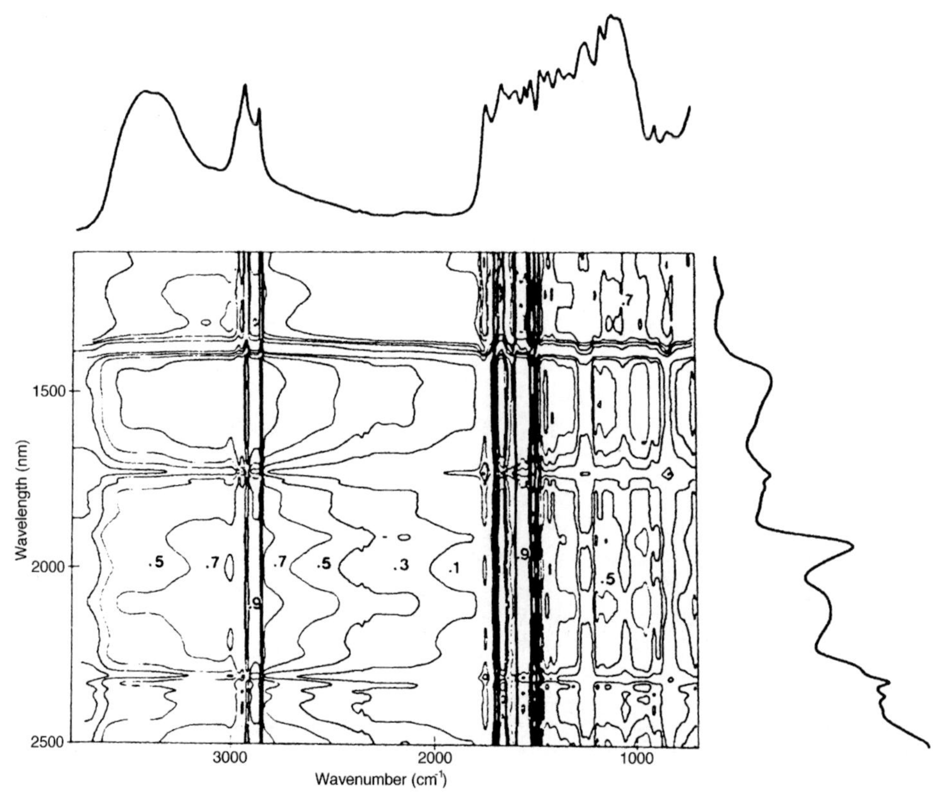

Fig. 8.3 A contour map plot of the NIR versus MIR against the coefficient of determination (R^2). The numerals on the contours are R^2 rounded to the nearest tenth. The map in this figure is a broad-range map which depicts the general shape of correlation over the entire regions (NIR and MIR). The number of contours in this case is 5, which shows the effects without appearing overly "busy". The R^2 values for the contours are 0.1, 0.3, 0.5, 0.7 and 0.9, respectively. (Reproduced from ref. [51] with permission. Copyright (1992) Society for Applied Spectroscopy)

$$r = n\sum xy - \left(\sum x\right)\left(\sum y\right) / \left\{\left[n\sum x^2 - \left(\sum x\right)^2\right]\left[n\sum y^2 - \left(\sum y\right)^2\right]\right\}^{1/2} \tag{8.7}$$

where $\boldsymbol{X}$ is an array of n measurements regressed upon an array $\boldsymbol{Y}$, containing n corresponding measurements. This equation yields a single correlation coefficient for the linear regression of the line described by the x, y pairs in the two arrays.

This process can be extended into another dimension by regressing the columns of two separate matrices with the same number of rows, but different numbers of columns. Let us consider two matrices $\boldsymbol{X}$ and $\boldsymbol{Y}$

$$X\begin{pmatrix} x_{11} & x_{12} & \cdots & x_{1px} \\ x_{21} & x_{22} & \cdots & x_{1px} \\ \cdots & \cdots & \cdots & \cdots \\ x_{n1} & x_{n2} & \cdots & x_{npx} \end{pmatrix} \quad (8.8)$$

$$Y\begin{pmatrix} y_{11} & y_{12} & \cdots & y_{1py} \\ y_{21} & y_{22} & \cdots & y_{2py} \\ \cdots & \cdots & \cdots & \cdots \\ y_{n1} & y_{n2} & \cdots & y_{npy} \end{pmatrix} \quad (8.9)$$

We can regard the matrices $\boldsymbol{X}$ and $\boldsymbol{Y}$ as an $n \times p_x$ matrix of MIR measurements and an $n \times p_y$ matrix of corresponding NIR measurements, respectively. Here, n represents the number of spectra in each matrix, and the rows of the $\boldsymbol{X}$ and $\boldsymbol{Y}$ matrices contain the MIR and NIR spectra of the individual samples, respectively. The numbers of data points in the MIR and NIR spectra are given by p_x and p_y, respectively. A column of the $\boldsymbol{X}$ or $\boldsymbol{Y}$ matrix represents the spectral responses at a single wavelength in the MIR or NIR, respectively. Note that p_x and p_y are not required to be equal and, in most cases, will be different. The matrices have the same number of elements in each column (n), so that a column of $\boldsymbol{X}$ can be regressed upon any column of $\boldsymbol{Y}$ to calculate a correlation coefficient. Regressing all possible combinations of the columns of $\boldsymbol{X}$ against the columns of $\boldsymbol{Y}$ creates a new matrix $\boldsymbol{R}$, which has the dimensions $p_x \times p_y$, contains the correlation coefficients for each combination and is calculated by:

$$\boldsymbol{R}(i,j) = n\sum x_j y_i - \left(\sum x_j\right)\left(\sum y_i\right) / \left\{\left[n\sum x_j^2 - \left(\sum x_j\right)^2\right]\left[n\sum y_i^2 - \left(\sum y_i\right)^2\right]\right\}^{1/2} \quad (8.10)$$

The subscripts i and j are the column indices of the $\boldsymbol{X}$ and $\boldsymbol{Y}$ matrices and range from 1 to p_x and 1 to p_y, respectively. They also indicate the corresponding element in the $\boldsymbol{R}$ matrix of correlation coefficients. Therefore, the calculated matrix element $\boldsymbol{R}(i,j)$ represents the correlation of the MIR responses at wavenumber i with the NIR responses at wavelength j. One can construct the correlation spectrum of all MIR wavenumbers with the individual NIR wavelength at index i by extracting a single row i from the $\boldsymbol{R}$ matrix. Likewise, by extracting a single column j from the $\boldsymbol{R}$ matrix one can obtain the correlation spectrum of all NIR wavelengths with the individual MIR wavenumber at index j.

Let us discuss the results in Fig. 8.3 in more detail. It can be seen from Fig. 8.3 that R^2 at 2130 nm (0.5) increases across the MIR spectrum from about 3700 cm^{-1} to a maximum at 2915–2850 cm^{-1}, then decreases to a minimum (0.1) at around 1850 cm^{-1}. This covers the spectral regions for OH and CH stretching modes in the MIR. Particularly striking is that although the OH stretching bands appear broader and stronger in the spectrum (see the top spectrum in Fig. 8.3), the C-H stretching bands are correlated more intensely. The fingerprint region of the MIR spectrum shows the same phenomena, but the overall correlations are smaller below 1500 cm^{-1}.

The NIR regions where most correlation activity is seen are 1385, 1730 and 2200–2450 nm. The MIR regions correlating with the above regions are 2900–2700 cm^{-1} (CH stretch), 1700–1500 cm^{-1} (C=O and C=C stretch) and 1100–800 cm^{-1} (C–C, C–O and C–N stretching bands).

In 2D statistical correlation spectroscopy "2D slices" play important roles in interpreting the correlations between the two spectral regions [51]. The slices, obtained by holding a wavelength in one region constant and letting the wavelengths in the other vary, can give a lot of information about the relationship between an absorber in one region and many absorbers in another.

Barton et al. [51] employed the CH stretching modes of the "waxy" material in the samples as the most prominent example in order to show how the correlation patterns from one region can be used to interpret the other regions of the spectrum. Fig. 8.4(a) and (b) shows NIR and MIR spectra, respectively, of bees' wax [51]. The three bands at 2307, 1726 and 1396 nm in the NIR spectrum are assigned to the CH combination mode, the first overtone of the CH stretching mode and its second overtone. These bands have major correlations to the MIR bands at 2919 and 2854 cm^{-1} due to the CH stretching modes, as shown in the correlation slices in Fig. 8.5(a) and (b) [51]. The correlation pattern is virtually identical for the two MIR slices, as it is for the three NIR slices [Fig. 8.6(a), (b) and (c)]. It is clear from Fig. 8.6 that the patterns at all three wavelengths 1390, 1729 and 2312 nm, respectively, are identical to the major correlations at 2919 and 2850 cm^{-1}.

The 2D approach of Barton et al. [51] has the following two advantages. Firstly, it is possible to elucidate which component in a sample contributes to a particular NIR band on the basis of an NIR correlation slice of a spectrum in another region (for example, a MIR spectrum). Secondly, 2D correlation spectroscopy enhances the understanding of chemometric models. With the aid of the MIR correlation slice of an NIR spectrum, it may also be possible to predict useful wavelengths for chemometric models. Examples of the applications of this 2D approach are found in references [51–54].

8.4
Sample-Sample Correlation Spectroscopy

Sample-sample correlation spectroscopy was derived from a new insight into the calculation of synchronous and asynchronous spectra [57]. According to Šašic et al. [57], synchronous and asynchronous spectra can be calculated by using linear algebra. Let us explain this point first. In general, spectral data sets can be expressed as the matrix product of the spectral characteristics and concentrations of the species present in the system,

$$X = CP \tag{8.11}$$

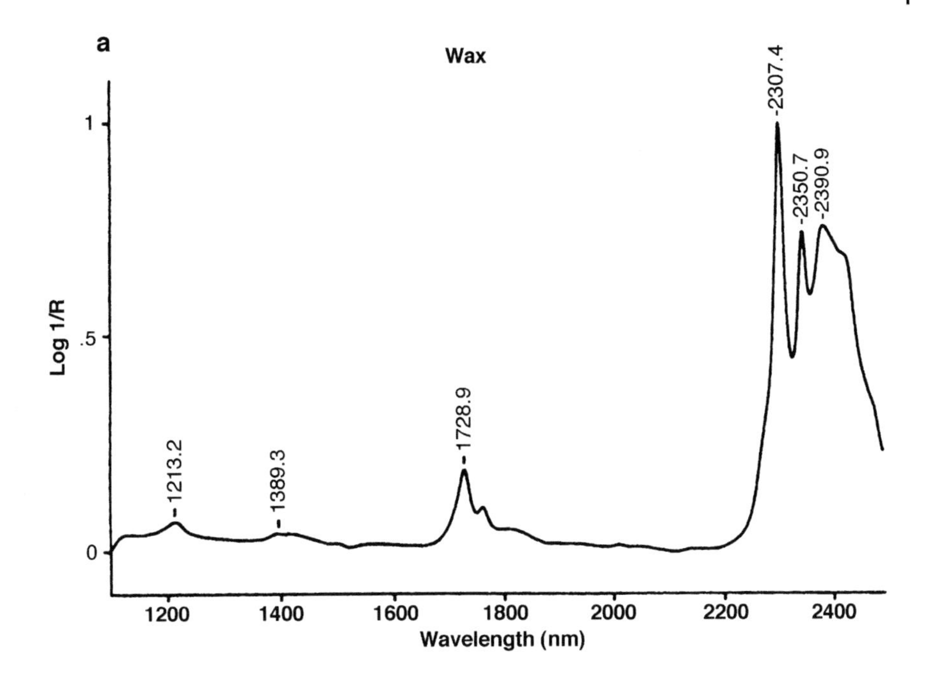

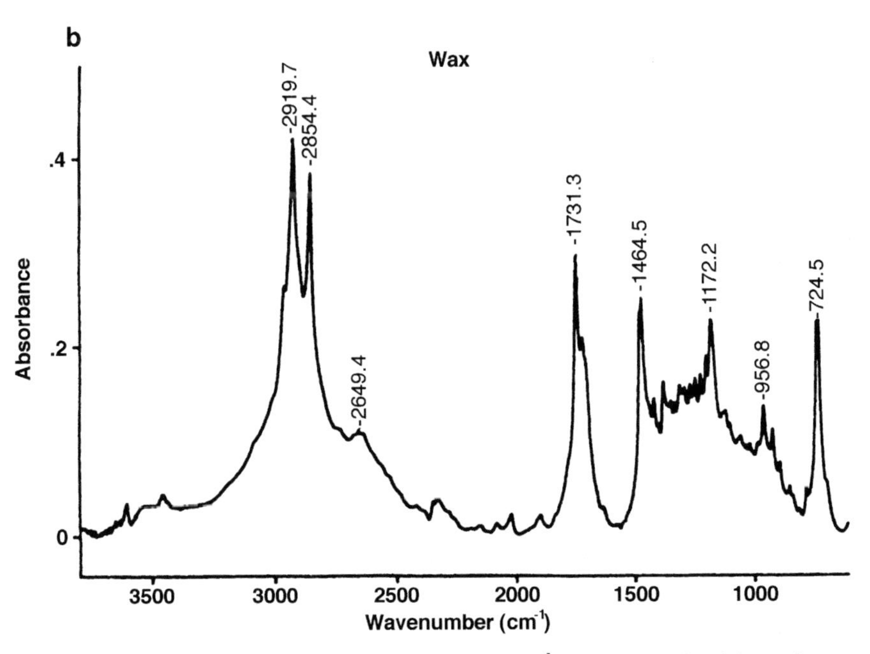

Fig. 8.4 (**a**) A NIR spectrum of bees' wax from the 1100 to 2500 nm region. (**b**) A MIR spectrum of the same sample from the 4000 to 600 cm^{-1} region. (Reproduced from ref. [51] with permission. Copyright (1992) Society for Applied Spectroscopy)

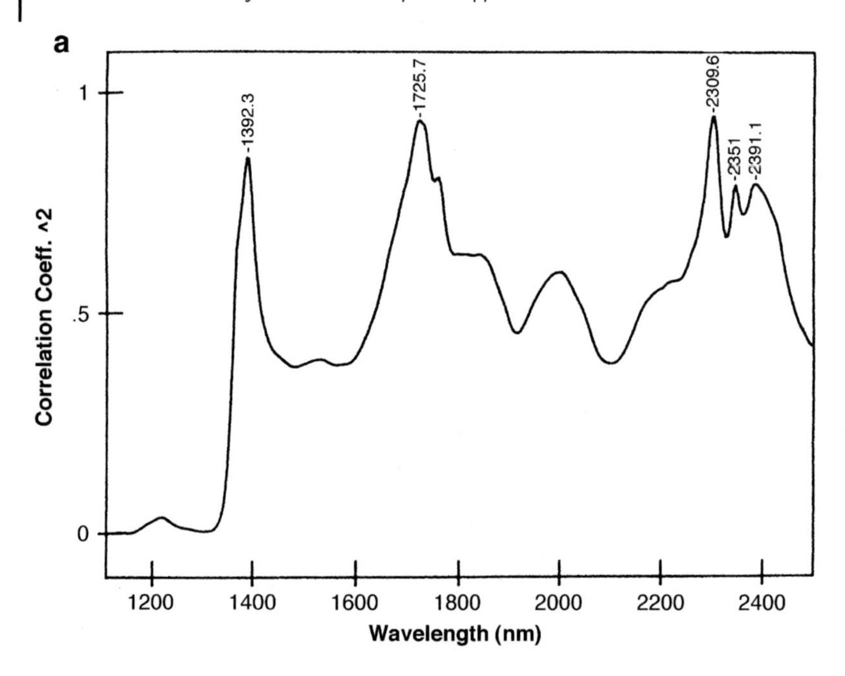

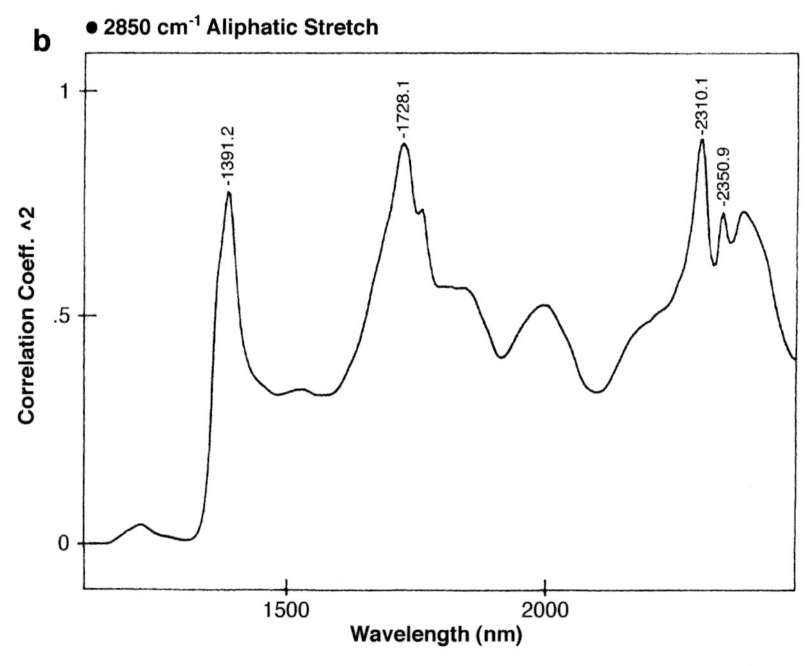

Fig. 8.5 (**a**) A 2919 cm^{-1} MIR correlation slice of the NIR spectrum in the 1100–2500 nm region. (**b**) A 2850 cm^{-1} correlation slice of the NIR spectrum in the 1100–2500 nm region. (Reproduced from ref. [51] with permission. Copyright (1992) Society for Applied Spectroscopy)

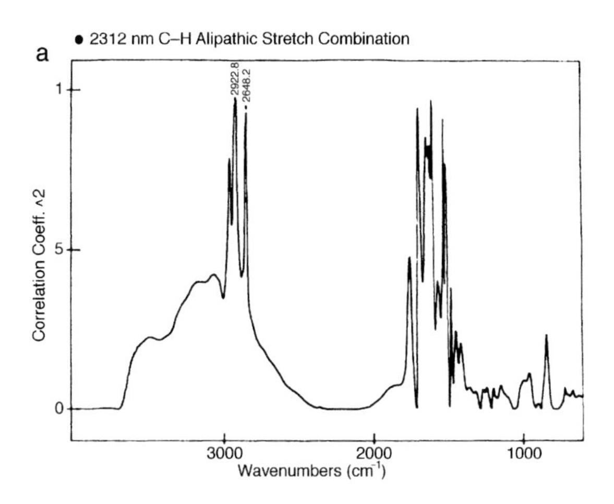

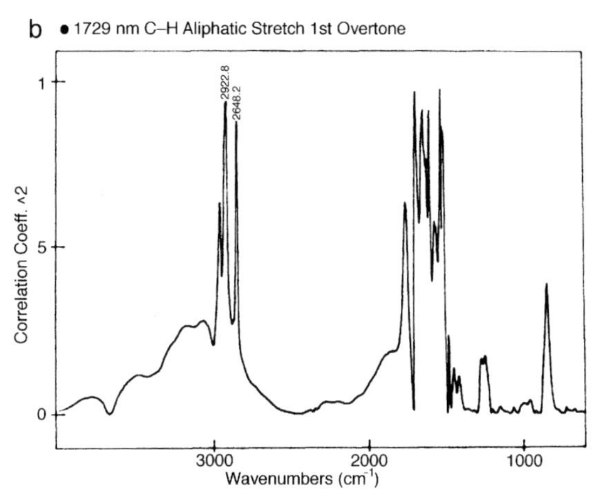

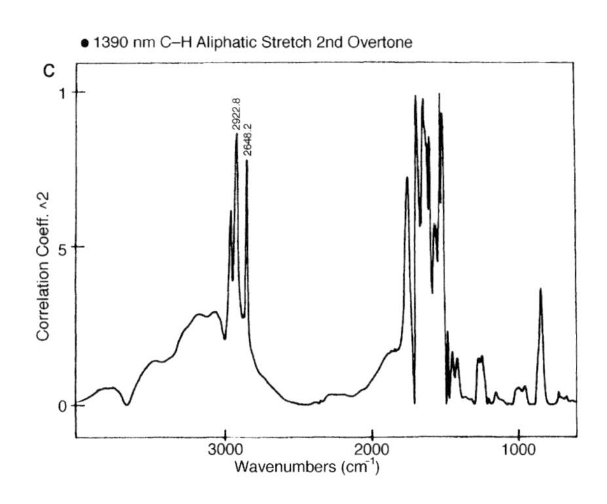

Fig. 8.6 (**a**) A 2312 nm NIR correlation slice of the MIR spectrum in the 4000–600 cm^{-1} region. (**b**) A 1729 nm NIR correlation slice of the MIR spectrum in the 4000–600 cm^{-1} region. (**c**) A 1390 nm NIR correlation slice of the MIR spectrum in the 4000–600 cm^{-1}. (Reproduced from ref. [51] with permission. Copyright (1992) Society for Applied Spectroscopy)

where X is an experimental matrix, $C(w \times n)$ is a matrix with n pure spectra (w data points) of pure components in columns, and $P(n \times s)$ is a concentration matrix with n concentration profiles of components in rows. Every row of the data matrix can be viewed as a vector of the spectral intensity changes at a given wavenumber. If one wants to examine the correlation between intensity changes for any particular pair of wavenumbers ν_1 and ν_2, then one should calculate the so-called rows-cross-product matrix $\mathbf{Z}$ (covariance matrix).

$$\mathbf{Z} = 1/(s-1)\boldsymbol{X}\boldsymbol{X}^{\mathrm{T}} \tag{8.12}$$

$\mathbf{Z}(w \times w)$ is simply a synchronous spectrum in variable-variable correlation spectroscopy. The corresponding asynchronous spectrum can be calculated as an array of the scalar products of the dynamic vectors in the original data matrix and the matrix that is orthogonal to the original one.

$$\mathbf{Z} = 1/(s-1)\boldsymbol{X}\boldsymbol{H}\boldsymbol{X}^{\mathrm{T}} \tag{8.13}$$

Here, H is the Hilbert-Noda transformation matrix [62]. Conventional generalised 2D correlation spectra present correlations among spectral features of the components, but one should note that the data matrix contains information about the concentrations as well as the spectra (Eq. (8.11)). Therefore, one can create a correlation between the concentrations as in the case of the variable-variable correlation:

$$\mathbf{Z} = 1/(w-1)\boldsymbol{X}^{\mathrm{T}}\boldsymbol{X} \tag{8.14}$$

$\mathbf{Z}$ in Eq. (8.14) is a synchronous spectrum in sample-sample correlation spectroscopy. The corresponding asynchronous sample-sample correlation spectrum is given by:

$$\mathbf{Z} = 1/(w-1)\boldsymbol{X}^{\mathrm{T}}\boldsymbol{H}\boldsymbol{X} \tag{8.15}$$

The synchronous and asynchronous correlation spectra in sample-sample correlation spectroscopy have samples on the axes instead of wavelengths or wavenumbers, and thus they provide straightforward data about concentration dynamics. Sample-sample correlation spectroscopy and variable-variable correlation spectroscopy are complementary to each other. As in the case of variable-variable correlation spectroscopy, sample-sample correlation spectroscopy can set any kind of perturbation variables such as temperature, concentration, pressure and time. Sample-sample correlation spectroscopy has been applied to temperature-dependent NIR spectra of neat oleic acid [56] and water [60], concentration-dependent short-wave NIR spectra of untreated milk [58], and MIR spectra of the polycondensation of bis(hydroxyethyl terephthalate) (BHET) measured on-line [59, 61].

8.5 References

[1] W. P. Aue, E. Bartholdi, R. R. Ernst, *J. Chem. Phys.* **1976**, *64*, 2229.

[2] K. Nagayama, A. Kumar, K. Wüthrich, R. R. Ernst, *J. Magn. Reson.* **1980**, *40*, 321.

[3] A. Bax, *Two Dimensional Nuclear Magnetic Resonance in Liquids.* Reidel, Boston, **1982**.

[4] D.L. Turner, *Prog. Nucl. Magn. Reson. Spectrosc.* **1985**, *17*, 281.

[5] I. Noda, *Bull. Am. Phys. Soc.* **1986**, *31*, 520.

[6] I. Noda, *J. Am. Chem. Soc.* **1989**, *111*, 8116.

[7] I. Noda, *Appl. Spectrosc.* **1990**, *44*, 550.

[8] M. Satkowski, J. T. Grothaus, S. D. Smith, A. Ashraf, C. Marcott, A. E. Dowrey, I. Noda in *Polymer Blends, Solutions, and Interfaces* (Eds. I. Noda D. N. Rubingh), Elsevier, New York, **1992**, pp. 89–108.

[9] I. Noda, A. E. Dowrey, C. Marcott, *Polym. News* **1993**, *18*, 167.

[10] I. Noda, A. E. Dowrey, C. Marcott, *Appl. Spectrosc.* **1993**, *47*, 1317.

[11] N. Reynolds, *Adv. Mater.* **1991**, *3*, 614.

[12] C. Marcott, A. E. Dowrey, I. Noda, *Anal. Chem.* **1994**, *66*, 1065A.

[13] P. J. Hendra, W. F. Maddams in *Polymer Spectroscopy* (Ed. A. H. Fawcett) Wiley, New York, **1996**, pp. 173–202.

[14] R. A. Palmer, C. J. Manning, J. L. Chao, I. Noda, A. E. Dowrey, C. Marcott, *Appl. Spectrosc.* **1991**, *45*, 12.

[15] C. Marcott, I. Noda, A. E. Dowrey, *Anal. Chim. Acta* **1991**, *250*, 131.

[16] V. G. Gregoriou, J. L. Chao, H. Toriumi, R. A. Palmer, *Chem. Phys. Lett.* **1991**, *179*, 491.

[17] I. Noda in *Modern Polymer Spectroscopy* (Ed. G. Zerbi), Wiley-VCH, Weinheim, **1999**, pp. 1-32.

[18] I. Noda, *Appl. Spectrosc.* **1993**, *47*, 1329.

[19] I. Noda, A. E. Dowrey, C. Marcott, Y. Ozaki, G. M. Story, *Appl. Spectrosc. A* **2000**, *54*, 236.

[20] Y. Ozaki, I. Noda, *Two-Dimensional Correlation Spectroscopy*, American Institute of Physics, New York, **2000**.

[21] M. Osawa, K. Yoshii, Y. Hibino, T. Nakano, I. Noda, *J. Electroanal. Chem.* **1997**, *426*, 11.

[22] A. Nabet, M. Pezolet, *Appl. Spectrosc.* **1997**, *51*, 466.

[23] S. J. Gadalleta, A. Gericke, A. L. Boskey, R. Mendelsohn, *Biospectroscopy* **1996**, *2*, 353.

[24] C. Roselli, J.-R. Burie, T. Mattioli, A. Boussa, *Biospectroscopy* **1995**, *1*, 329.

[25] T. Gustafson, D. L. Morris, L. A. Huston, R. M. Bustler, I. Noda, *Time-Resolved Vibrational Spectroscopy, Vol. V* (Eds. A. Lau, F. Siebert), Springer, Berlin, **1994**, pp. 131-135.

[26] I. Noda, Y. Liu, Y. Ozaki, M. A. Czarnecki, *J. Phys. Chem.* **1995**, *99*, 3068.

[27] Y. Liu, Y. Ozaki, I. Noda, *J. Phys. Chem.* **1996**, *100*, 7326.

[28] I. Noda, Y. Liu, Y. Ozaki, *J. Phys. Chem.* **1996**, *100*, 8665.

[29] I. Noda, Y. Liu, Y. Ozaki, *J. Phys. Chem.* **1996**, *100*, 8674.

[30] Y. Ozaki, Y. Liu, I. Noda, *Macromolecules* **1997**, *30*, 2391.

[31] Y. Ozaki, Y. Liu, I. Noda, *Appl. Spectrosc.* **1997**, *51*, 526.

[32] Y. Ozaki, Y. Liu, I. Noda, *Macromol. Symp.* **1997**, *119*, 49.

[33] Y. Ozaki, I. Noda, *J. Near Infrared Spectrosc.* **1996**, *4*, 85.

[34] M. A. Czarnecki, H. Maeda, Y. Ozaki, M. Suzuki, M. Iwahashi, *Appl. Spectrosc.* **1998**, *52*, 994.

[35] M. A. Czarnecki, P. Wu, H. W. Siesler, *Chem. Phys. Lett.* **1998**, *283*, 326.

[36] Y. Wang, K. Murayama, Y. Myojyo, R. Tsenkova, N. Hayashi, Y. Ozaki, *J. Phys. Chem. B* **1998**, *102*, 6655.

[37] S. V. Shilov, S. Okretic, H. W. Siesler, M. A. Czarnecki, *Appl. Spectrosc. Rev.* **1996**, *31*, 125.

[38] M. Muler, R. Buchet, U. P. Fringeli, *J. Phys. Chem.* **1996**, *100*, 10810.

[39] I. Noda, G. M. Story, A. E. Dowrey, R. C. Reeder, C. Marcott, *Macromol. Symp.* **1997**, *119*, 1.

[40] N.L. Sefara, N. P. Magtoto, H. H. Richardson, *Appl. Spectrosc.* **1997**, *51*, 563.

[41] P. Pancoska, J. Kubelka, T. A. Keiderling, *Appl. Spectrosc.* **1999**, *53*, 655.
[42] L. Smeller, K. Heremans, *Vib. Spectrosc.* **1999**, *19*, 375.
[43] Y. Ren, T. Murakami, T. Nishioka, K. Nakashima, I. Noda, Y. Ozaki, *J. Phys. Chem. B* **2000**, *104*, 679.
[44] D. G. Craff, B. Pastrana-Rios, S. Venyaminov, F. G. Prendergast, *J. Am. Chem. Soc.* **1997**, *119*, 11282.
[45] M. A. Czarnecki, H. Maeda, Y. Ozaki, M. Suzuki, M. Iwahashi, *J. Phys. Chem. B* **1998**, *46*, 9117.
[46] Y. Ren, M. Shimoyama, T. Ninomiya, K. Matsukawa, H. Inoue, I. Noda, Y. Ozaki, *Appl. Spectrosc.* **1999**, *53*, 919.
[47] Y. Ren, A. Matsushita, K. Matsukawa, H. Inoue, Y. Minami, I. Noda, Y. Ozaki, *Vib. Spectrosc.* **2000**, *23*, 207.
[48] B. Czarnik-Matusewicz, K. Murayama, R. Tsenkova, Y. Ozaki *Appl. Spectrosc.* **1999**, *53*, 1582.
[49] Y. Wu, B. Czarnik-Matusewicz, K. Murayama, Y. Ozaki, *J. Phys. Chem. B* **2000**, *104*, 5840.
[50] C. P. Schultz, H. Fabian, H. H. Mantsch, *Biospectrosc.* **1998**, *4*, 19.
[51] F. B. Barton II, D. S. Himmelsbach, J. H. Duckworth, M. J. Smith, *Appl. Spectrosc.* **1992**, *46*, 420.
[52] F. B. Barton II, D. S. Himmelsbach, *Appl. Spectrosc.* **1993**, *47*, 1920.
[53] W. F. McClure, H. Maeda, J. Dong, Y. Liu, Y. Ozaki, *Appl. Spectrosc.* **1996**, *50*, 467.
[54] F. B. Barton II, D. S. Himmelsbach, D. E. Akin, A. Sethuraman, K.–E. L. Eriksson, *J. NIR Spectrosc.* **1995**, *3*, 25.
[55] S. Šašic, A. Muszynski, Y. Ozaki, *J. Phys. Chem. A* **2000**, *104*, 6380.
[56] S. Šašic, A. Muszynski, Y. Ozaki, *J. Phys. Chem. A* **2000**, *104*, 6388.
[57] S. Šašic, A. Muszynski, Y. Ozaki, *Appl. Spectrosc.* **2001**, *55*, 343.
[58] S. Šašic, Y. Ozaki, *Appl. Spectrosc.* **2001**, *55*, 163.
[59] S. Šašic, Y. Ozaki, *Anal. Chem.* **2001**, *173*, 2294.
[60] V. Segtnan, S. Šašic, T. Isaksson, Y. Ozaki, *Anal. Chem.* **2001**, *73*, 3153.
[61] S. Šašic, T. Amari, Y. Ozaki, *Anal. Chem.* in press.
[62] I. Noda, *Appl. Spectrosc.* **2000**, *54*, 994

9
Applications in Chemistry

Yukihiro Ozaki

9.1
Introduction

NIR spectroscopy has been applied extensively to various practical problems, such as those in the agricultural and food industries, polymer industries and biomedical sciences [1–3]. However, it is also a powerful tool in studies of basic science. For example, it is very useful for the investigation of hydrogen bonds, hydration and self-association of molecules. This chapter describes the applications of NIR spectroscopy to basic problems in chemistry.

The chapter is divided into two parts. The first section is concerned with NIR studies of hydrogen bonds, hydration and self-association of basic compounds. This section has three purposes: One is to learn the spectral pattern and band assignments of basic compounds, for which the NIR spectra of water, fatty acids and alcohols are presented and analysed. Another is to show how NIR spectroscopy is powerful in investigating hydrogen bonds and self-association. In the last part of the first section, an NIR study of hydration in proteins will be introduced. Although proteins are not simple compounds, the study described is a basic example for the application of NIR spectroscopy to the investigation of hydration.

The second section describes the applications of chemometrics to chemical problems. Determination of the physical and chemical properties of water, discrimination of various alcohols and resolution enhancement of NIR spectra by loadings plots will be discussed.

9.2
NIR Studies of Hydrogen Bonds, Hydration and Self-Association of Basic Compounds

NIR spectroscopy has been employed to investigate the hydrogen bonds, hydration and self-association of a variety of compounds from simple molecules, such as water and alcohols, to complicated ones, such as polymers and proteins. The reason why NIR spectroscopy is so suitable for such studies is that bands due to overtones and combinations associated with OH and NH groups appear strongly in the NIR region (14000–4000 cm^{-1}). The anharmonicity of XH bonds is very large and, in addition, the fundamentals of XH stretching modes appear in the high frequency region of IR spectra. Therefore, an NIR spectrum is, in general, dominated by bands arising from XH groups.

The advantages of NIR spectroscopy over IR spectroscopy in studying hydrogen bonds and the dissociation of self-associated molecules may be summarised as follows:

- OH and NH stretching bands of monomeric and polymeric species are better separated in the NIR region than in the IR region. Even bands ascribed to free, terminal OH and NH groups of the polymeric species can be clearly identified.
- Because of the larger anharmonicity, bands ascribed to the first overtones of OH and NH stretching modes of monomeric species are enhanced compared with the corresponding bands arising from polymeric species. Therefore, it may be easier to monitor the dissociation process from polymeric species into monomeric ones in the NIR region rather than in the MIR region by using the first overtone of the OH or NH stretching mode of the monomeric species.
- NIR bands have much weaker absorption intensities relative to MIR bands, so a more convenient cell pathlength can be used and the exact volume of the sample can be evaluated. In MIR spectroscopy, one must use a very thin cell or attenuated total reflection (ATR) prism and, therefore, one often encounters adsorption problems.

The NH and OH bands, which are key bands for the investigations of self-association and hydrogen bonds, are often severely overlapped with CH bands and, thus, we have to investigate how we can select or discriminate the bands due to NH or OH groups from those due to CH groups. In order to pick out spectral information about the NH and OH bands from rather complicated NIR spectra, one can employ difference spectra, second derivatives, chemometrics and generalised 2D correlation spectroscopy (Chapter 8). In the studies described below these techniques are utilised.

9.2.1 Water

Studies of water by NIR spectroscopy stretch back more than thirty years because they are very important, not only in basic science, but also in a variety of applications [4–10, 35–37]. Water is involved in almost all kinds of substances, and water content and the structure of the water in them are often key factors in determining their functions and structure. NIR spectroscopy has been employed to investigate water content, hydrogen bonds of water and hydration in various fields such as the agricultural and food industries, medical and pharmacological sciences and the polymer and textile industries. Studies of water have always been a matter of keen interest in NIR spectroscopy [1–10, 35–37]. However, it is still not easy to understand the NIR spectra of water completely because water does not exist as a single species, and water molecules form various cluster structures.

Fig. 9.1 shows NIR spectra of water in the 900–2500 nm (11100–4000 cm^{-1}) region. The intense tail near 2500 nm is due to the fundamentals of the OH stretching modes (ν_1, ν_3; see Fig. 9.2). Bands near 1910, 1430 and 942 nm (5235, 6900 and 10613 cm^{-1}) are assigned to the $\nu_2+\nu_3$, $\nu_1+\nu_3$ and $2\nu_1+\nu_3$ modes of water, respectively. As shown in this figure, the intensities of water bands decrease stepwise with decreasing wavelength. This means that one can select the spectral region to be used or the pathlength of a cell. This is a characteristic feature of NIR spectroscopy. The three spectra in Fig. 9.1 were measured by use of cells with pathlengths of 10 and 1.0 mm should be 10 (a), 1.0 (b) and 0.05 (c) mm.

The intensities of water bands are much weaker in the NIR region than in the MIR region, so it is much easier to measure spectra of aqueous solutions in the NIR region. In the NIR spectra of water, a number of weak bands are observed; Tab. 9.1 summarises the frequencies and assignments for the water bands.

Let us examine the 9000–5500 cm^{-1} (1110–1820 nm) region in more detail. Fig. 9.3 shows an NIR spectrum (a) and its second derivative (b) of water in the 9000–5500 cm^{-1} region measured at 5 °C [9]. The second derivative indicates that the band between 6000 and 7500 cm^{-1} consists of at least six components, marked by arrows. A weak shoulder near 7440 cm^{-1} is assigned to a $\nu_1+\nu_2+\nu$ (ν; rotational

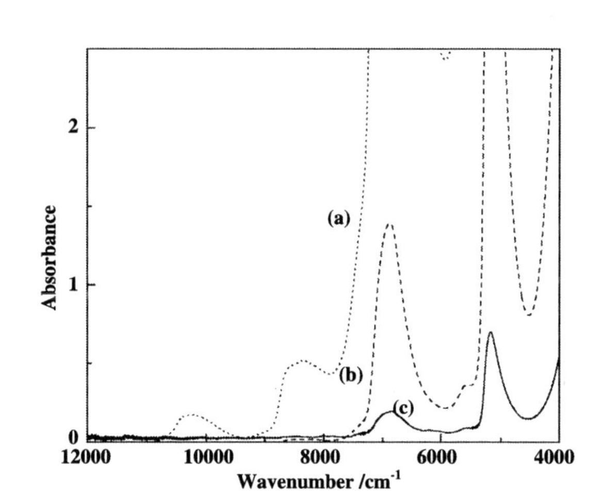

Fig. 9.1 NIR spectra of water in the 900–2500 nm (11100–4000 cm^{-1}) region

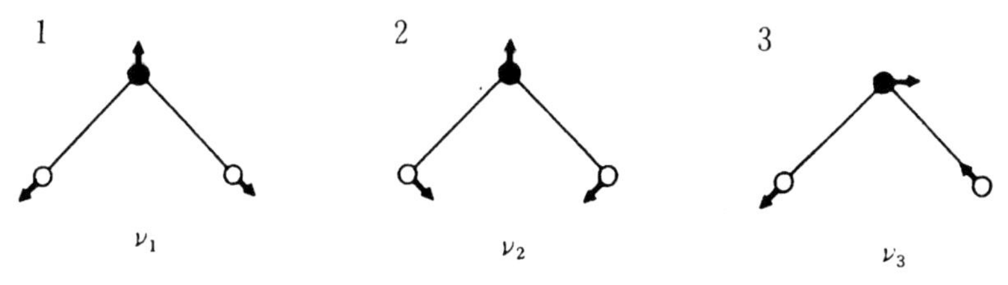

Fig. 9.2 Vibrational modes of water: ν_1 symmetric OH stretching mode, ν_2 OH bending mode, ν_3 antisymmetric OH stretching mode

Table 9.1 Frequencies and assignments for the bands due to water

Vibrational quantum number of upper level			***Wavenumber (cm^{-1}) of band center***
ν_1	ν_2	ν_3	H_2O
0	1	0	1594.59
1	0	0	3656.65
0	0	1	3755.79
0	2	0	3151.4
0	3	0	4667
1	1	0	5235
0	1	1	5332
1	2	0	6775
0	2	1	6874
2	0	0	7201
1	0	1	7250
0	0	2	7445
2	1	0	8762
1	1	1	8807
0	1	2	9000
3	0	0	10600
2	0	1	10613
1	0	2	10869
0	0	3	11032

mode) mode while the remaining five absorptions probably correspond to bands due to the $\nu_1+\nu_3$ modes of the water species.

The NIR spectra of water are very sensitive to temperature and to the ions involved. Fig. 9.4(a) presents NIR spectra of water measured over a temperature range of 5–85 °C at increments of 5 °C [9]. Note that the spectrum of water changes gradually with temperature. In Fig. 9.4(b), a series of difference spectra of water is shown, calculated by subtracting the spectrum at 5 °C, taken as a reference spectrum, from the other spectra measured at various temperatures. A band at 7089 cm^{-1} becomes stronger while that at 6718 cm^{-1} becomes weaker with temperature. Recently, Segtnan et al. [10] investigated the structure of water using 2D correlation spectroscopy and PCA. Fig. 9.5(a) and (b) shows a synchronous 2D correlation spectrum generated from temperature-dependent spectral variations of water measured over a temperature range of 6 to 80 °C at 2 °C increments, and loadings and scores of the first-principal component constructed from the same data, respectively. The results of the difference spectra, 2D correlation spectroscopy and PCA all suggest that there are two major water species giving the $\nu_1+\nu_3$ band at 7089 (1412 nm) and 6718 (1491 nm) cm^{-1}. Segtnan et al. [10] also found that the wavelengths 1412 and 1491 nm account for more than 99% of the spectral variations. Based upon these results, they concluded that according to its tem-

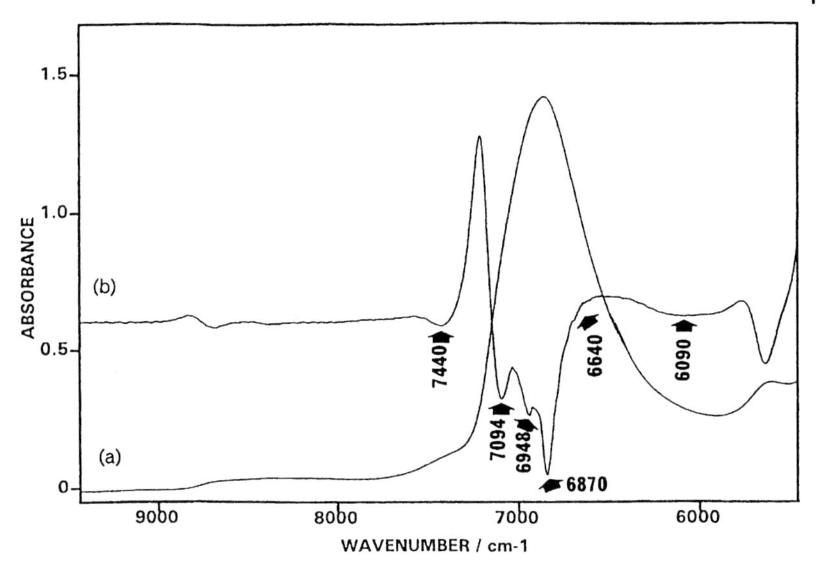

Fig. 9.3 A NIR spectrum in the 9000–5500 cm^{-1} region of water (a) and its second derivative (b) measured at 5°C. (Reproduced from ref. [9] with permission. Copyright 1995) NIR Publications)

perature-dependent NIR spectra, water can be portrayed as a quasi-two-component mixture.

Fig. 9.6(a) shows the effect of increasing sodium salt concentrations of 0–10% m/m on the NIR absorption of aqueous solutions [7]. The effect can be seen more clearly in the difference spectra shown in Fig. 9.6(b). Note that there are considerable differences between the effect of either hydroxide or carbonate and chloride ions. The sodium chloride solutions show much stronger absorption near 1900 nm than the other two compounds; this indicates that in chloride solutions there is a lower percentage of hydrogen bonds in the solvent [7]. Therefore, sodium chloride appears to be a structure-breaking reagent.

Several research groups have employed NIR spectroscopy for ice research [11, 12]. Fig. 9.7 depicts NIR spectra of Ice VII, Ice VI and high pressure water (10.2 kbar and 1 bar) measured at 25 °C. Kato et al. analysed a band near 5100 cm^{-1} in these spectra by use of a Gauss-Lorentzian product function [12]. They assumed two components for the band of the ices and three components for that of high-pressure water and found that each component shows a downward shift with pressure. Therefore, the hydrogen bonds in water and ice become stronger and the distances between neighbouring molecules become shorter with pressure. If the shift depends largely upon a change in the O···O distance between the neighbouring water molecules, one can calculate the change in distance from the shift. Tab. 9.2 summarises the relations between the frequency shift $d\nu/dp$ (p; pressure) and the change in the O···O distance ($d(\mathrm{O}\cdots\mathrm{O})/dp$).

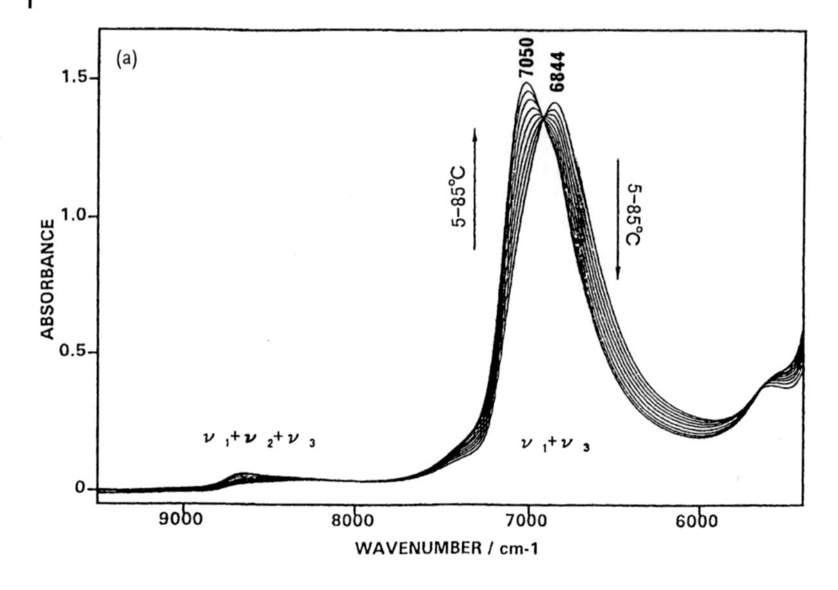

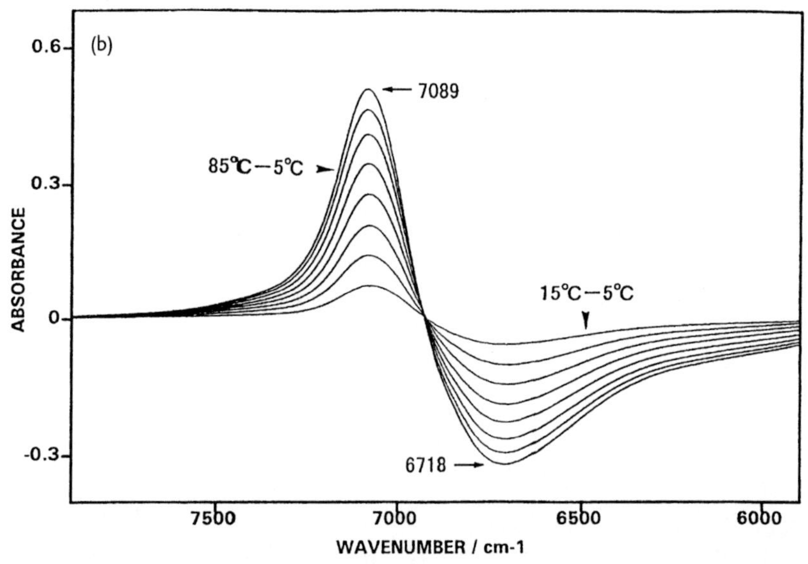

Fig. 9.4 (a) NIR spectra of water measured over a temperature range of 5–85 °C at increments of 5 °C. (b) Difference spectra of water calculated by subtracting the spectrum at 5 °C, taken as a reference spectrum, from the other spectra measured at various temperatures. (Reproduced from ref. [9] with permission. Copyright (1995) NIR Publications)

Fig. 9.5 (a) A synchronous 2D correlation spectrum generated from the temperature-dependent spectral variations of water measured over a temperature range of 6 to 80 °C at 2 °C increments. (b) Loadings and scores of the first principal component constructed from the same data as those for (a). (Reproduced from ref. [10] with permission. Copyright (2001) American Chemical Society)

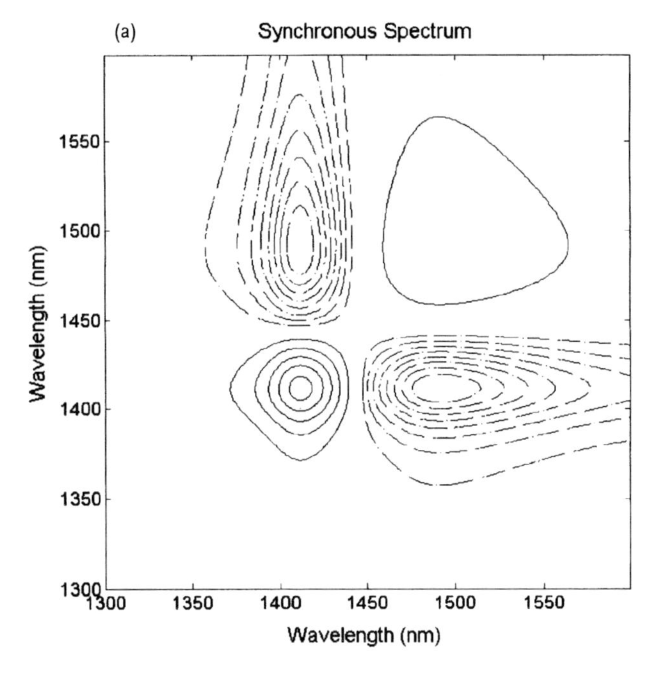
(a)
Synchronous Spectrum
1550
1500
1450
1400
1350
1300
Wavelength (nm)
1300
1350
1400
1450
1500
1550
Wavelength (nm)

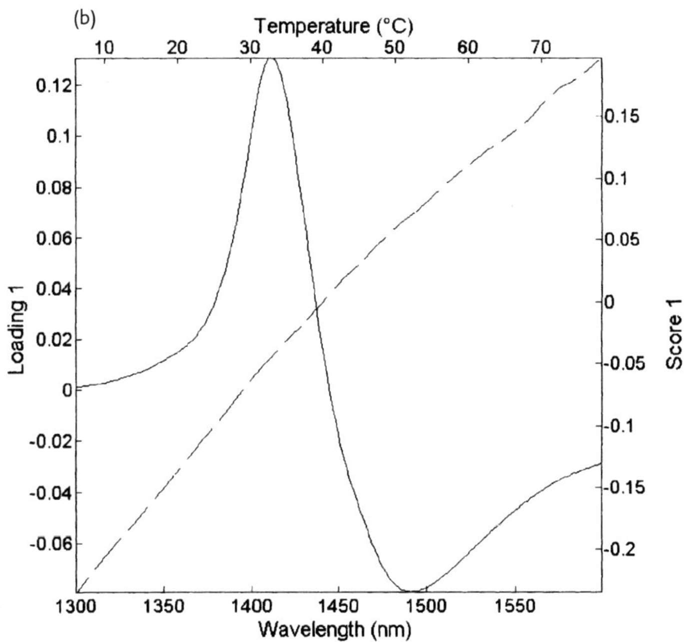
(b)
Temperature (°C)
10
20
30
40
50
60
70
0.12
0.1
0.08
0.06
0.04
0.02
0
-0.02
-0.04
-0.06
Loading 1
0.15
0.1
0.05
0
-0.05
-0.1
-0.15
-0.2
Score 1
1300
1350
1400
1450
1500
1550
Wavelength (nm)

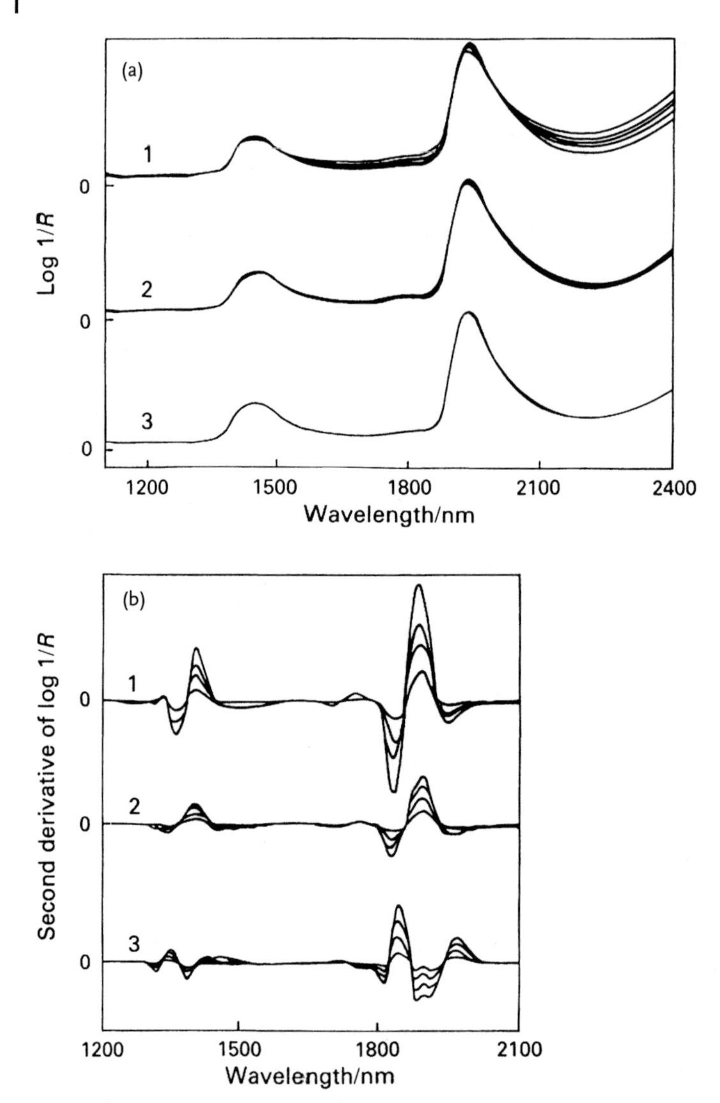

Fig. 9.6 (a) NIR spectra of aqueous solutions of NaOH (1), Na_2CO_3 (2) and NaCl (3) with concentrations of 0–10% (m/m). (b) Difference spectra obtained by subtracting second derivative spectra of distilled water from those of the aqueous sodium salt solutions (2.5–10% m/m). (Reproduced from ref. [7] with permission. Copyright (1989), Royal Society of Chemistry)

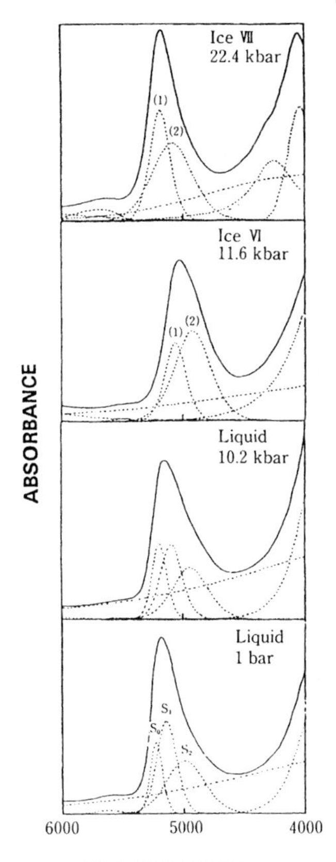

Fig. 9.7 NIR spectra of Ice VII, Ice VI and high pressure water (10.2 kbar and 1 bar) measured at 25°C. (Reproduced from ref. [12] with permission. Copyright (1992) Hokkaido University Press)

Table 9.2 Relations between the frequency shift ($d\nu/dp$, p; pressure) and the change in the O···O distance ($d(O\cdots O)/dp$)

	Water			*Ice VI*		*Ice VII*		*Ice Ih*
	S_0	S_1	S_2	S_1	S_2	S_1	S_2	
$d\nu/dp$ cm^{-1} $kbar^{-1}$	−3.1	−3.9	−4.1	−4.9	−6.5	−2.0	−4.3	
$d(O\cdots O)/dp$ pm $kbar^{-1}$	−0.40	−0.50	−0.53	−0.63	−0.83	−0.26	−0.55	−1.00

9.2.2 Fatty Acids

It is well known that fatty acids form so called “ring dimers” in the liquid phase and in solutions, and that NIR spectroscopy is suitable for monitoring the dissociation of these dimers [13–16]. Iwahashi et al. [14] investigated the dissociation of dimers of *cis*-9-octadecenoic acid (oleic acid) into the monomeric species in the pure liquid state by use of NIR spectroscopy. Fig. 9.8(a) shows NIR spectra of neat oleic acid at various temperatures. The spectra were normalised with respect to the 1209 nm standard band at 15.2 °C [14]. An intense band at 1209 nm is assigned to the second overtone of the CH stretching mode of the methylene groups. The intensity of this band changes little with temperature. Bands in the 1380–1500 nm region contain contributions from the overtones of CH vibrations and the first overtone of an OH stretching mode of the monomer. Note that a band at 1445 nm involving the first overtone of the OH stretching mode increases with temperature. This reveals that the dimer dissociates into the monomer as the temperature is increased. In order to clarify the temperature-dependent intensity change in the band at 1445 nm, difference spectra were calculated by using the spectrum at 15.2 °C as a standard. In Fig. 9.8(b) the difference spectra of oleic acid at various temperatures are shown [14]. It can be clearly seen that the band at 1445 nm increases with temperature.

Iwahashi et al. [14] plotted the absorbance at 1445 nm versus temperature and then calculated the degree of dissociation as a function of temperature. If the process of dissociation of a dimer into monomers is given by

$$D \rightleftharpoons 2M \tag{9.1}$$

the degree of dissociation, a, can be calculated by:

$$a = MA_{\mathrm{M}}/(1000d\varepsilon_{\mathrm{OH}}) \tag{9.2}$$

where M, d, A_{M} and $\varepsilon_{\mathrm{OH}}$ are the molar weight of the sample, its density, the absorbance and the molar absorption coefficient of the monomer, respectively. The molar absorption coefficient at 1445 nm was obtained through the absorbance measurements in carbon tetrachloride solutions at very low concentrations [14].

Fig. 9.9 shows the temperature dependence of the degree of dissociation, a, for oleic acid in the pure liquid state [14]. Interestingly, the a-T curve has two break points at 30 and 55 °C. The break-point temperatures correspond to the transition temperatures in the liquid structures of oleic acid [15]; 30 °C corresponds to the transition from a quasi-smectic liquid crystal to a more disordered liquid crystal, and 55 °C from disordered liquid crystal to isotropic liquid. Fig. 9.10 illustrates the arrangement proposed for the dimers of oleic acid molecules in the clusters of a quasi-smectic liquid crystal [16].

Czarnecki et al. [16, 17] introduced the Fourier-transform (FT) NIR technique to the study of self-associated molecules. The technique enables highly accurate spec-

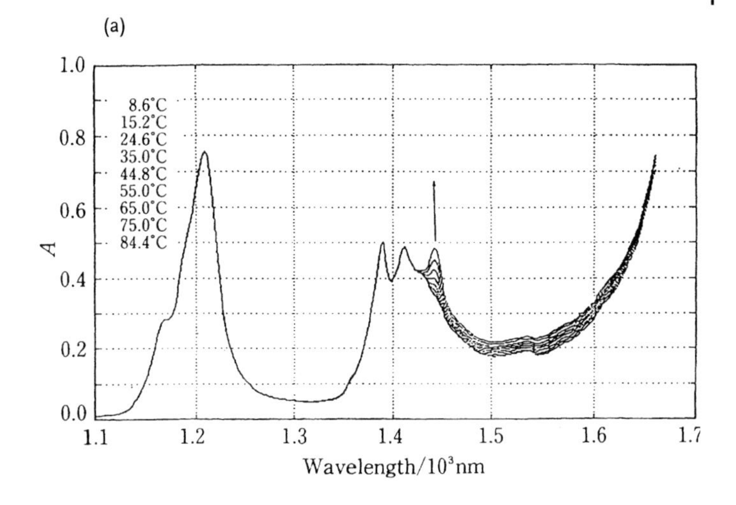

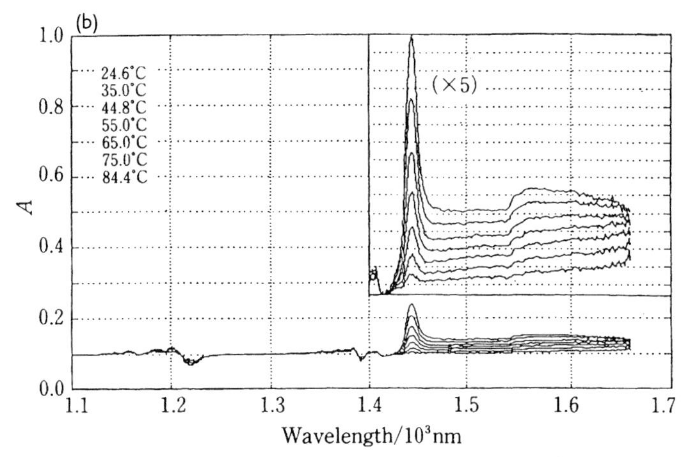

Fig. 9.8 (a) NIR spectra of oleic acid in the pure liquid state over a temperature range of 8.6-84.4 °C. The spectra were normalised with respect to the 1209 nm standard band at 15.2 °C. The normalised spectrum for a supercooled sample at 8.3 °C is identical to the spectrum at 15.2 °C throughout all the wavelength region. (b) Difference spectra for oleic acid at various temperatures. (Reproduced from ref. [14] with permission. Copyright (1993) American Chemical Society)

tra in terms of both wavenumber and absorbance scale to be obtained, even at very low concentrations. Fig. 9.11 shows FT-NIR spectra of octanoic acid in the pure liquid state over a temperature range of 15 to 92 °C after density correction [16]. It can be seen from this figure that the intensity of the band at 6920 cm^{-1}, which is due to the first overtone of an OH stretching mode of the free acid (monomer), increases with temperature. This observation reveals that the proportion of the monomer increases as the temperature is raised [16]. In order to estimate the intensity of the band at 6920 cm^{-1}, difference spectra were calculated by

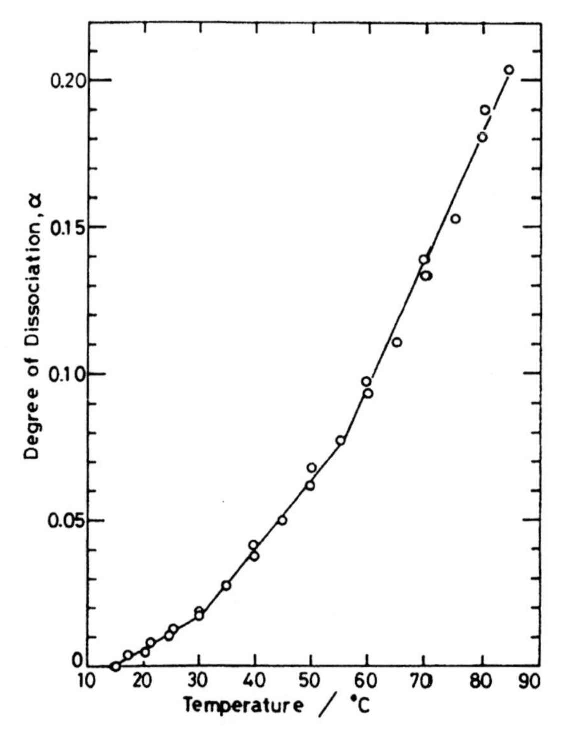

Fig. 9.9 Temperature dependence of the degree of dissociation, α, for oleic acid in the pure liquid state. (Reproduced from ref. [14] with permission. Copyright (1993) American Chemical Society)

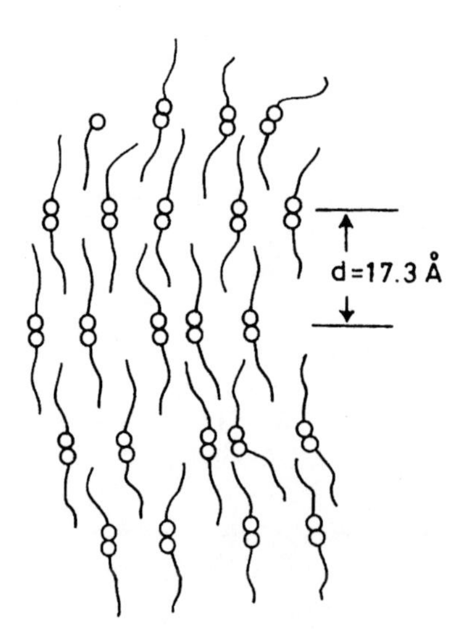

Fig. 9.10 Proposed structure for the dimers of oleic acid molecules in clusters taking the structure of a quasi-smectic liquid crystal. (Reproduced from ref. [15] with permission. Copyright (1991) American Chemical Society)

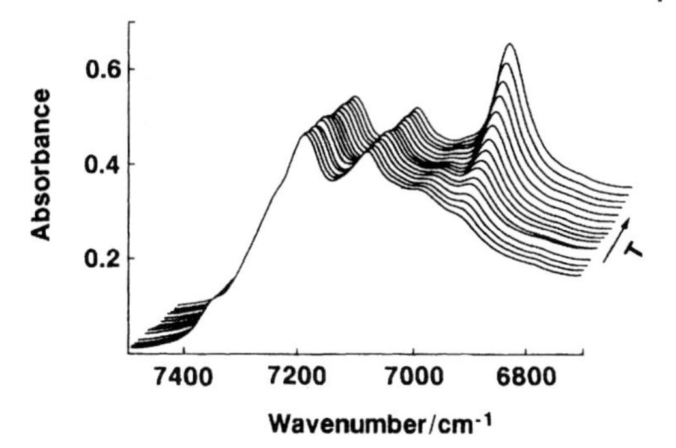

Fig. 9.11 FT-NIR spectra of octanoic acid in the pure liquid over a temperature range of 15 to 92 °C (after density correction). (Reproduced from ref. [16] with permission. Copyright (1993) Society for Applied Spectroscopy)

subtracting a reference spectrum from the spectra in Fig. 9.11. The spectrum measured at the temperature closest to the melting point, 15.1 °C was taken as the reference spectrum because the concentration of the monomer at that temperature should be negligible.

The difference spectra thus obtained are presented in Fig. 9.12 [16]. Note that, even after the calculation of the difference spectra, the monomer band is still partially overlapped. Fortunately, however, the high-frequency side of the monomer band is nearly unperturbed, as shown in Fig. 9.13, assuming a symmetric band shape [16]. Thus, half of the band area was calculated by using this unperturbed wing and then the result was doubled. This method enables the band area and peak height for the band due to the first overtone of the OH stretching mode of the free acid to be estimated.

In order to determine the degree of dissociation and the thermodynamic parameters of octanoic acid in the liquid phase, the molar absorption coefficients (integrated and peak height) of the first overtone of the OH stretching mode of the monomeric acid were calculated based upon a series of NIR spectra of carbon tetrachloride solutions of octanoic acid with concentrations ranging from 0.0117 to 0.4502 mol dm^{-3} at 15 °C [16].

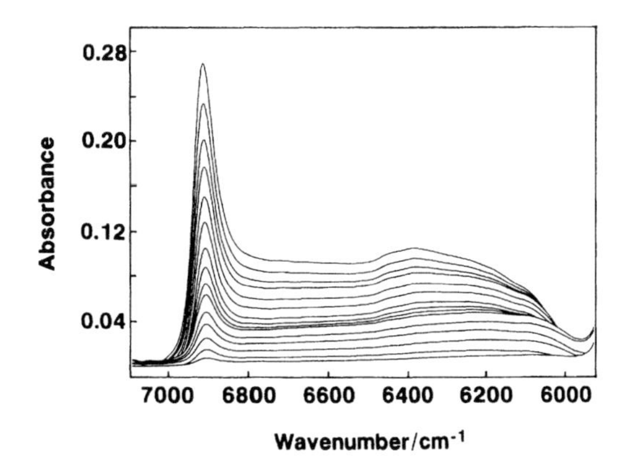

Fig. 9.12 Difference spectra of neat octanoic acid over a temperature range of 20 to 92 °C. The spectrum at 15.1 °C was taken as a reference spectrum. (Reproduced from ref. [16] with permission. Copyright (1993) Society for Applied Spectroscopy)

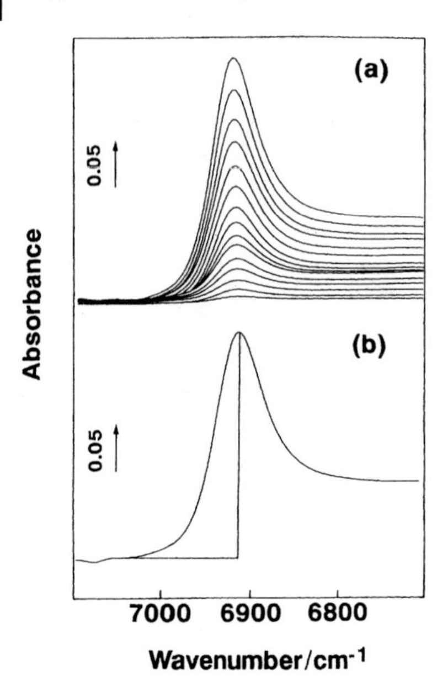

Fig. 9.13 (a) Expanded difference spectra in the 7100-6700 cm^{-1} region over the temperature range of 20 to 92 °C. (b) The difference spectrum at 92 °C, illustrating the idea of evaluation of the band area and peak height. (Reproduced from ref. [16] with permission. Copyright (1993) Society for Applied Spectroscopy)

Fig. 9.14 shows the temperature dependence of the degree of dissociation of neat octanoic acid calculated by use of band area (△) and peak height (◇) [16]. On the basis of the values of the degree of dissociation, the thermodynamic parameters (ΔH and ΔS) for the dissociation process of the dimer into monomers could be evaluated. Plots of the logarithm of the equilibrium constant against the inverse of the absolute temperature for the values calculated by use of the band areas and peak heights, respectively, give straight lines and from the slopes of the lines the enthalpy and entropy for the process of dissociation were determined.

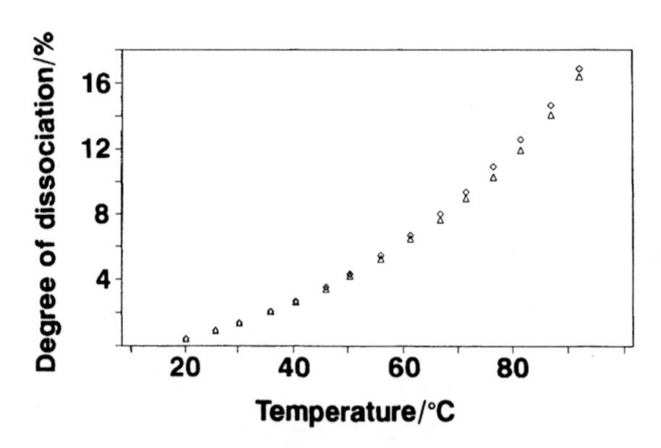

Fig. 9.14 Temperature dependence of the degree of dissociation of neat octanoic acid in the pure liquid state calculated by use of band area (△) and peak height (◇). (Reproduced from ref. [16] with permission. Copyright (1993) Society for Applied Spectroscopy)

The values obtained by using the band areas are $\Delta H=33.21$ kJ mol^{-1} and $\Delta S=86.95$ J K^{-1} mol^{-1} [16]. The corresponding values determined from the peak heights are $\Delta H=33.51$ kJ mol^{-1} and $\Delta S=88.18$ J K^{-1} mol^{-1} [16].

9.2.3 Alcohols

Alcohols form a variety of self-associated species as shown in Fig. 9.15. The hydrogen bonds and self-association of alcohols have been investigated extensively by means of NIR spectroscopy [17–21]. It may be assumed that bonds *e, f, g* and *h* have nearly the same strength: this gives the same OH stretching frequency (ν_4) in the NIR region and bonds *b* and *d* are nearly the same in strength, having the same frequency (ν_2). In earlier literature [22, 23], it was assumed bond *a* as well as *b* and *d* showed the same OH stretching frequency. Recent FT-NIR [20] and generalised 2D NIR correlation spectroscopy [21] studies have clearly discriminated the band due to *a* from *b* and *d*. Here, the FT-NIR and 2D NIR correlation spectroscopy studies of oleyl alcohol in the pure liquid are explained.

Fig. 9.16 shows FT-NIR spectra in the 11500–6000 cm^{-1} region of oleyl alcohol in the pure liquid state measured over a temperature range of 6.5–90 °C after density correction [17]. Bands at 10753, 8555 and 8254 cm^{-1} are assigned to the third overtones of CH stretching modes, the second overtone of a CH stretching mode of the HC=CH group and the second overtones of the CH stretching modes of CH_3 and CH_2 groups, respectively. A weak band at 10375 cm^{-1} and an intense feature at 7090 cm^{-1} are due to the second and first overtones of an OH stretch-

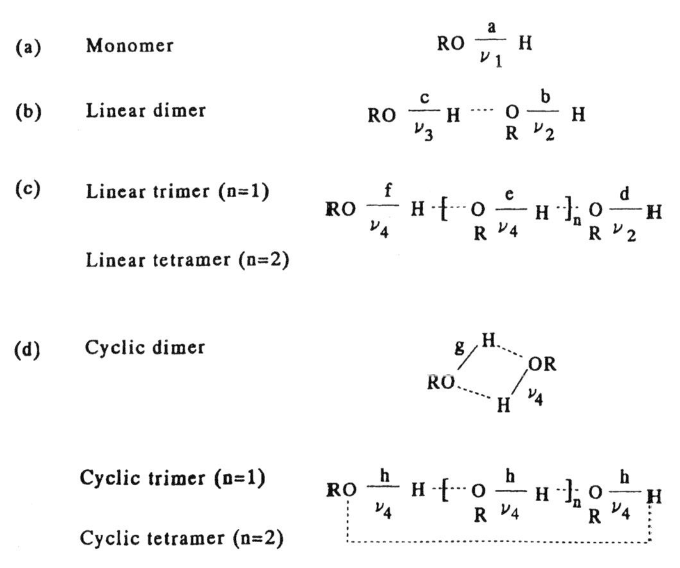

Fig. 9.15 Possible structure and hydrogen-bond types for alcohols (Reproduced from ref. [20] with permission. Copyright (1995) Society for Applied Spectroscopy)

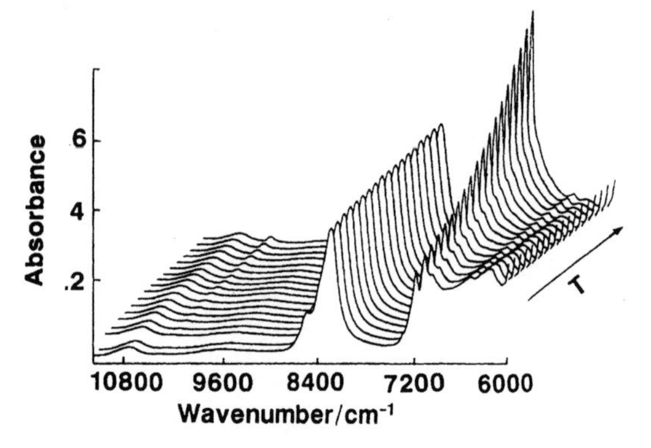

Fig. 9.16 FT-NIR spectra of oleyl alcohol in the pure liquid state measured over a temperature range of 6.5–90 °C. (Reproduced from ref. [17] with permission. Copyright (1993) Society for Applied Spectroscopy)

ing mode of the monomer, respectively. The latter band heavily overlaps with bands due to combinations of CH vibrations. A broad feature in the 6600–6200 cm^{-1} region is ascribed to the first overtone of OH stretching modes of polymeric forms of the alcohol. The temperature-dependent changes in the first and second overtones of the OH stretching modes show that the hydrogen-bonded species dissociate into the monomeric species with increasing temperature [17].

In order to extract useful information from the temperature-dependent spectral variations in Fig. 9.16, one can calculate the difference spectra, as in the case of fatty acids [17, 20]. However, here, the analysis of the temperature-dependent NIR spectra by generalised 2D correlation spectroscopy (Chapter 8) is described [21]. Fig. 9.17 shows the full view of the pseudo three-dimensional *stacked-trace* representation of the 2D NIR synchronous correlation spectrum; this represents the temperature-dependent spectral intensity variations in the 11 500–6000 cm^{-1} region of oleyl alcohol between 6.5 and 90 °C [21]. A dominant autopeak at the diagonal position near 7090 cm^{-1} corresponds to the temperature-induced intensity variation of the first overtone of the OH stretching mode of the monomeric species. Although such a stacked representation offers the best overall view of a 2D correlation spectrum, it is usually easier to use a contour map representation to inspect the detailed peak shapes and positions.

Fig. 9.18(a) shows the contour map representation of the synchronous 2D NIR correlation spectrum in the 8000–6000 cm^{-1} region [21]. Autopeaks at 7090 and around 6300 cm^{-1} are ascribed to the first overtones of the OH stretching mode of the monomer and the polymeric forms, respectively. The negative cross peaks between the two bands mean that the directions of the intensity changes at these wavenumbers are opposite. The simultaneous appearance and disappearance of the OH stretching bands of the monomeric and polymeric forms suggest a population shift of oleyl alcohol from the polymeric to the monomeric form with temperature. However, the asynchronous 2D NIR spectrum shown in Fig. 9.18(b) clearly indicates

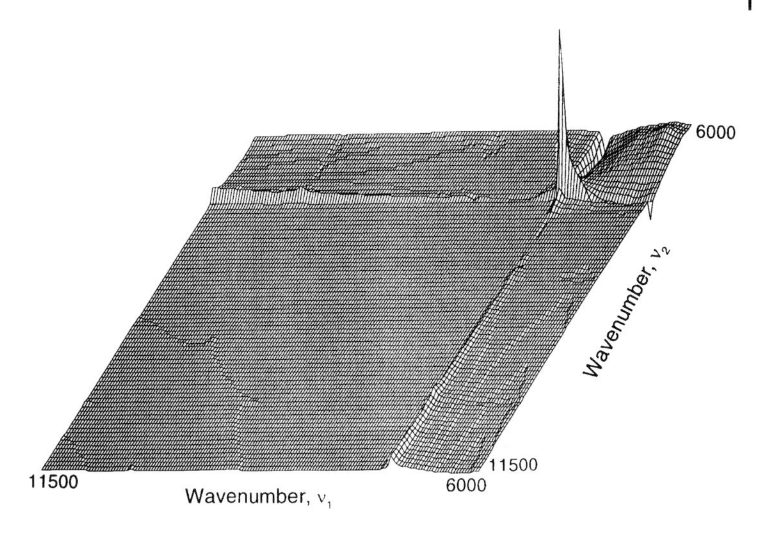

Fig. 9.17 Full view of the pseudo-three-dimensional stacked-trace representation of synchronous 2D FT-NIR correlation spectra of oleyl alcohol obtained from 6.5 to 90 °C. (Reproduced from ref. [21] with permission. Copyright (1995) American Chemical Society)

that the pattern of the temperature dependence for the peak intensity change at 7090 cm^{-1} is different from those of other bands [21]. The signs of the cross peaks suggest that the change in the peak intensity at 7090 cm^{-1} occurs at higher temperature than those at other spectral coordinates. In other words, the disappearance of the polymeric form does not simultaneously result in the formation of the monomeric form; this indicates that there may be intermediate species such as the dimer [21]. The cross peaks near 6840 and 6200 cm^{-1} in the asynchronous 2D NIR correlation spectrum probably correspond to the dimer and polymer bands, respectively (Fig. 9.18(b)). Thus, it is concluded that the alcohol in the pure liquid state consists of the polymeric, intermediate (e.g., dimer and trimer) and monomeric species [21].

The synchronous and asynchronous spectra in the 7300–6800 cm^{-1} region are shown in Fig. 9.19(a) and (b), respectively [21]. Note that the autopeak at 7090 cm^{-1} is significantly extended into the off-diagonal area of the spectrum (Fig. 9.19(a)); this indicates the existence of additional cross peaks obscured by the strong autopeak. The additional cross peaks probably arise from a band due to free terminal OH groups of the linear hydrogen-bonded species (bonds *b* and *d* in Fig. 9.15) [21].

The asynchronous spectrum (Fig. 9.19(b)) reveals that there are several different types of spectral responses of oleyl alcohol to the temperature change; bands at 7115, 7090, 7070 and 6840 cm^{-1} are clearly differentiated. The splitting of the two bands at 7115 and 7090 cm^{-1} may be ascribed to a rotational isomerism of the free OH group [21]. The band at 6840 cm^{-1} is assigned to the dimer as described above. In this way, the bands due to the rotational isomers, the free terminal OH groups (7070 cm^{-1}) and the dimer can be identified separately in the 2D spectra [21].

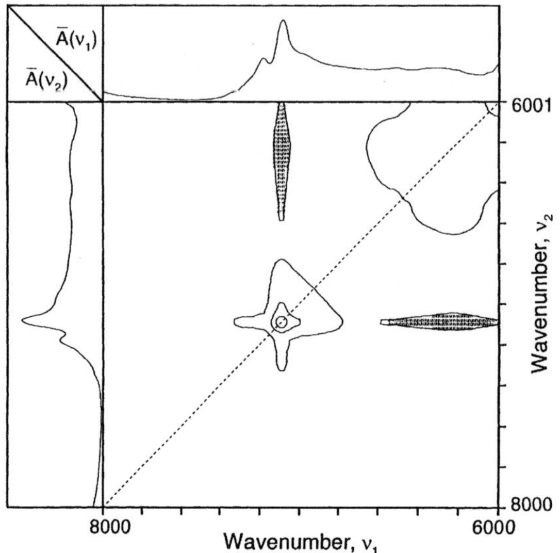

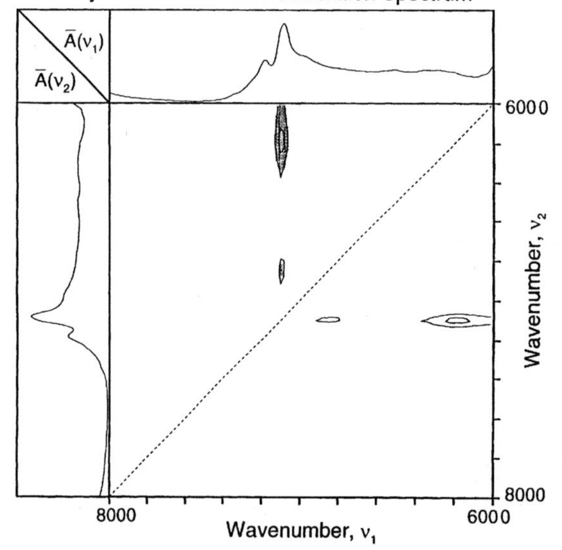

Fig. 9.18 Synchronous (a) and asynchronous (b) 2D NIR correlation spectra of oleyl alcohol in the 8000–6000 cm^{-1} region. (Reproduced from ref. [21] with permission. Copyright (1995) American Chemical Society)

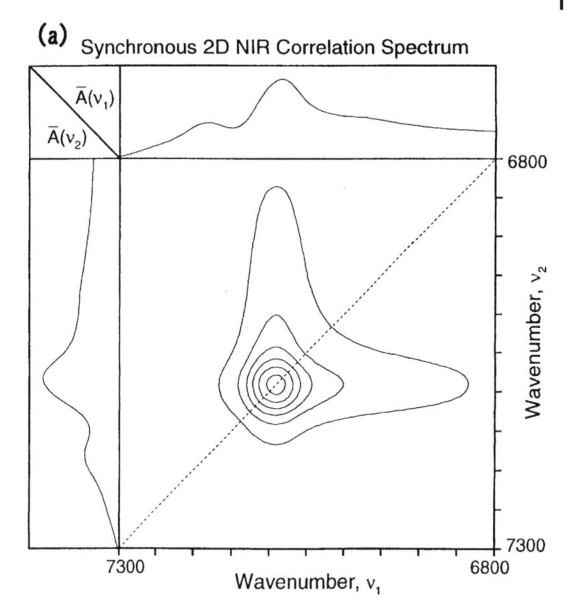

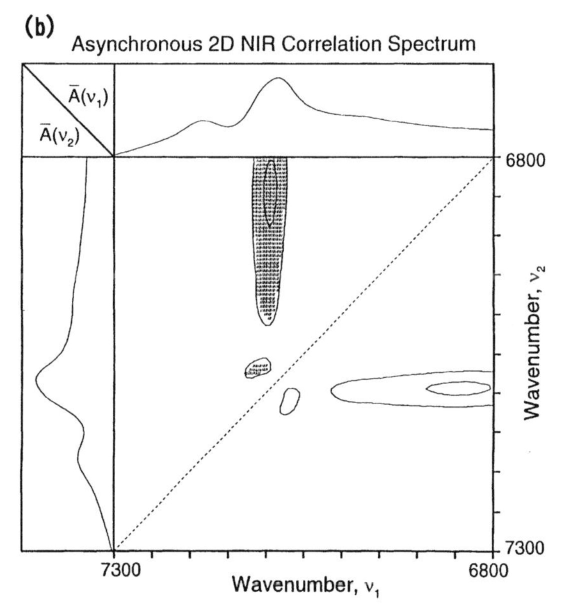

Fig. 9.19 Synchronous (a) and asynchronous (b) 2D NIR correlation spectra of oleyl alcohol in the 7300–6800 cm^{-1} region. (Reproduced from ref. [21] with permission. Copyright (1995) The American Chemical Society)

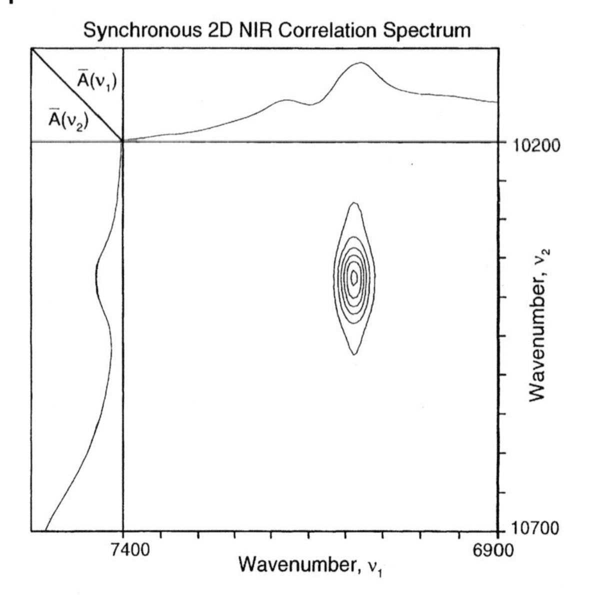

Fig. 9.20 Localised view of the off-diagonal position of a synchronous 2D FT-NIR spectrum of oleyl alcohol. (Reproduced from ref. [21] with permission. Copyright (1995) American Chemical Society)

The 2D NIR correlation analysis of oleyl alcohol also suggests the possibility of correlating various overtone and fundamental bands to establish unambiguous assignments in the NIR region [21]. A synchronous 2D NIR spectrum, shown in Fig. 9.20, compares the temperature-dependent intensity changes of two bands at 10380 and 7090 cm^{-1} due to the second and first overtones of the OH stretching mode of the monomeric form of the alcohol, respectively. A strong cross peak appears between the two bands; this indicates that they share the same temperature-dependent pattern, as expected. The correlation between the bands arising from the same group of the same species suggests that one can make correlations for various overtones and fundamental bands to establish assignments in the rather complicated NIR region.

Similar 2D correlation spectroscopy studies have been carried out for butanol [24, 25] and *N*-methylacetamide [26], polyamides [27, 28] and Nylon [29].

9.2.4
Proteins

NIR spectroscopy is useful for studying hydration and hydrogen bonds in proteins. Wang et al. [30] and Wu et al. [31] demonstrated the potential of generalised 2D NIR correlation spectroscopy in protein research. Wang et al. [30] measured temperature-dependent NIR spectra for ovalbumin solutions over a temperature range of 45–80 °C, which covers the whole heat denaturation process. A series of dynamic NIR spectra modulated by the concentration at various temperatures were used to construct 2D synchronous and asynchronous correlation spectra [30].

Fig. 9.21(a) and (b) shows NIR spectra in the 4900-4550 cm^{-1} region of ovalbumin solutions with concentrations of 0, 2, 5 and 8 wt% measured at 45 and 80 °C, respectively, at which temperatures the protein is in the natively folded and partially unfolded states [30]. It is impossible to identify any band in the spectra except for a tail of the strong absorption band of water near 5200 cm^{-1}. Thus, the second derivatives of the spectra were calculated (Fig. 9.22(a) and (b)). The calculation of the second derivative makes some bands observable. Two bands at 4850 and 4600 cm^{-1} in Fig. 9.22(a) and (b) are assigned to combinations of free NH stretching and amide II (amide A/II) and intramolecular hydrogen-bonded NH stretching and amide II (amide B/II) of ovalbumin, respectively [30].

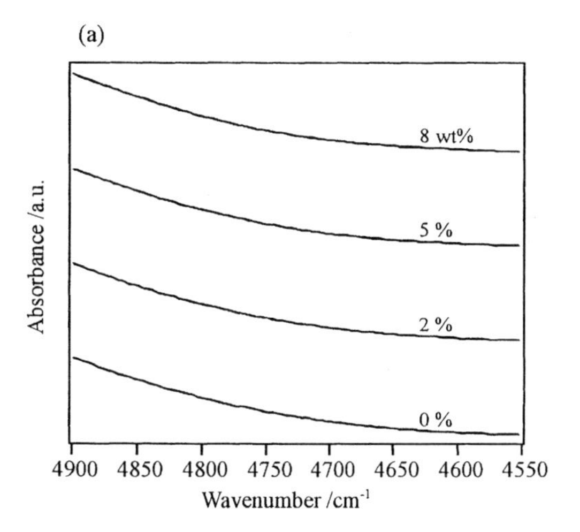

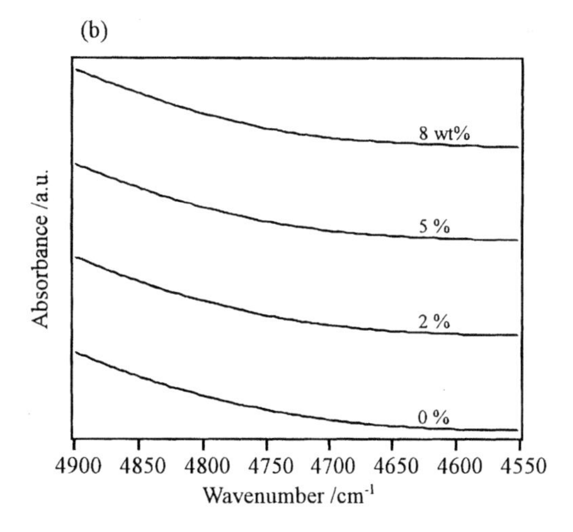

Fig. 9.21 NIR spectra in the 4900–4550 cm^{-1} region of ovalbumin solutions with a concentration of 0, 2, 5 and 8 wt%, respectively. (a) 45 °C, (b) 80 °C. (Reproduced from ref. [30] with permission. Copyright (1998) American Chemical Society)

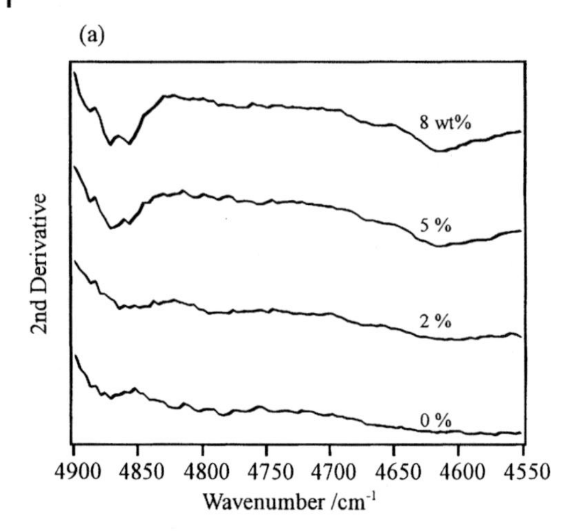

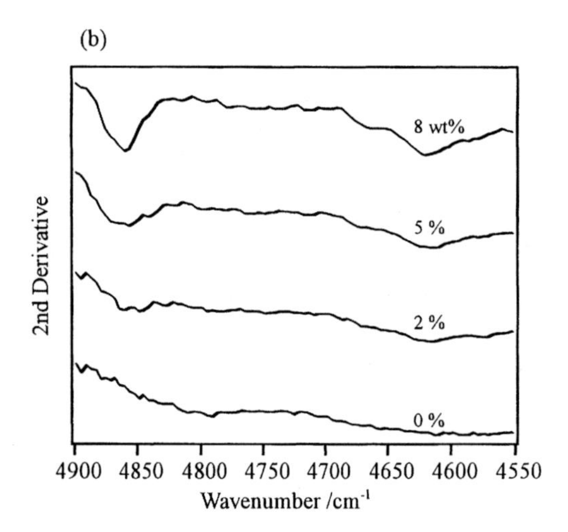

Fig. 9.22 The second derivative of the NIR spectra shown in Figure 9.21. (a) 45 °C, (b) 80 °C. (Reproduced from ref. [30] with permission. Copyright (1998) American Chemical Society)

Fig. 9.23(a), (b) and Fig. 9.24(a), (b) show contour map representations of synchronous and asynchronous 2D correlation spectra, which are constructed from the concentration-perturbed NIR spectra of ovalbumin solutions at 45 and 80 °C [30]. Note that two strong autopeaks are clearly observed at 4850 and 4800 cm^{-1} in the synchronous spectrum at 45 °C, although the peak at 4800 cm^{-1} could hardly be identified even in the second derivative spectra (cf. Fig. 9.22(a)). The two peaks probably correspond to the combination band of the amide A/II and the second overtone of OH bending ($3\nu_2$) of water, respectively. The appearance of the autopeaks means that the intensities of these two bands change most significantly with the increase in the concentration in the folded state. In addition, positive cross peaks (4850 vs. 4800 cm^{-1}) are seen in the synchronous spectrum; this indicates that their band intensities increase in phase (simultaneously). NIR spectros-

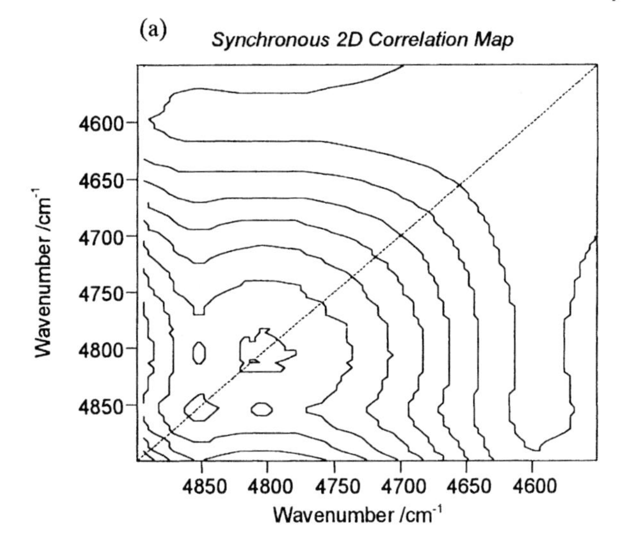

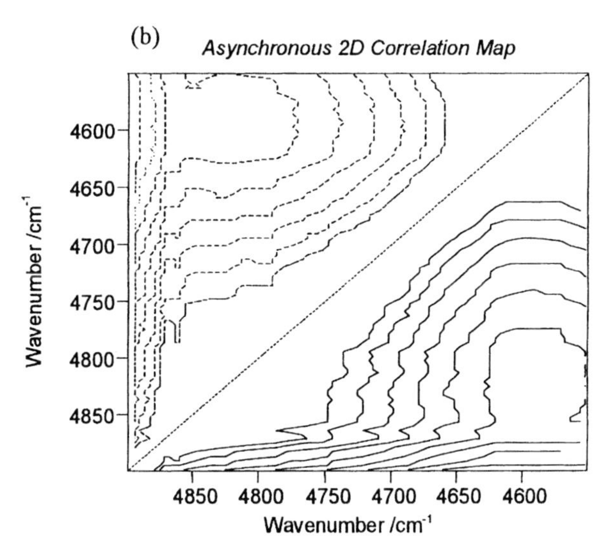

Fig. 9.23 (a) Synchronous and (b) asynchronous 2D NIR correlation spectra in the 4900–4550 cm^{-1} region, constructed from concentration-perturbed spectra of ovalbumin solutions at 45 °C. Solid and dashed lines in the contour maps denote positive and negative correlation peaks, respectively. (Reproduced from ref. [30] with permission. Copyright (1998) American Chemical Society)

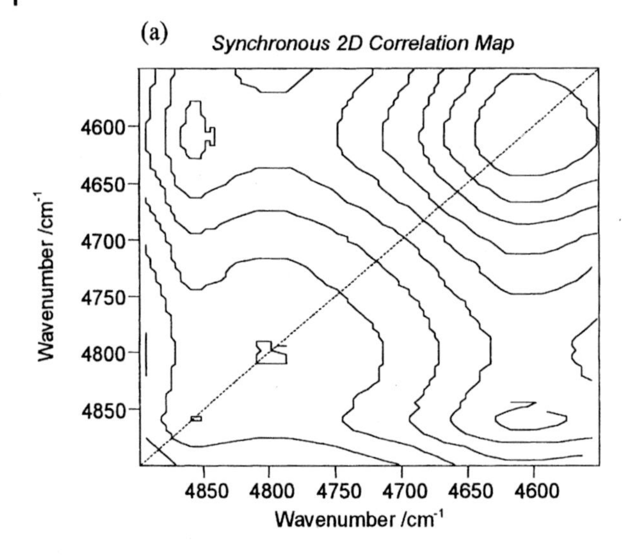

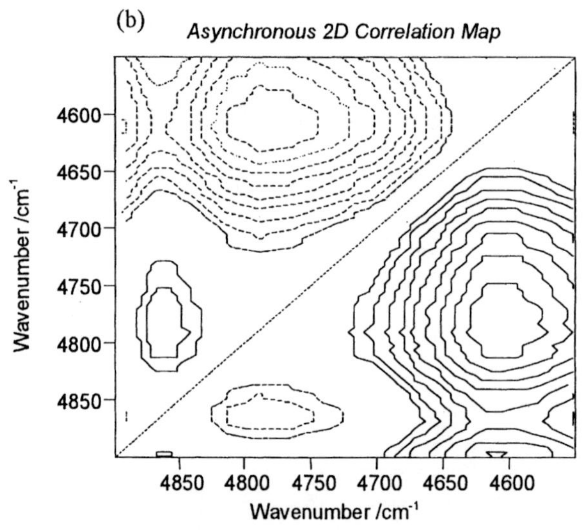

Fig. 9.24 Synchronous (a) and asynchronous (b) 2D NIR correlation spectra in the 4900–4550 cm^{-1} region, constructed from concentration-perturbed spectra of ovalbumin solutions at 80°C. (Reproduced from ref. [30] with permission. Copyright (1998) American Chemical Society)

copy has one unique advantage for exploring protein structures in solution: in the NIR region the absorption of water becomes relatively weak, so that one can monitor spectral changes due to protein and water simultaneously. Therefore, it is possible to investigate the hydration of proteins directly.

The asynchronous 2D map of Fig. 9.23(b) is dominated by asymmetric cross peaks (4850 vs. 4600 cm^{-1} and 4800 vs. 4600 cm^{-1}). The asynchronous correlation spectrum reveals that at 4600 cm^{-1}, the amide B/II band shows an out-of-phase variation with the amide A/II and $3\nu_2$ bands. The sign of the cross peaks in the asynchronous spectrum gives additional useful information about relative temporal relationships for different bands. According to the rule proposed by Noda [32] for determining the sequential relationships between different bands, the negative peaks (4850 and 4800 vs. 4600 cm^{-1}) in Fig. 9.23(b) indicate that the intensity of the amide B/II band varies at a lower concentration than those of the amide A/II and $3\nu_2$ bands.

A strong autopeak is observed at 4610 cm^{-1} in the synchronous spectrum at 80 °C (Fig. 9.24(a)); this shows that the intensity of the amide B/II band changes most significantly in the spectra of the unfolded state. The two autopeaks at 4850 and 4800 cm^{-1} in the synchronous spectrum of Fig. 9.23(a) become weak in Fig. 9.24(a). Instead, two new positive cross peaks (4860 vs. 4610 cm^{-1}) appear in the synchronous map. This indicates the in-phase intensity variation of the amide A/II band with the amide B/II band. The in-phase intensity variation between the amide A/II band and the $3\nu_2$ band in the folded state vanishes completely. In addition, the $3\nu_2$ band changes out-of-phase with the amide A/II and amide B/II bands as verified by the corresponding asynchronous spectrum (Fig. 9.24(b)).

The asynchronous map of Fig. 9.24(b) shows two pairs of new cross peaks (4860 vs. 4780 cm^{-1} and 4780 vs. 4610 cm^{-1}). These peaks confirm the out-of-phase intensity variation of the $3\nu_2$ band with the amide A/II and amide B/II bands. Note that the amide A/II and amide B/II bands show a high frequency shift of about 10 cm^{-1} while the $3\nu_2$ band shifts downward by some 20 cm^{-1} in the 2D spectra as the temperature increases from 45 to 80 °C. These shifts may be due to the thermally-induced unfolding of the secondary structure of ovalbumin [30].

The corresponding 2D correlation spectra at 76 °C, the midpoint temperature during the unfolding process of ovalbumin, are similar to those at 80 °C shown in Fig. 9.24(a) and (b); this indicates that the intermolecular interaction between ovalbumin and water, that is the hydration of ovalbumin, is changed substantially even in the intermediate state of the unfolding process. Moreover, the $3\nu_2$, the amide A/II and amide B/II bands have already shifted to some extent at 76 °C.

In order to investigate how the correlation patterns vary with temperature during the heat denaturation process, a series of the 2D correlation spectra were calculated over a temperature range from 45 to 80 °C at intervals of 2 °C [30]. The correlation patterns below 65 °C and above 71 °C are very close to those of 45 and 80 °C, respectively. Of particular note is that the correlation patterns change in a very narrow temperature range of 67–69 °C. Therefore, it was concluded from these results that the hydration of ovalbumin remains in the native state below 65 °C, then undergoes the sudden change from the native to the denatured state in the critical temperature region of 67–69 °C and is unchanged again above 71 °C. The unfolding of secondary structures begins at about 69 °C and continues progressively until 80 °C [30]. Therefore, the change in the hydration occurs earlier than the unfolding process. It seems very likely that the change in the hydration in the region of 67-69 °C initiates the unfolding process from 69 to 80 °C.

9.3 Chemometric Approach to Basic Chemical Problems

In NIR spectroscopy, chemometrics has been employed mainly for practical problems [1–3]. However, it is also useful for basic research in chemistry. Three examples of the applications of chemometrics to basic problems in chemistry are discussed below.

9.3.1 Determination of the Physical and Chemical Properties of Water

Lin and Brown [8] applied NIR spectroscopy as a universal approach to determine the physical and chemical properties of water and their functions of temperature. PCR and MLR models were developed and employed to correlate the spectra with the properties of water at temperatures between 5 and 65 °C. The properties examined were density, refractive index, dielectric constant, relative viscosity, surface tension, vapour pressure, sound velocity, isothermal compressibility, thermal expansivity, thermal capacity, thermal conductivity, enthalpy, free energy, entropy and ionisation constant. They demonstrated that these fifteen properties of water can be determined simultaneously simply by measuring a set of NIR spectra of water [8].

In principle, all the physical and chemical properties of water depend upon its structure; the spatial relations between water molecules and the hydrogen bonding between OH groups. Since an NIR spectrum of water reflects its packing and hydrogen bonds, there should be relations between the properties and spectra of water as a function of temperature. Therefore, it is expected that these properties can be determined from the NIR spectra of water. In other words, a universal spectroscopic approach can be developed.

Fig. 9.25 shows linear correlations between the PCR predicted and actual properties [8]. Very good linear regressions were found between them. All of the squared coefficients (R^2) are larger than 0.9978 and most of the slopes are in the range of 0.987 to 1.000, except for isothermal compressibility, thermal capacity and vapour pressure [8]. (Liu and Brown compared the PCR models with MLR models. For a more detailed discussion, see ref. [8].)

9.3.2 Discrimination of 24 Kinds of Alcohols by PCA

Twentyfour kinds of low-molecular-weight alcohols with straight and branched chains can be easily identified by PCA of their NIR spectra [33]. The NIR spectra in the 1100–2200 nm region were measured for the 24 kinds of primary, secondary and tertiary alcohols listed in the legend to Fig. 9.26. The spectra were subjected to PCA and it was found that PC 1 and PC 2 can construct a model in which all the alcohols are 100% correctly identified simultaneously. The sample pattern of the learning set in the 2-factor plot is shown in Fig. 9.26 [33]. The sam-

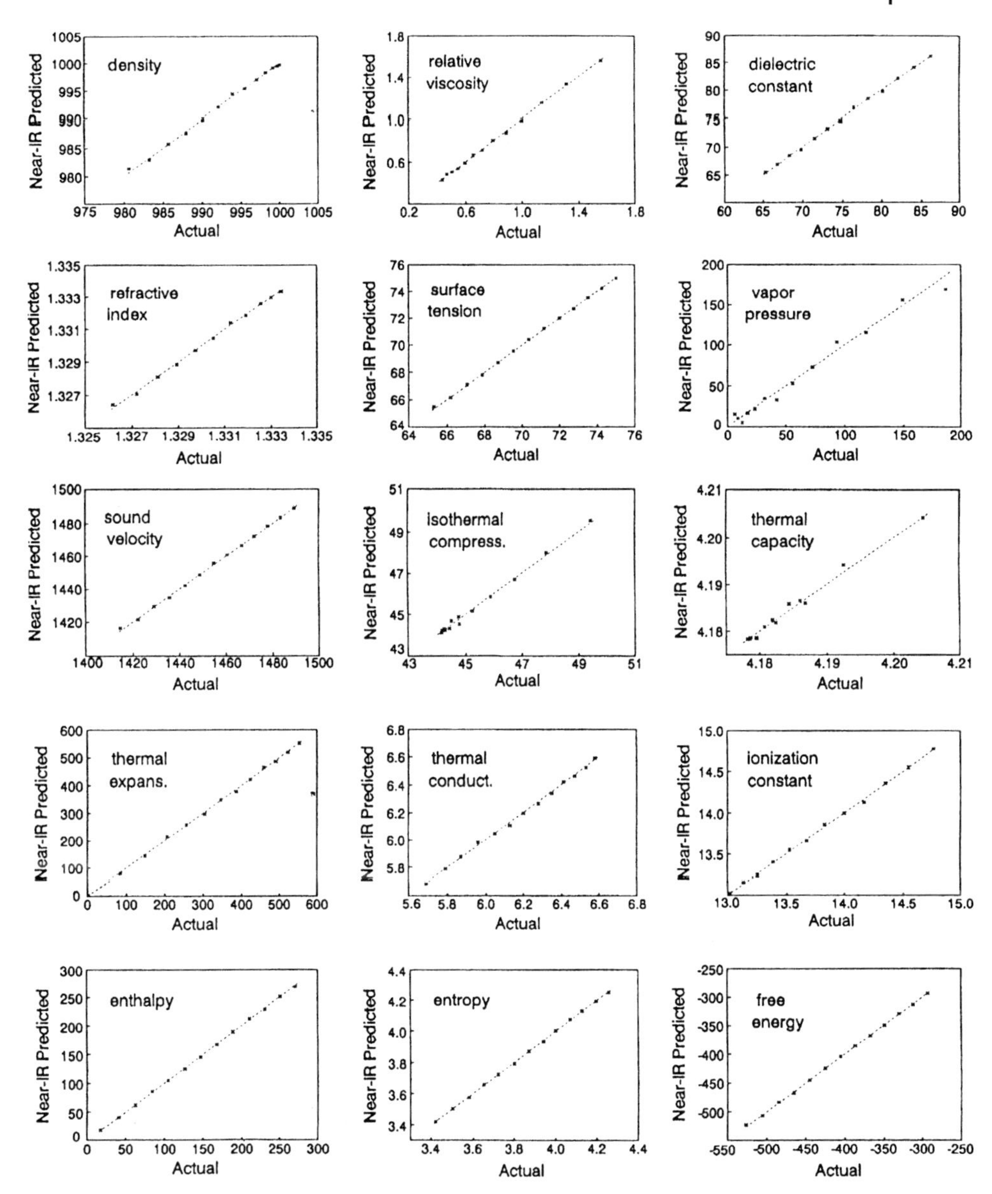

Fig. 9.25 Linear correlations between the PCR predicted properties and actual properties. The dashed lines are perfect linear regressions (1:1 lines) with zero intercepts. (Reproduced from ref. [8] with permission. Copyright (1993) Society for Applied Spectroscopy)

ple set contains some alcohols whose spectra are very similar to each other. For example, the spectra of 1-octanol and 1-nonanol are so close in the whole spectral region that one might misidentify them in a conventional procedure such as superimposing the spectra. In the two-factor plot, however, the difference between

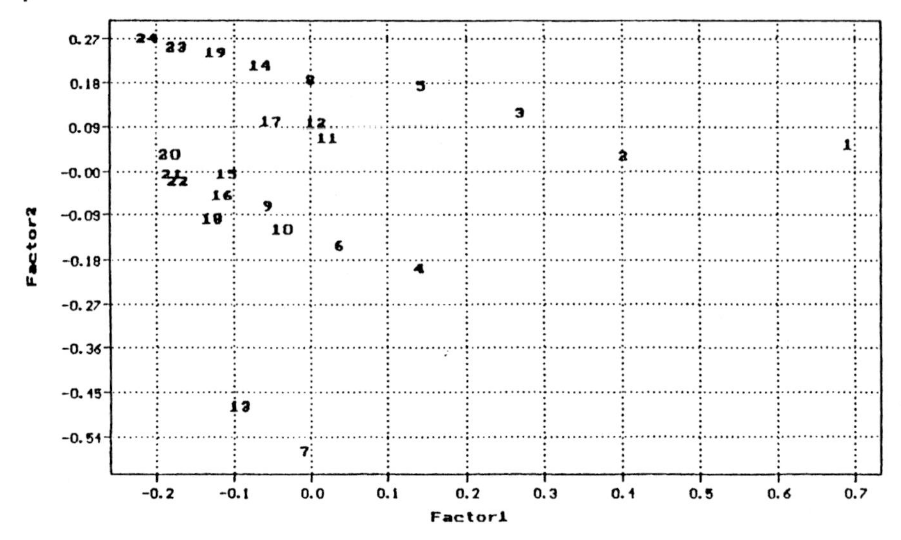

Fig. 9.26 1.2-factor plot of PC2 versus PC1. Primary alcohol [straight]: 1) methanol, 2) ethanol, 3) propan-1-ol, 5) butan-1-ol, 8) pentan-1-ol, 14) hexan-1-ol, 19) heptan-1-ol, 23) 1-octan-1-ol, 24) nonan-1-ol; [branched]: 11) 2-methylbutan-1-ol, 12) 3-methylbutan-1-ol, 17) 2-methylpentan-1-ol; Secondary alcohol [straight]. 4) propan-2-ol, 6) butan-2-ol, 9) pentan-2-ol, 15) hexan-2-ol, 20) heptan-2-ol, 21) heptan-3-ol, 22) heptan-4-ol; [branched]: 18) 4-methylpentan-2-ol; Tertiary alcohol [straight]: 7) 2-methylpropan-2-ol, 13) 2-methylbutan-2-ol. (Reproduced from ref. [33] with permission. Copyright (1995) Impact Printing (Vic.) Pty Ltd)

them can be clearly seen. Note that in Fig. 9.26 the 24 kinds of alcohol make up three major groups consisting of primary, secondary and tertiary alcohols. For the primary and secondary alcohols, even subgroups of the straight and branched structures can be recognised.

Interpretation of the PC is very important in order to understand the model shown in Fig. 9.26. Fig. 9.27(a) depicts the eigenvector of PC 1 [33]. Note that peaks around 1580 and 2040 nm, corresponding to the first overtone of the OH stretching mode and the combination of the OH vibrations, respectively, change greatly in intensity. Significant changes are also seen for features around 1200 and 1700 nm ascribed to the second and first overtones of the CH stretching modes, respectively. The intensity changes in the OH bands are negative in sign; this means that as the number of carbon atoms increases, the band intensity decreases. This is in good agreement with the fact that the increase in the number of carbon atoms decreases the relative molar concentration of the OH group. A peak at 1410 nm and a broad feature in the 1450–1600 nm region arise from the free and hydrogen-bonded alcohols, respectively, The relative intensity of these two peaks suggests that the proportion of the free OH group increases in the larger alcohols due to steric hindrance.

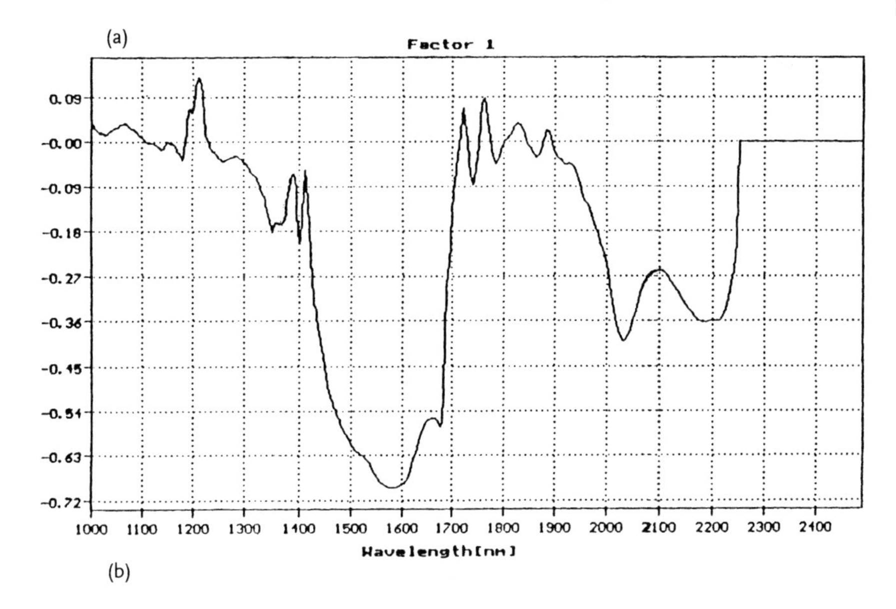

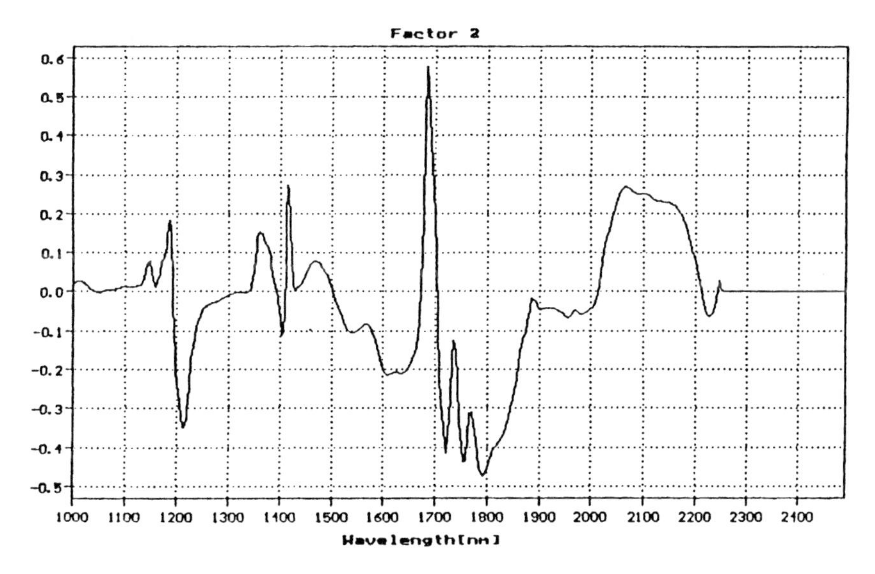

Fig. 9.27 (a) Eigenvector of PC1. (b) Eigenvector of PC2. (Reproduced from ref. [33] with permission. Copyright (1995) Impact Printing (Vic.) Pty Ltd)

For the CH absorption bands, the sign of the PC 1 eigenvector changes with the groups (methyl, methylene and methine groups). The bands around 1180, 1355, 1680 and 1740 nm are ascribed to the methyl groups because of the negative sign, while those around 1210, 1390, 1722 and 1762 nm are assigned to the methylene groups because of their positive sign.

The eigenvector of PC 2 shown in Fig. 9.27(b) has a different aspect. It can be seen from this figure that the contributions from the CH groups are larger in PC 2 than in PC 1 and that, for the OH group, the first overtone and the combination bands have opposite signs. The sample set investigated includes not only alcohols with different numbers of carbon atoms, but also those of different classes. While PC 1 seems to be attributed mainly to the effect of the number of carbon atoms, PC 2 may reflect the variation due to structural isomerism.

9.3.3
Resolution Enhancement of NIR Spectra by Loadings Plots

Chemometrics is useful for the resolution enhancement of NIR spectra [34], and loadings plots or regression coefficients play important roles in it. Fig. 9.28 shows the NIR spectra of a solution of *sec*-butanol (0.3 M) in carbon tetrachloride measured over a temperature range of 10–60 °C [34]. An intense band at 7081 cm^{-1} is due to the first overtone of an OH-stretching mode of the monomeric species. As described above, alcohols form various kinds of self-associated species (Fig. 9.15). In carbon tetrachloride they also form a weak hydrogen bond with the solvent.

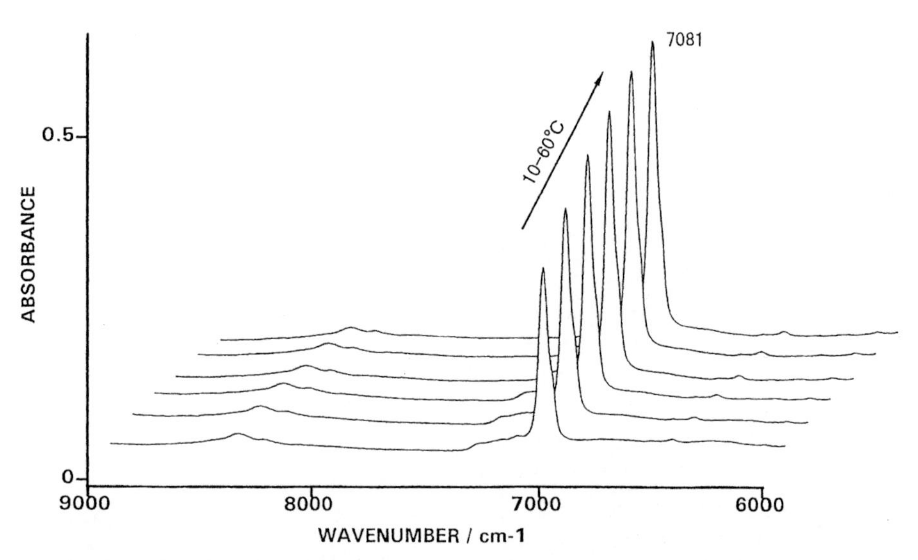

Fig. 9.28 NIR spectra of *sec*-butanol in carbon tetrachloride (0.3 M) measured over a temperature range of 10–60 °C. (Reproduced from ref. [34] with permission. Copyright (1998) Elsevier)

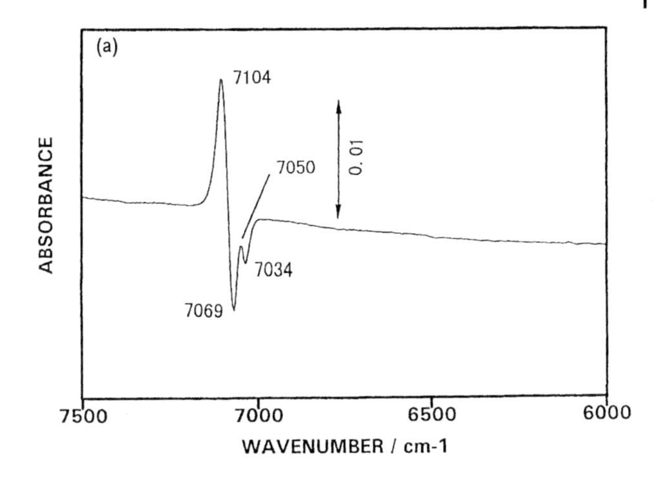

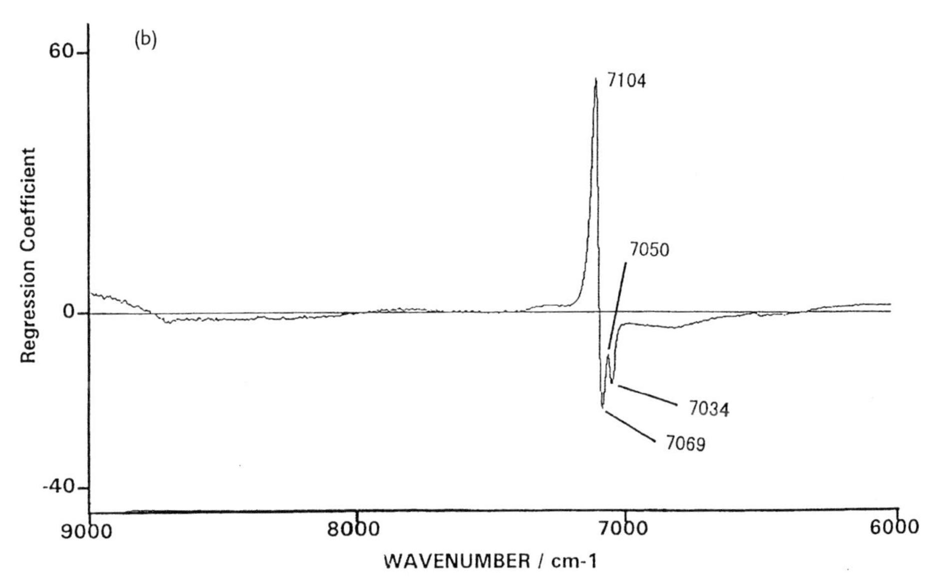

Fig. 9.29 (a) A difference spectrum of *sec*-butanol in carbon tetrachloride at 40 °C. (b) Regression coefficients for the model predicting temperature of *sec*-butanol based upon the NIR spectra measured over a temperature range of 10–60 °C. (Reproduced from ref. [34] with permission. Copyright (1998) Elsevier)

The band at 7081 cm^{-1} should consist of at least four OH-stretching features; two bands arising from rotational isomerism of the alcohol, a band due to the terminal free OH group of the hydrogen-bonded species and one attributed to the OH group hydrogen bonding with carbon tetrachloride.

In order to make these component bands observable, difference spectra were calculated by subtracting the spectrum measured at 10 °C from the spectra obtained at higher temperatures. Fig. 9.29(a) shows the difference spectrum of *sec*-

butanol in carbon tetrachloride at 40 °C [34]. Upward peaks at 7104 and 7050 cm^{-1} are assigned to the free OH groups of the monomer on the basis of the temperature-dependent behaviour [34]. The splitting is attributed to the rotational isomerism. The bands at 7069 and 7034 cm^{-1} are due to the OH group engaging in the hydrogen bond with carbon tetrachloride and to the terminal OH groups.

Fig. 9.29(b) depicts regression coefficients (RCs) for the model predicting temperature of *sec*-butanol in carbon tetrachloride based on the NIR spectra measured over the temperature range of 10–60 °C [34]. The model was developed by use of partial least squares (PLS) regression [34]. Particularly striking is that the RCs are so close to the difference spectrum. Not only the difference spectrum but also RCs can clearly show the existence of many bands. In this way, chemometrics can be used to analyse rather complicated NIR spectra. In other words, chemometrics is powerful even for structural studies of molecules.

9.4 References

[1] *Near-Infrared Technology in the Agricultural and Food Industries*, 2nd ed., (Ed. P. Williams, K. Norris), American Association of Cereal Chemists, St. Paul, **1990**.

[2] *Handbook of Near-Infrared Analysis* (Eds. D. A. Burns, E. W. Ciurczak), Marcel Dekker, New York, **2001**.

[3] B. G. Osborne, T. Fearn, P. H. Hindle, *Practical Near Infrared Spectroscopy with Applications in Food and Beverage Analysis*, Longman Scientific & Technical, Essex, **1993**.

[4] K. Buijs, G. R. Choppin, *J. Chem. Phys.* **1963**, *39*, 2035.

[5] T. Hirschfeld, *Appl. Spectrosc.* **1985**, *39*, 740.

[6] M. Iwamoto, J. Uozumi, K. Nishinari, *Proceedings of the International NIR/NIT Conference* (Ed. J. Hollo, K. J. Kaffka, J. L. Gonczy), Akademiai Kiado, Budapest, **1987**, 3–1.

[7] A. Grant, A. M. C. Davies, T. Bilverstone, *Analyst* **1989**, *114*, 819.

[8] J. Lin, C. W. Brown, *Appl. Spectrosc.* **1993**, *47*, 1720.

[9] H. Maeda, Y. Ozaki, M. Tanaka, N. Hayashi, T. Kojima, *J. Near Infrared Spectrosc.* **1995**, *3*, 191.

[10] V. H. Segtnan, S. Šašic, T. Isaksson, Y. Ozaki, *Anal. Chem.* **2001**, *73*, 3153.

[11] V. Fornes, J. Chaussidon, *J. Chem. Phys.* **1978**, *68*, 4667.

[12] M. Kato, Y. Taniguchi, S. Sawamura, K. Suzuki, *Physics and Chemistry of Ice* (Ed. N. Maeno, T. Hondoh), Hokkaido University Press, Sapporo, Japan, **1992**, p. 83.

[13] A. N. Fletcher, C. A. Heller, *J. Phys. Chem.* **1967**, *71*, 3742.

[14] M. Iwahashi, N. Hachiya, Y. Hayashi, H. Matsuzawa, M. Suzuki, Y. Fujimoto, Y. Ozaki, *J. Phys. Chem.* **1993**, *97*, 3129.

[15] M. Iwahashi, T. Kato, T. Horiuchi, I. Sakurai, M. Suzuki, *J. Phys. Chem.* **1991**, *95*, 445.

[16] M. A. Czarnecki, Y. Liu, Y. Ozaki, M. Suzuki, M. Iwahashi, *Appl. Spectrosc.* **1993**, *47*, 2162.

[17] Y. Liu, M. A. Czarnecki, Y. Ozaki, M. Suzuki, M. Iwahashi, *Appl. Spectrosc.* **1993**, *47*, 2169.

[18] A. N. Fletcher, C. A. Heller, *J. Phys. Chem.* **1967**, *71*, 3742.

[19] M. Iwahashi, Y. Hayashi, N. Hachiya, H. Matsuzawa, H. Kobayashi, *J. Chem. Soc. Faraday Trans.* **1993**, *89*, 707.

[20] Y. Liu, H. Maeda, Y. Ozaki, M. A. Czarnecki, M. Suzuki, M. Iwahashi, *Appl. Spectrosc.* **1995**, *49*, 1661.

[21] I. Noda, Y. Liu, Y. Ozaki, M. A. Czarnecki, *J. Phys. Chem.* **1995**, *99*, 3068.

[22] F. S. Parker, K. R. Bhaskar, *Biochemistry* **1966**, *7*, 1286.

[23] H. C. Van Ness, J. V. Winkle, H. H. Richtol, H. B. Hollinger, *J. Phys. Chem.* **1967**, *71*, 1483.

[24] M. A. Czarnecki, H. Maeda, Y. Ozaki, M. Suzuki, M. Iwahashi, *Appl. Spectrosc.* **1998**, *52*, 994.

[25] M. A. Czarnecki, B. Czarnik-Matusewicz, Y. Ozaki, M. Iwahashi, *J. Phys. Chem. A*, **2000**, *104*, 4906.

[26] Y. Liu, Y. Ozaki, I. Noda, *J. Phys. Chem.* **1996**, *100*, 7326.

[27] M. A. Czarnecki, P. Wu, H. W. Siesler, *Chem. Phys. Lett.* **1998**, *283*, 326.

[28] P. Wu, H. W. Siesler in *Two-Dimensional Correlation Spectroscopy* (Ed. Y. Ozaki, I. Noda), American Instutute of Physics, New York, **2000**, p. 18.

[29] Y. Ozaki, Y. Liu, I. Noda, *Macromolecules*, **1997**, *30*, 2391.

[30] Y. Wang, K. Murayama, Y. Myojo, R. Tsenkova, N. Hayashi, Y. Ozaki, *J. Phys. Chem. B* **1998**, *102*, 6655.

[31] Y. Wu, B. Czarnik-Matusewicz, K. Murayama, Y. Ozaki, *J. Phys. Chem. B*, **2000**, *104*, 5840.

[32] I. Noda, *Appl. Spectrosc.* **1993**, *47*, 1329.

[33] K. Sakurai, T. Miura, Y. Liu, Y. Ozaki in *Leaping Ahead with Near Infrared Spectroscopy; Proceedings of NIR-94* (Ed. G. D. Batten, P. C. Flinn, L. A. Welsh, A. B. Blakeney), Impact Printing (Vic.) Pty Ltd., Brunswick, Australia, **1995**, p. 75.

[34] H. Maeda, Y. Wang, Y. Ozaki, M. Suzuki, M. A. Czarnecki, M. Iwahashi, *Chemometrics and Int. Lab. Syst.* **1998**, *45*, 121.

[35] W. A. P. Luck, W. Ditter, *J. Mol. Struct.* **1967–1968**, *1*, 261.

[36] W. A. P. Luck, W. Ditter, *J. Phys. Chem.* **1970**, *74*, 3687.

[37] C. Bourderon, C. Sandorfy, *J. Chem. Phys.* **1973**, *59*, 2527.

10
Applications to Polymers and Textiles

H. W. Siesler

10.1
Introduction

Synthetic polymers have become an integral part of our every-day life and this chapter will demonstrate that NIR spectroscopy has developed over the last few years into one of the most important characterisation and control techniques for the whole life cycle of a polymeric product. Thus, in many polymer production facilities, NIR spectroscopy is used for the quality control of raw materials and, increasingly, also as a process-monitoring technique. For example, modern polymer processing plants have NIR spectroscopy sensors integrated into extruder operations to monitor blending quality and additive quantification. Last, but not least, although not realised on a large commercial scale, NIR spectroscopy has been shown to be an extremely useful tool for sorting polymeric waste in recycling operations (see Chapter 11).

The main reason for the NIR-specific response of polymers is based on their comparatively simple, "NIR-favourable" structure: most polymer backbones contain aliphatic, olefinic or aromatic CH functionalities, and, frequently, NIR-sensitive functionalities, such as -NH, -OH, -COOH and -C=O, or combinations thereof (e.g. -NH-CO- or -NH-COO-), make up important side-group or main-chain structural characteristics. Based on these structural differences, some authors have studied the NIR spectra of the most common synthetic polymers and assigned certain absorption bands to functional groups [1–10]. Thus, characteristic overtone and combination bands in the NIR spectra of polyolefins, polyesters, polyamides, polyurethanes, polyamic acids and urea and phenol formaldehyde resins have been shown to be of great value for identification purposes, the study of polymerisation kinetics and polymer physical investigations [3, 11–21].

In previous chapters of this book, we have addressed the specific differences of mid-infrared and near-infrared spectroscopy. With the exception of the higher structural specificity of mid-infrared spectroscopy, most of these differences will turn out as advantages to the NIR technique primarily with reference to the nondestructive analysis of polymers and textiles. Thus, the reduced intensities of NIR absorptions necessitate larger optical pathlengths for this spectral region (about 0.1–5 mm for undiluted samples or up to 100 mm for solutions, depending on the wavelength region of interest). This facilitates the testing of solid polymers such as films or sheets, granulates or powders, fibres, fibre fleeces and textiles and turns out to have a considerable advantage over MIR spectroscopy for sample handling of polymers.

As an example, the MIR and NIR spectra of poly(dimethylsiloxane) (PDMS) are shown in Fig. 10.1. Whereas most of the fundamental absorptions in the MIR spectrum of a 30 μm thick film are beyond useful absorbance values, NIR spectroscopy of a 200 μm film yields an excellent spectrum. On top of this, most of the absorption bands in the NIR spectrum can be assigned to their vibrational origin (Tab. 10.1). From the absorbance scale in Fig. 10.1(b), it is also evident that even a 1 mm film thickness would yield a useful spectrum in the NIR region.

Polymer solutions can be studied in quartz or glass cells with pathlengths from 1 mm up to several centimetres. Similarly to mid-infrared spectroscopy, however, the choice of solvents is limited. Solvents that have been used are chloroform, dichloromethane and dioxane. Water-soluble polymers may be studied complementarily in H_2O and D_2O solutions [3]. For specific applications, the choice of a suitable solvent has to be adjusted to the wavenumber window required for the particular analytical NIR method.

The restricted response of NIR spectroscopy to only a few chemical functionalities has, unfortunately, led to the underestimation of the information content of an NIR spectrum. In actual fact, in analogy to a conventional mid-infrared spectrum, an NIR spectrum also contains information on the following parameters of the investigated polymer [11]:

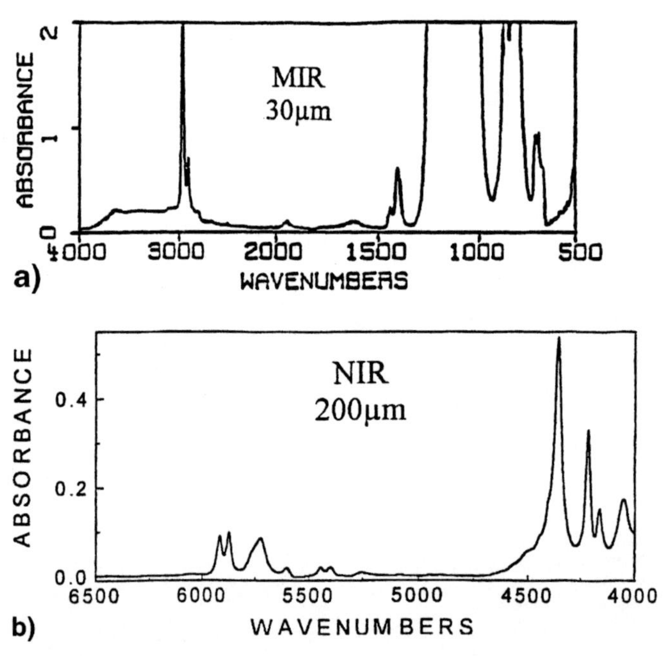

Fig. 10.1 (a) MIR and (b) NIR spectra of a 30 μm (a) and a 200 μm (b) film of poly(dimethylsiloxane) (PDMS)

Tab. 10.1 Band assignment of the overtone and combination vibrations of PDMS (for absorption bands >4000 cm^{-1}, see Fig. 10.1b) (private communication, L. Bokobza, ESPCI, Paris)

Wavenumber [cm^{-1}]	*Assignment*	*Polarisation*
1262	$\delta_s(CH_3)A_1$	σ
1413	$\delta_{as}(CH_3)E$	σ
2499	$2\times\delta_s(CH_3)$	σ
2905	$\nu_s(CH_3)A_1$	π
2964	$\nu_{as}(CH_3)E$	π
4052	$2\times\delta_{as}(CH_3) + \delta_s(CH_3)$	π
4164	$\nu_s(CH_3) + \delta_s(CH_3)$	σ
4216	$\nu_{as}(CH_3) + \delta_s(CH_3)$	π
4358	$\nu_{as}(CH_3) + \delta_{as}(CH_3)$	π
5405	$\nu_s(CH_3) + 2\times\delta_s(CH_3)$	σ
5448	$\nu_{as}(CH_3) + 2\times\delta_s(CH_3)$	π
5605	$\nu_{as}(CH_3) + \delta_{as}(CH_3) + \delta_s(CH_3)$	σ
5725	$2\times\nu_s(CH_3)$	σ
	$\nu_s(CH_3) + \nu_{as}(CH_3)$	
	$\nu_s(CH_3) + 2\times\delta_{as}(CH_3)$	
5872	$2\times\nu_{as}(CH_3)A_1 + E$	σ
5915	$2\times\nu_{as}(CH_3)A_1 + E$	π

- constitution, configuration and conformation
- crystallinity
- anisotropy (from polarisation measurements)
- inter- and intramolecular interactions (e.g. hydrogen bonding)
- thermal and mechanical pretreatment
- number of end groups/molar mass/viscosity
- particle size/fibre diameter (from diffuse-reflection measurements).

It is interesting to note that many of the earlier NIR polymer studies were very fundamental in nature, focussing on the characterisation of structural details such as hydrogen bonding, crystallinity or anisotropy [19, 20]. For this purpose, more sophisticated techniques such as polarisation and variable-temperature measurements or isotope exchange [3, 10] were applied, which then resulted in detailed assignments of the NIR spectra and their correlation with the investigated structural phenomenon. Examples of recent applications of NIR spectroscopy in fundamental research will be discussed in some detail at the end of this chapter.

With reference to the application of diffuse-reflection measurements, the smaller absorptivities and larger scattering coefficients at the shorter wavelengths lead to a larger scattering/absorption ratio in the NIR spectral region [22]. These effects make diffuse reflection a very linear phenomenon in the NIR and support its application for the quantitative analysis of powders and granular materials.

10.2 Selected Analytical Applications

A large amount of literature is available on applications of near-infrared spectroscopy to polymer and textile analysis; it ranges from the determination of the OH-number to copolymer composition to the residual monomer content in curing reactions [1–6, 11–15, 78–80]. Only some selected examples, which are representative of different phenomena and parameters, will be outlined here.

The spectral separation of the $\nu(CH_2)$ and $\nu(CH_3)$ overtones and of the $(\nu+\delta)$ combination bands of the organic OH and water OH absorptions in the NIR region, for example, opened up the possibility of the simultaneous determination of the ethyleneoxide/propyleneoxide content, the OH number and the water content of aliphatic copolyethers [3] (Fig. 10.2(a) and (b)). In fact, a more detailed analysis (Fig. 10.3) shows that even primary- and secondary-OH functionalities can be differentiated by NIR spectroscopy. Such investigations have gained large industrial importance due to the tremendous amount of time that may be saved once a calibration model has been established.

The sensitivity of NIR spectroscopy to CH functionalities involved in carbon-carbon rings or open chains and double or single bonds has led to numerous polymer-analytical applications in the field of ring-opening and curing reactions [3, 11]. Thus, Fig. 10.4(a) and (b) demonstrates the monitoring of the UV-induced polymerisation of a diacrylate by MIR and NIR spectroscopy. In both cases, information on the number of residual carbon-carbon double bonds in the cured resin can be readily derived from the decrease of the $\nu(C{=}C)$ absorption (1600 cm^{-1}) in the MIR and the $2{\times}\nu({=}C{-}H)$ absorption (1625 nm) in the NIR, with the advantage of an almost 400-fold increase in sample thickness and a concomitant ease of sample handling for NIR spectroscopy.

In a series of publications, Buback et al. [23–26] have demonstrated the application of quantitative near-infrared spectroscopy to investigations of the high-pressure polymerisation of ethylene (Fig. 10.5). Thus, kinetic parameters have been derived from intensity changes in the ethylene and polyethylene $\nu(CH_2)$ overtone and combination bands at 8970, 8740 and 8260 cm^{-1}. In this application, full advantage is taken of the NIR transmittance and the mechanical stability of sapphire windows in the construction of a high-pressure cell for experiments at up to 3.5 kbars.

With the introduction of chemometric evaluation procedures [27, 28] in combination with diffuse-reflection measurements [29] for the multicomponent analysis of solid samples, NIR spectroscopy has been launched into a new era of analytical applications not only for the agricultural and food but also for the chemical and, primarily, the polymer industries [22, 30–33, 80]. The potential of NIR diffuse-reflection measurements has been demonstrated for the analysis of polymers of widely varying morphology [3, 6, 11, 34, 35] including the determination of the characteristic chemical and physical parameters of synthetic fibres. Here, some experimental results obtained on as-received poly(acrylonitrile) (PAN) fibres will be presented. The calibration was based on a multilinear wavelength regression

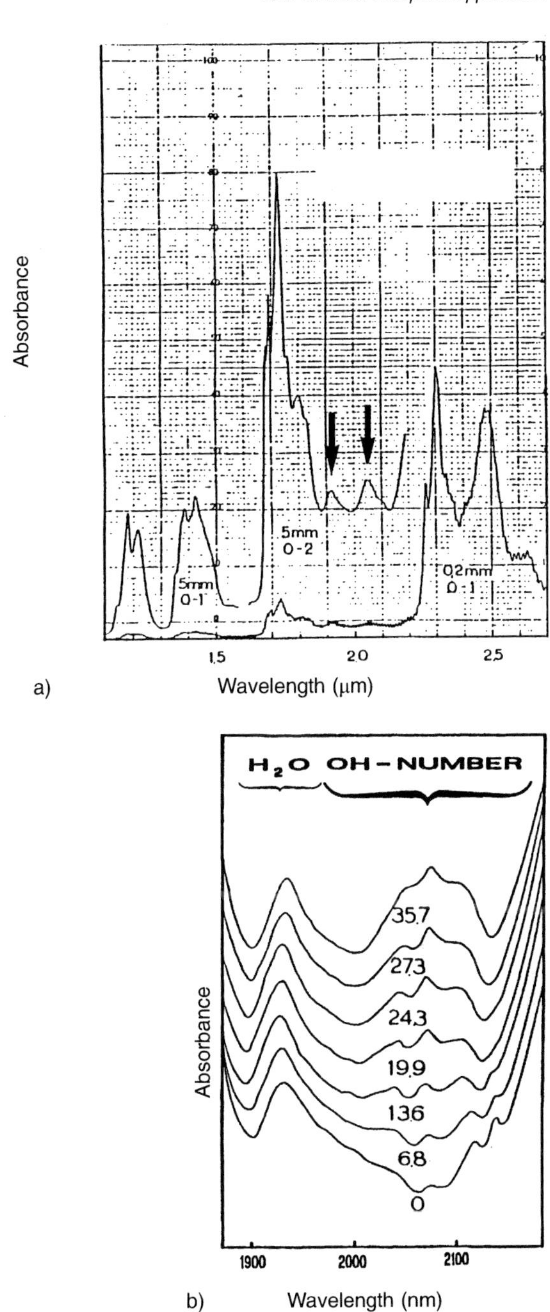

Fig. 10.2 NIR transmission spectrum of an ethyleneoxide/propyleneoxide copolymer measured with different cell pathlengths (a) and NIR transmission spectra of poly(propylene-oxide)mixtures with different OH numbers (b) (the arrows in (a) mark the expanded region of (b))

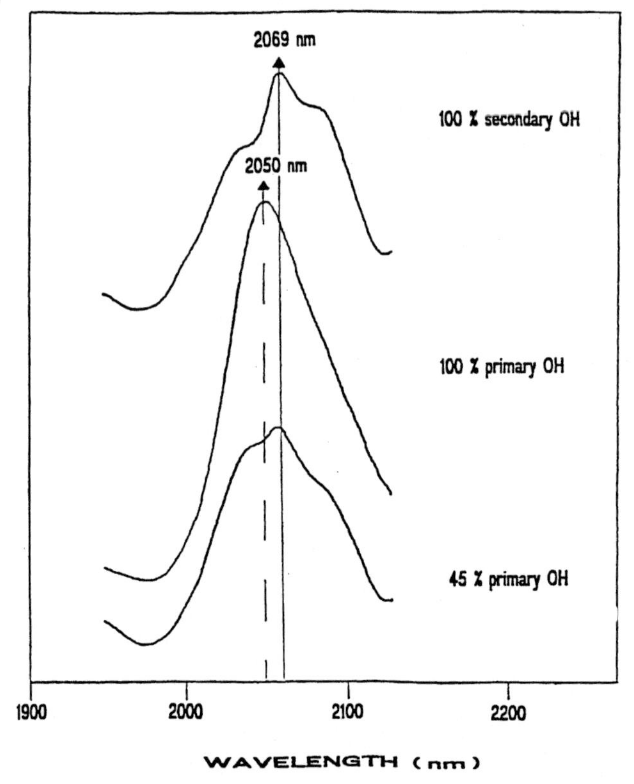

Fig. 10.3 NIR spectra of the ν(OH)+δ(OH) region of polyethers with different primary and secondary OH content

(MLWR) of the log $1/R$ spectra of 70 reference samples. The spectrum of such a reference sample is shown in Fig. 10.6. In Tab. 10.2 the actual (*A*), the NIR-predicted (*P*) and the residual (*R*) values ($R = P - A$) of the chemical and physical parameters under investigation are shown for three "unknown" test samples and reflect the good agreement between the actual and predicted values. It should be pointed out that physical parameters such as strain also have their "signature" in the spectrum of the polymeric specimen under examination. It has been shown in numerous examples of mid-infrared spectroscopy that the mechanical pretreatment of a polymer is clearly reflected in band shifts and band distortions [3] apart from the intensity changes in the polarisation spectra of such a polymer [3, 18, 36, 37]. Additionally, the fibre diameter or particle size can be differentiated by diffuse reflection measurements. Thus, based on a suitable calibration, not only chemical and physical parameters, but also morphological effects can be determined.

An interesting commercial application of qualitative, diffuse-reflection NIR spectroscopy has recently been reported for the rapid sorting of carpet waste. In fact, dedicated instruments have been developed for this purpose [38].

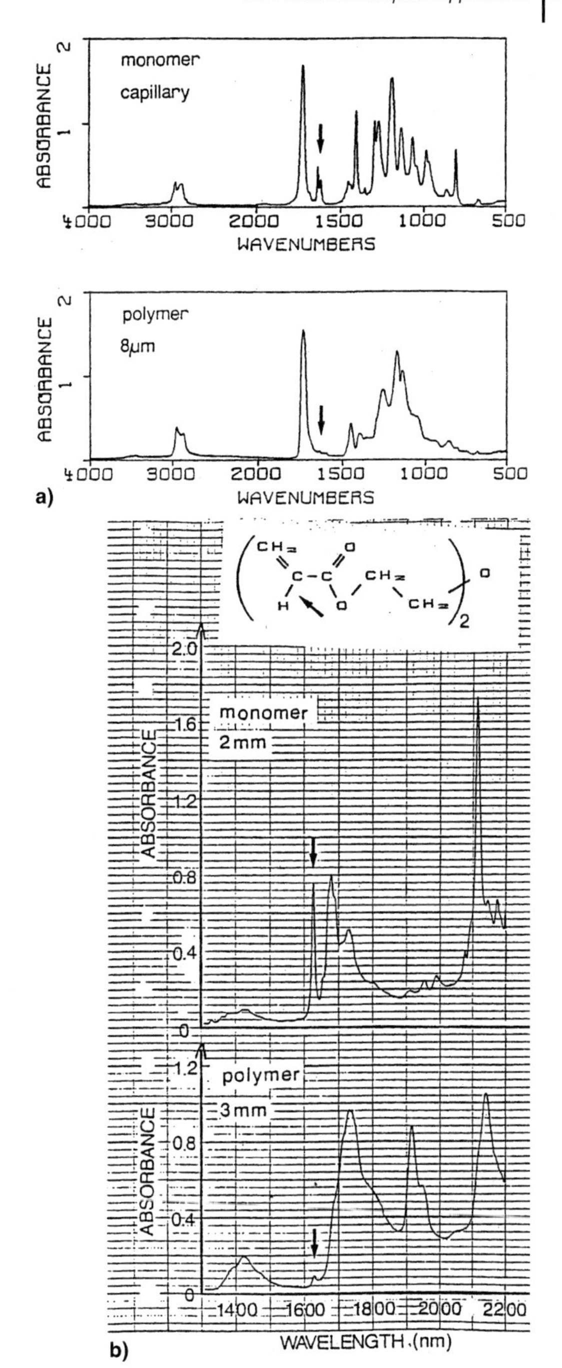

Fig. 10.4 The UV-induced polymerisation of a diacrylate: (a) mid-infrared spectra of the monomer and polymer, (b) near-infrared spectra of the monomer and polymer (→ marks the absorption bands which are characteristic of the residual monomer content)

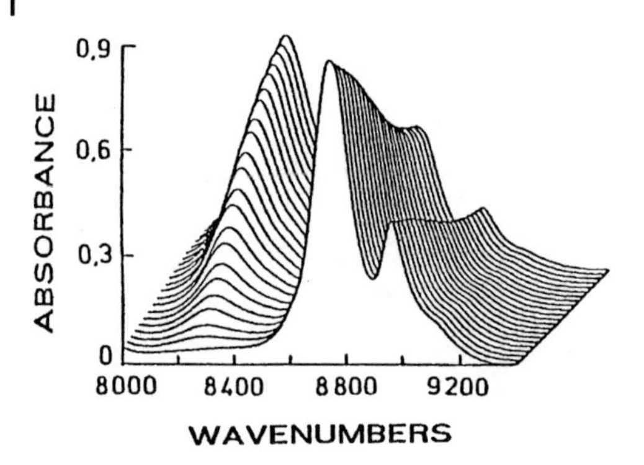

Fig. 10.5 NIR spectra recorded during high-pressure polymerisation of polyethylene (from M. Buback [26])

In another example, the diffuse-reflection technique was applied to the nondestructive, quantitative NIR analysis of the OH and COOH numbers and the molar mass of poly(ethyleneterephthalate) (PET) granulate [39]. Taking into account that the individual parameters usually have to be determined separately by time-consuming titration (OH- and COOH-numbers) or gel-permeation chromatographic (molar mass) procedures, the results of the three test samples shown in Tab. 10.3, which were obtained within a few minutes (based on a previous calibration of the first derivative spectra) by transflection measurements of monogranular layers, demonstrate the outstanding potential of NIR spectroscopy for rational multiparameter analysis.

While in conventional analytical spectroscopy the sample of interest is measured in the spectrometer, the principle of light-fibre spectroscopy is based on the transfer of light from the spectrometer through the light fibre to the measurement point and back to the spectrometer after transmission of or reflection from the sample (see also Chapter 11). The obvious advantage of the light-fibre technique is the possibility of separating and varying the location of measurement from the spectrometer within the limits given by the length of the light fibre

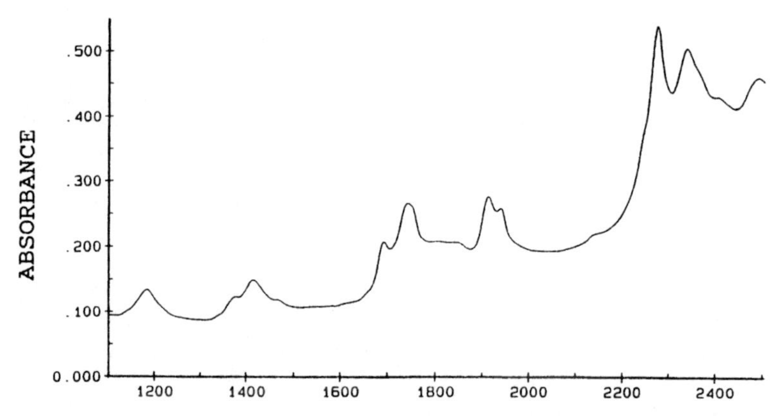

Fig. 10.6 NIR diffuse-reflection spectrum of a poly(acrylonitrile) fibre bundle

Tab. 10.2 Actual (*A*) and NIR-predicted (*P*) values and their residuals (*R*=*P*–*A*) of the content of dimethylformamide (DMF) solvent residues, the surface preparation, shrinkage and strain of 3 PAN test samples measured as fibre bundles in diffuse reflection

Sample		DMF %	Preparation %	Shrinkage %	Strain %
	P	0.9342	0.2715	4.7389	22.9778
1	A	0.9200	0.2800	5.0000	22.8000
	R	0.0142	–0.0085	–0.2611	0.1778
	P	0.6661	0.2655	4.0705	20.4069
2	A	0.7000	0.2600	4.0000	21.0000
	R	–0.0339	0.0055	0.0705	–0.5931
	P	0.7107	0.2797	3.9485	22.8066
3	A	0.7300	0.2700	3.9000	23.0000
	R	0.0193	0.0097	0.0485	–0.1934

Tab. 10.3 Actual (*A*) and NIR-predicted (*P*) values and their residuals (*R*=*P*–*A*) of the molar mass (M_n), specific viscosity (η_{spez}) and the OH- and COOH-numbers of 2 PET granular test samples measured in triplicate (SEP: standard error of prediction)

Sample		Molar mass (M_n)	Viscosity (η_{spez})	COOH mmol/kg	OH mmol/kg
1.1	P	17215	0.589	12.6	111.6
	A	15968	0.570	11.7	112.1
	R	1247	0.019	0.9	–0.5
1.2	P	16777	0.590	13.3	106.0
	A	15968	0.570	11.7	112.1
	R	809	0.020	1.6	–6.1
1.3	P	17949	0.610	15.3	103.4
	A	15968	0.570	11.7	112.1
	R	1981	0.040	3.6	–8.7
2.1	P	28657	1.066	18.8	47.8
	A	29936	1.030	19.4	45.9
	R	–1279	0.036	–0.6	1.90
2.2	P	28549	1.048	19.0	46.0
	A	29936	1.030	19.4	45.9
	R	–1387	0.018	–0.4	0.1
2.3	P	29076	1.073	17.9	46.9
	A	29936	1.030	19.4	45.9
	R	–860	0.043	–1.5	1.0
SEP		1318	0.031	1.8	4.4

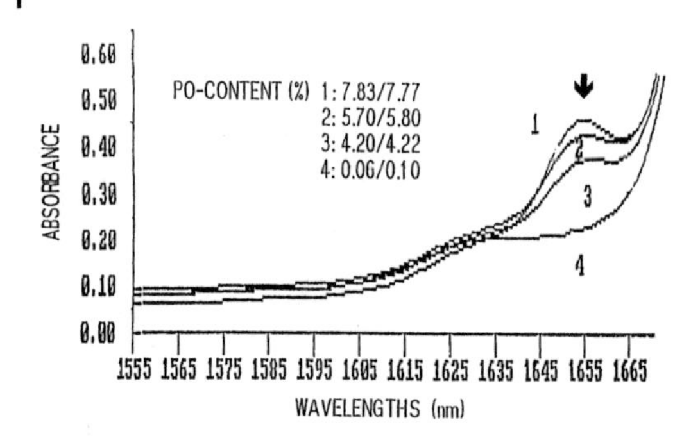

Fig. 10.7 NIR-spectroscopic determination of propyleneoxide (PO) in poly(propyleneoxide) (PO content in % (w/w): 1: 7.80, 2: 5.75, 3: 4.21 and 4: 0.08; sample thickness: 10 mm, light-fibre length: 50 m)

(presently up to several hundred meters) [40]. This has led to the development of a multiplicity of probes that can be integrated into different reactors or by-passes for remote-sensing in-line process-control applications. This has proved of great importance, primarily for the analysis of hazardous materials in chemical plants, and will be further detailed in Chapter 11. However, even for the laboratory analysis of toxic compounds, a difficult situation may be significantly alleviated by the reduction of sample-handling steps through the implementation of an NIR method with a suitable light-fibre probe. The determination of monomeric residues of propyleneoxide in poly(propyleneoxide) by quantitative evaluation of the first overtone of the ν(C–H) epoxy-ring vibration at 1655 nm in NIR transmission spectra (Fig. 10.7) acquired with an insertion light-fibre probe may serve as an example here.

10.3 Specific Features of NIR Spectroscopy

Most current NIR work focuses primarily on analytical aspects by exploiting the advantages of recent progress in instrumentation and multivariate data treatment. Thus, the emphasis is placed on the nondestructive analysis of polymers with a broad range of morphologies (granulates, fibres, films, chips or pellets) and on the on-line monitoring of parameters in a processing environment. However, the uncritical evaluation of NIR spectroscopic data by chemometric methods caused by neglecting the complex relationship between the chemical composition of a polymer and its physical structure (as, for example, in the case of polyurethanes) may lead to serious misinterpretations of the obtained results. Unfortunately, over-optimistic "all-round" applications of NIR spectroscopy on the one hand, and a too conservative attitude towards chemometric data treatment on the other, have delayed the general acceptance and technology transfer of an otherwise extremely valuable analytical tool.

As an example of a possible pitfall, the phenomenon of hydrogen bonding and its consequences for the absorption intensities of fundamental and overtone vibrations will be discussed in some detail here. Hydrogen bonding is an extremely important phenomenon for polymers with -OH, -NH, -C=O, amide and urethane functionalities due to its impact on the thermal and mechanical properties of these polymers and their phase separation. Fig. 10.8(a) shows the mid-infrared spectrum of the ν(NH) region of a polyamide 11 (PA11) film of about 30 µm thickness measured at room temperature. Under these conditions, the majority of -NH groups (>99%) occur in the associated, hydrogen-bonded form. This is directly reflected in the very low intensity of the $\nu(NH)_{free}$ absorption at 3450 cm^{-1} relative to the dominating $\nu(NH)_{ass}$ absorption at 3300 cm^{-1}. If the same polymer is

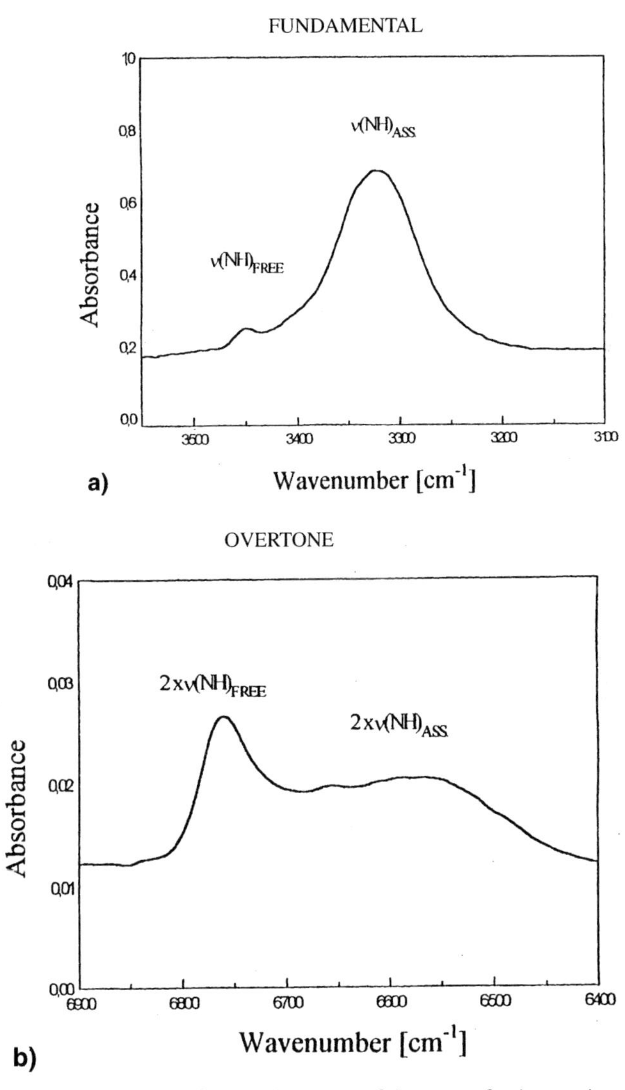

Fig. 10.8 (a) MIR and (b) NIR spectra of PA11 films in the region of the ν(NH) fundamental (a) and $2\times\nu$(NH) overtone (b) absorption region

investigated at room temperature with a film thickness of about 750 µm in the near-infrared, the spectrum shown in Fig. 10.8(b) is obtained. Here, the intensity ratio of the $2\times\nu(NH)_{free}$ at 6760 cm^{-1} relative to the broad $2\times\nu(NH)_{ass}$ absorption at about 6510 cm^{-1} is almost reversed, although the state of order of the polymeric material has not been changed [10]. The solution to this anomaly is that the intensity of the first overtone of the $\nu(NH)_{free}$ increases drastically due to the much larger anharmonicity of this vibration relative to the corresponding vibration of the associated NH-groups [41, 42]. Hydrogen bonding is equivalent to increasing the mass of the vibrating H-atom thereby making the vibration mechanically less anharmonic and therefore less intense in the near-infrared. The uncontrolled use of intensities without proper care for the corresponding absorptivities would lead to dramatic errors in the evaluation of the extent of hydrogen bonding. For a detailed discussion of these aspects with reference to the MIR and NIR spectra of polyamides the reader is referred to the pertinent literature [10, 21, 64].

10.4 Polymer Optical Fibres

The use of polymers as light-fibre material is an indirect example of NIR and visible spectroscopy of polymers. Two commercial polymers, polycarbonate (PC) and poly(methylmethacrylate) (PMMA), are primarily used in light-fibre applications in windows of a few wavelengths at the border of the NIR and the visible regions (Fig. 10.9). However, due to the unfavourable attenuation characteristics compared with quartz [43–46] (Fig. 10.10) radiation can only be transported over short distances of a few meters in a polymer optical fibre (POF) as, for example, in automotive applications. To improve the transmission characteristics of PMMA, selected C-H protons can be exchanged for fluorine because C-F functionalities do not absorb in the NIR region [43].

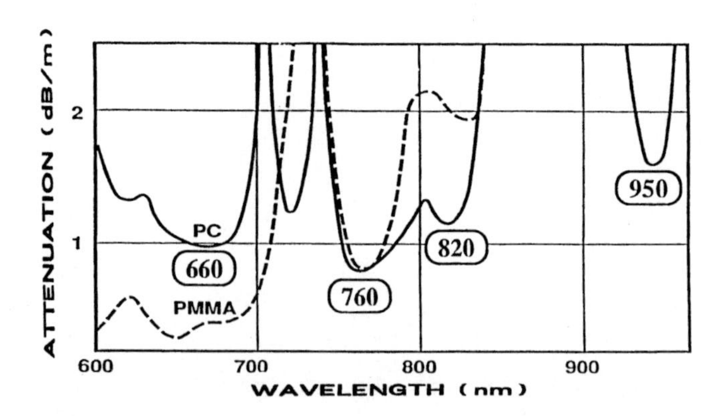

Fig. 10.9 The transmission characteristics of PMMA and PC in the Vis/NIR border wavelength region

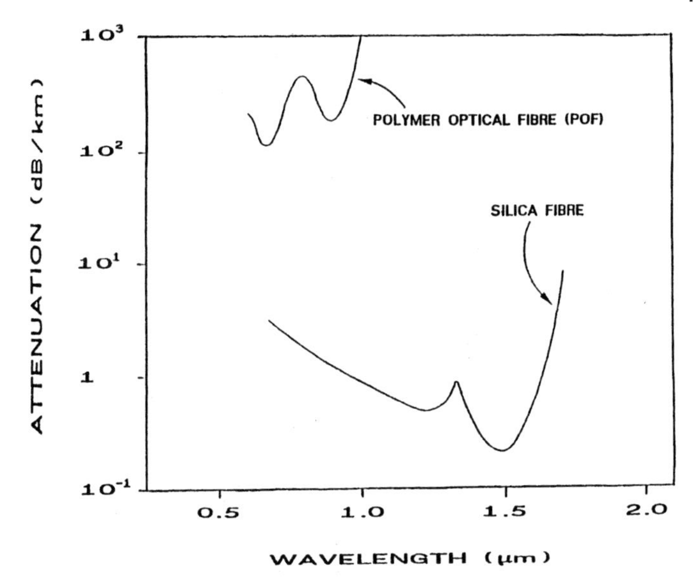

Fig. 10.10 Comparison of quartz and polymer optical fibre attenuation characteristics

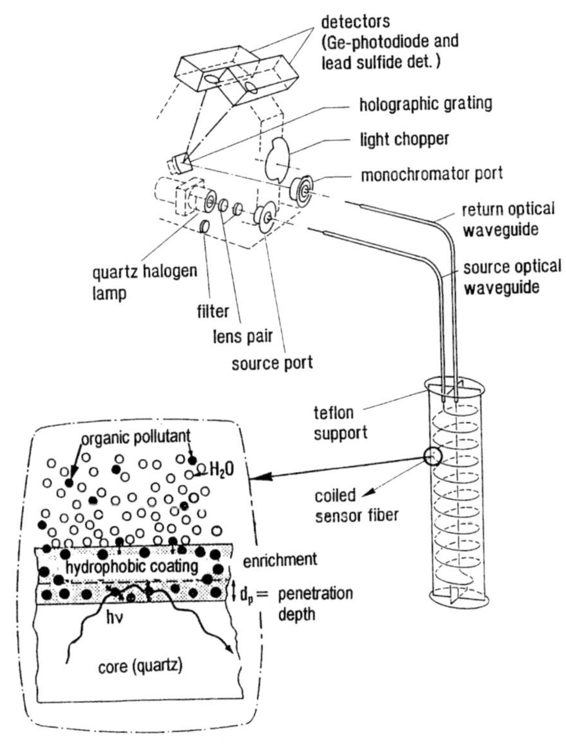

Fig. 10.11 The principle of an NIR evanescent-field absorbance sensor [48, 81]

Some researchers have described a fibre-optic NIR evanescent-field absorbance sensor [47, 48, 81], in which the polymer cladding is an integral part of the analytical tool. Basically, a light fibre based on a quartz core, a polymer cladding [e.g. poly(dimethylsiloxane) (PDMS)] and a polyamide coating are modified to act as an immersion sensor. Thus, upon removal of the polyamide coating by dissolution (over distances of several meters), the function of the remaining light-fibre is based on the effect that the compound to be determined can be enriched in the PDMS polymer cladding (thereby eliminating the interference from water) and detected by internal reflection of the NIR beam at the core/cladding interface (Fig. 10.11). To compensate for the absorptions of the polymer cladding, a background has to be measured against air. Trace amounts of chlorinated hydrocarbons in water have been measured quantitatively by this technique [81]. Similar applications have also been described for the MIR region with a polymer-coated internal reflection element [49] and a Teflon®-cladding fluoride glass fibre [50].

10.5 Fundamental Polymer Research by NIR Spectroscopy

The straightforward use of NIR spectroscopy has left many researchers with the impression of a black-box technique in the sense that, at best, it can be applied to simple routine problems. Unfortunately, this wrong impression has been nourished over the years by naive users who not only neglected the simplest spectroscopic know-how but, on top of that, applied chemometric routines in a sort of blind analysis without understanding the most basic background. In order to demonstrate that NIR spectroscopy certainly has the potential to contribute to the elucidation of sophisticated problems in polymer science, two case studies will be outlined in some detail at the end of this chapter.

10.5.1 The Study of Diffusion of Alcohols and Water in Polyamide 11 [21]

The diffusion of small molecules into polymeric materials is of importance in many areas such as food packaging, protective clothing or separation membranes. The physical properties of polyamides depend on a series of factors, one of the most important being their ability to absorb small molecules such as water and alcohols. The absorbed molecules change the melting and glass-transition temperatures of the polyamides [51] and strongly influence their mechanical [52] and electrical [53] properties. Therefore, the measurement of the diffusion coefficient is crucial to both a better understanding of the material transport mechanism and for the design of materials with optimised permeation properties.

A number of experimental techniques have been used to measure the diffusion coefficients of small molecules in a polymer film by monitoring either the release from, uptake into or permeation through the polymeric material. Gravimetric techniques [54, 55] are the most frequently used methods. With this technique, polymer samples are mounted on a quartz spring microbalance and the change in mass is recorded as a function of time. The recorded weight is corrected for buoyancy to obtain the mass of the sample. If a sensitive microbalance is used, this technique can be very accurate for determining diffusion coefficients for gases and vapours. However, if the density of the liquid is close to the density of the polymer, correction for buoyancy leads to considerably less accurate values of the sorbed mass; therefore, the quartz spring microbalance cannot be used to determine diffusion coefficients for liquids. In this case, the "pat-and-weigh" technique [56] is applied. This technique involves immersing the polymer sample in the liquid penetrant and periodically removing the sample, blotting the surfaces to remove excess liquid, and then weighing the sample on a laboratory balance.

The main difficulties with semicrystalline polymers arise from the inhomogeneity of the medium in which diffusion takes place. Usually semicrystalline polymers are regarded as composite materials consisting of impermeable microcrystalline domains imbedded in a matrix of permeable amorphous substance. The diffusion of penetrant molecules into perfectly crystalline regions of the polymer is expected to be negligible [57, 58] and the diffusion coefficient may be considered to be zero. In the amorphous regions, the diffusion basically follows the free-volume model proposed by Cohen [59]. For polymers with labile hydrogen atoms in NH or OH functionalities, the H/D-isotope exchange (deuteration), in combination with infrared spectroscopy, is a convenient method of determining the accessible regions [3, 60, 61]. This method is based on the fact that deuteration only takes place in the amorphous regions, therefore, the proportion of the deuterated functional groups should be equal to the proportion of accessible volume. However, the use of MIR transmission spectroscopy to measure sorption kinetics suffers from many of the same limitations that are inherent in the "pat-and-weigh" technique. Due to the very strong absorptivity of the absorption bands in the MIR region (4000–400 cm^{-1}), only very thin films (approximately 10–20 μm) can be used. Apart from sample-handling problems, if the specimen is too thin, or the diffusion coefficient is too high, a significant amount of penetrant may desorb during the time the sample is not immersed. Therefore, the attenuated total reflection (ATR) [56, 62] and photo-acoustic techniques [63] have been proposed for this purpose. Here, we would like to demonstrate that near-infrared transmission spectroscopy is an even better alternative for measuring the sorption kinetics of small molecules in situ. With this technique, film specimens up to a few millimetres thick can be investigated. In such cases, the extent of deuterium reexchange during the spectroscopic measurements is insignificant and the sample handling is extremely simple.

In an attempt to better understand the state of order of polyamide 11 (PA11) and its implication on the mechanical properties of this polymer, extensive diffusion experiments with PA11 film specimens in D_2O and in a series of OD-deuterated alcohols of different molecular size and geometry have been performed [21]. To monitor the rate of diffusion, we exploited the NIR-spectroscopic isotope effects arising from the NH/ND-exchange induced by the penetrating deuteration agents. The idea behind this investigation was to study the mobility of the penetrant in the polymeric matrix as a function of its molecular size and geometry. To this end the diffusion coefficients for the different deuteration agents in PA11 have been determined.

Polyamide 11 films with a thickness ranging from 0.5–1 mm and a crystallinity of about 25%, as determined from differential scanning calorimetry (DSC), were cut into test specimens with dimensions of 40×4 mm^2. The molecular structures of the different deuterated alcohols utilised as diffusants are listed in Tab. 10.4. The deuteration agents were varied with the aim of systematically increasing their size and, additionally, to compare the diffusion behaviour of linear versus spherical structures. For a diffusion experiment, a PA11 test specimen was immersed in 5 mL of the deuteration agent at 50 °C and NIR transmission spectra were recorded before and at selected time intervals during the deuteration process.

The NIR spectrum of a PA11 film (500 μm) in the wavenumber range 10000-4000 cm^{-1} is shown in Fig. 10.12. The absorption bands observed in this region and their assignments are summarised in Tab. 10.5. These assignments were verified by different techniques and investigations [64]:

- by relating fundamental vibrational frequencies of various structural groups in PA11 to the frequency positions of possible overtone and combination bands
- by interpreting the spectrum of partially deuterated PA11
- by taking into account the dichroic effects in spectra of stretched PA11 films
- by comparing spectra measured at different temperatures.

Tab. 10.4 The molecular structures of the different deuteration agents of PA11

Deuteration agents	*Structure*
D_2O	D–O–D
methanol(OD)	H_3C–OD
ethanol(OD)	H_3C–CH_2–OD
n-propanol(OD)	H_3C–CH_2–CH_2–OD
n-butanol(OD)	H_3C–$(CH_2)_3$–OD
n-pentanol(OD)	H_3C–$(CH_2)_4$–OD
t-butanol(OD)	DO–C(CH_3)(CH_3)(CH_3)
3-ethyl-3-pentanol(OD)	DO–C(CH_2CH_3)(CH_2CH_3)(CH_2CH_3)

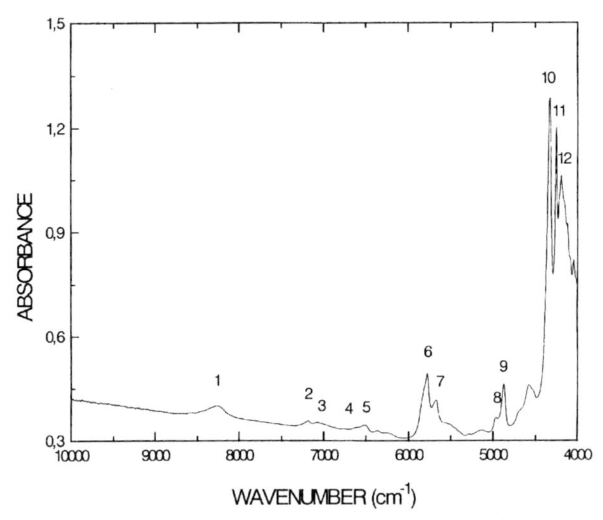

Fig. 10.12 NIR spectrum of a PA11 film (500 μm, 27 °C; the numbers correspond to the assignments in Tab. 10.5)

Fig. 10.13 shows the NIR spectra of a PA11 film (880 μm) before and after 48 h of deuteration with *n*-propanol(OD). While the $\nu(CH_2)$ absorptions at 5780 and 5680 cm^{-1} are not affected by the deuteration, significant intensity reductions upon deuteration are reflected by those absorption bands that belong to vibrations that involve the amide-hydrogen atom. Thus, the weak band at 6760 cm^{-1}, which has been assigned to the first overtone of the $\nu(NH)_{free}$ mode of the nonbonded

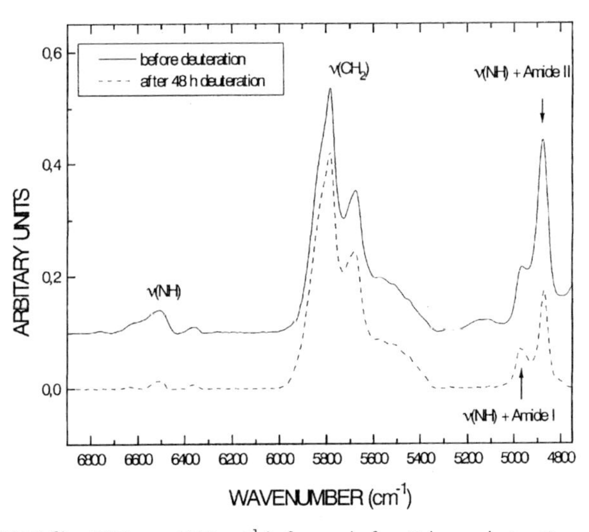

Fig. 10.13 NIR spectrum of a PA11 film (880 μm) in the wavenumber region 6800–4800 cm^{-1} before and after 48 hours deuteration with n-propanol(OD) at 50 °C

NH functionalities (Tab. 10.5), has completely disappeared; this indicates that the small percentage of NH groups which are not involved in hydrogen bonding are readily deuterated. Furthermore, the first overtone of the $\nu(NH)_{ass}$ absorption of the bonded NH groups at 6510 cm^{-1} and the two bands at 4970 and 4870 cm^{-1}, which have been attributed to combination bands of the $\nu(NH)_{ass}$ absorption and the Amide I and Amide II bands, respectively, were reduced in intensity after deuteration. In Fig. 10.14, the NIR spectra (5050–4750 cm^{-1}) of a PA11 film (850 μm) deuterated with ethanol(OD) are shown as a function of deuteration time. From this figure it becomes clear that the reduction in intensity of the band at 4970 cm^{-1} is much less significant than for the band at 4870 cm^{-1}. The reason is that, with progressing deuteration, the 4970 cm^{-1} combination band is increasingly superimposed by the evolving $2\times\nu(ND)_{ass}$ overtone [65]. In view of this overlap and the comparatively low intensity of the $2\times\nu(NH)_{ass}$ overtone at 6510 cm^{-1}, the isolated right wing of the 4870 cm^{-1} band was utilised for the evaluation of the deuteration progress as a function of time. Generally, the accessibility of a PA11 film in the deuteration process can be calculated quantitatively according to:

$$Z(\%) = \frac{A_{t=0}(NH) - A_t(NH)}{A_{t=0}(NH)} \times 100(\%) \tag{10.1}$$

Tab. 10.5 Assignment of the absorption bands in the NIR spectrum (10000–4000 cm^{-1}) of PA11 (free = not hydrogen bonded, ass = hydrogen bonded, as = antisymmetric, s = symmetric, δ = bending, γ_w = wagging; w, m, s: weak, medium and strong intensity) (see also Fig. 10.12)

	Wavenumber (cm^{-1})	*Intensity*	*Tentative assignment*
1	8265	m	$3\times\nu(CH_2)$
2	7184	m	$2\times\nu_{as}(CH_2)+\delta(CH_2)$
3	7070	m	$2\times\nu_s(CH_2)+\delta(CH_2)$
4	6760	w	$2\times\nu(NH)_{free}$
5	6510	m	$2\times\nu(NH)_{ass}$
	6368	w	$\nu(NH)_{ass}+2\times$Amide II
	6256	w	$\nu(NH)_{ass}+\nu_{as}(CH_2)$
	6180	w	$\nu(NH)_{ass}+\nu_s(CH_2)$
6	5780	s	$2\times\nu_{as}(CH_2)$
7	5680	s	$2\times\nu_s(CH_2)$
8	4970	m	$\nu(NH)_{ass}$+Amide I
9	4870	m	$\nu(NH)_{ass}$+Amide II
	4706	m	Amide I+2×Amide II
	4586	m	$\nu(NH)_{ass}$+Amide III
	4527	m	2×Amide I+Amide III
10	4337	s	$\nu_s(CH_2)+\delta(CH_2)$
11	4258	s	$\nu_s(CH_2)+\delta(CH_2)$
12	4196	s	$\nu_s(CH_2)+\gamma_w(CH_2)$

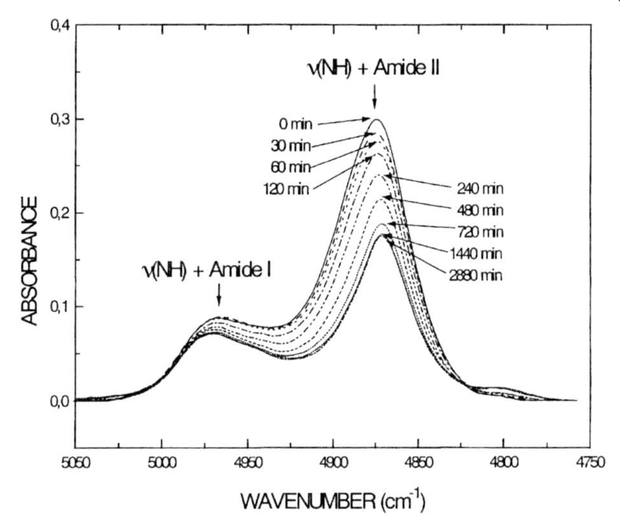

Fig. 10.14 NIR spectra of a PA11 film (850 μm) in the range 5050–4750 cm^{-1} recorded in different time intervals during deuteration with ethanol(OD)

where A_t(NH) and $A_{t=0}$(NH) are the absorbances of an NH-related band measured at deuteration time t and before the start (at $t=0$) of the deuteration, respectively [3].

The accessibilities of PA11 for the different deuterating agents as a function of immersion time at 50 °C are shown in Fig. 10.15. The graphs prove that the accessibility reaches a saturation level which is independent of the deuteration agent. Thus, after 50 hours of deuteration the linear alcohols have reached a plateau of

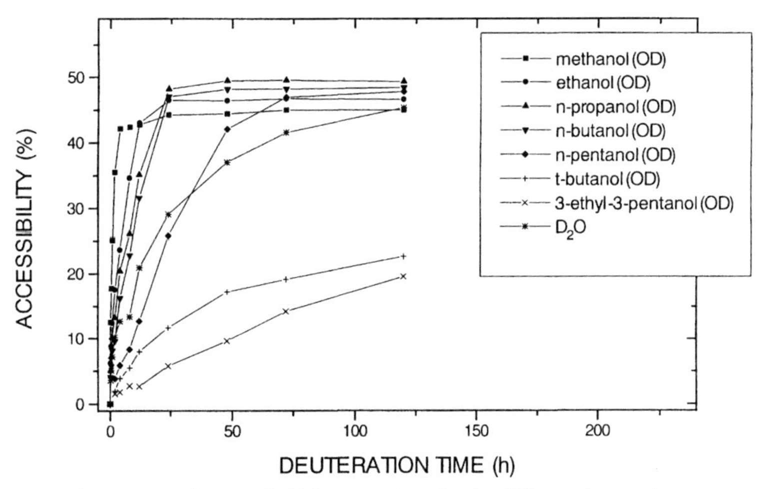

Fig. 10.15 Extent of NH/ND-exchange of PA11 versus time for the different deuteration agents

about 47±2% accessibility. Only *t*-butanol(OD), 3-ethylpentan-3-ol (OD) and D_2O had not reached their saturation level after this period due to their molecular size and their different structure.

The maximum accessibility of about 47%, however, is much lower than the amorphous fraction (about 75%) of the investigated PA11 samples. Mukai [66] has shown that the amorphous fraction in polyamides can be categorised into two regions: conventional amorphous and rigid amorphous. The latter is located at the interface between the amorphous and crystalline regions. At this interface, there are necessarily changes in density and chain mobility. These changes, from a disordered amorphous state to an ordered crystalline state, occur gradually. As a result, a transition zone exists between amorphous and crystalline domains wherein the mobility of molecules is restricted compared with that in fully amorphous regions. The deuteration experiments indicate that, like the crystalline regions, the so-called rigid amorphous regions in polyamides are not accessible for the deuteration agents at 50 °C.

When deuteration agents of the same type, for example the linear alcohols, are compared (Fig. 10.15) with reference to the deuteration progress, an inverse relationship between the size of the diffusant and the deuteration rate could be derived. The same trend can be observed for the alcohols with spherical structure (*t*-butanol(OD) and 3-ethylpentan-3-ol(OD)). However, if deuteration agents of the same size but of linear versus spherical geometry are compared, the alcohols with the spherical morphology exhibit a much slower deuteration progress [e.g. *t*-butanol(OD) $\ll$ *n*-butanol (OD) (Fig. 10.16)]. Obviously, the possibility of reptational motion [67] of the diffusant with a linear structure leads to a faster diffusion process.

With reference to the slow and anomalous deuteration progress of PA11 with D_2O (Fig. 10.15), however, it becomes obvious that the diffusive transport is not only governed by the molecular dimensions and geometry of the penetrant, but that other factors, such as the polarity and especially the hydrogen bonding of the penetrating molecules, play an important role. Firstly, the deuterium bonding between D_2O molecules is much stronger than that in the alcohols (OD). This is also reflected by the high boiling point (100 °C) of D_2O compared with methanol (OD) (65 °C), ethanol (OD) (79 °C) and *n*-propanol (OD) (97 °C). Due to the stronger intermolecular forces in D_2O, the molecules can associate into larger oligomers and, thereby, take on a larger spherical shape. Secondly, each D_2O molecule can form two deuterium bonds with amide groups in the polyamide chain. Therefore, the D_2O molecules are more strongly retarded in the polymer matrix compared with the alcohols(OD), which can only form one deuterium bond.

Generally, the one-dimensional molecular diffusion in a polymer film with constant diffusion coefficient can be described by the second Fickian law [68]:

$$\frac{\partial c}{\partial t} = D\frac{\partial^2 c}{\partial x^2} \tag{10.2}$$

where c is the concentration of the penetrant, D is the diffusion coefficient and x is the diffusion coordinate.

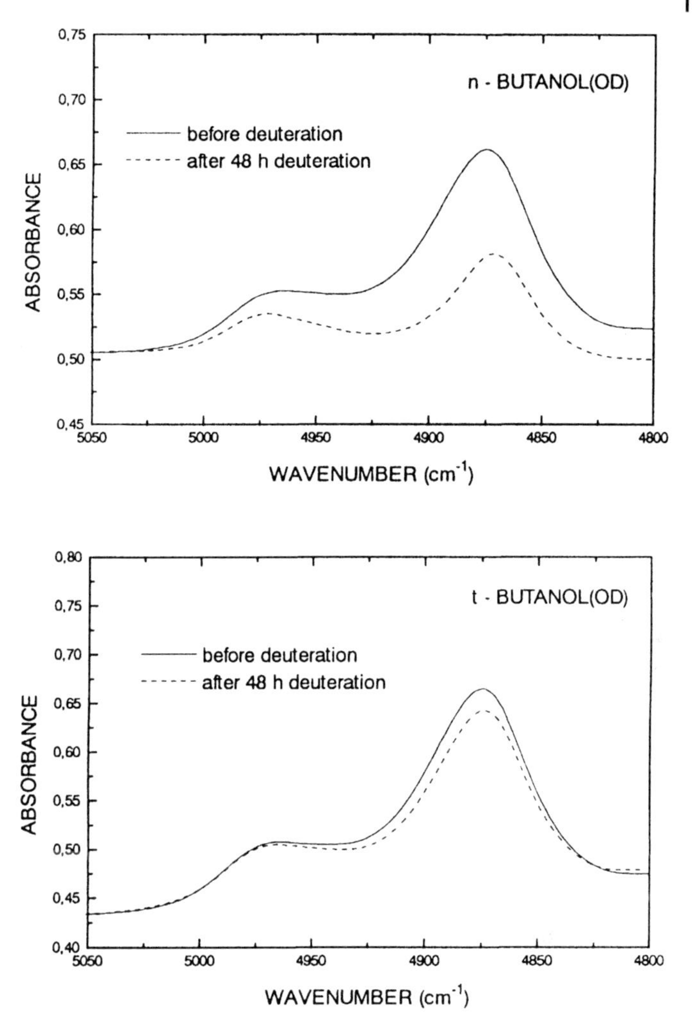

Fig. 10.16 Comparison of the extent of NH/ND-exchange of PA11 with *n*-butanol(OD) and *t*-butanol(OD) at equal deuteration time of 48 hours

If a polymer film is placed in an infinite bath of diffusant, it has been shown that, under certain boundary conditions (neglecting the effects if diffusion at the edges of the film), the sorption kinetics can be expressed by:

$$\frac{M_t}{M_{max}} = 4\left[\frac{D}{\pi}\right]^{0.5} \frac{t^{0.5}}{d} \tag{10.3}$$

where M_{max} is the mass uptake at saturation, M_t is the mass uptake at time t, d is the film thickness and D is the diffusion coefficient [68, 69]. It has been demon-

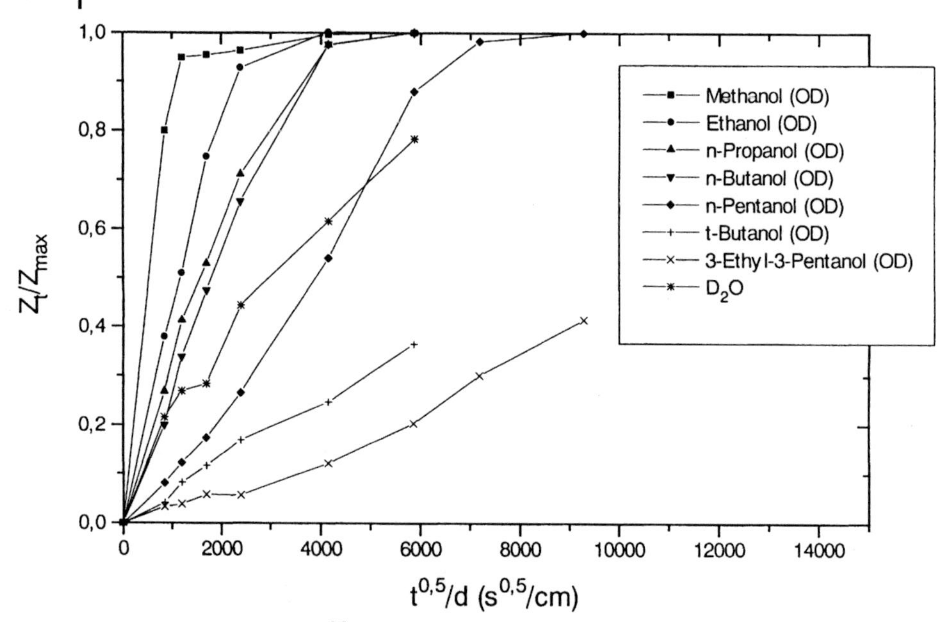

Fig. 10.17 Z_t/Z_{max} versus $t^{0.5}/d$ plots for the investigated deuteration agents

strated that at $M_t/M_{max} \leq 0.5$, the diffusion coefficient of the penetrant in the polymer can be derived according to:

$$D = \frac{\pi}{16} I_a^2 \tag{10.4}$$

where I_a is the initial slope in a M_t/M_{max} versus $t^{0.5}/d$ plot. Because we have determined the accessibility by the progress of the NH/ND exchange instead of the mass uptake, the y-axis in our terminology reads Z_t/Z_{max}. The Z_t/Z_{max} versus $t^{0.5}/d$ plots for the investigated deuteration agents are shown in Fig. 10.17. The diffusion coefficients of the different deuteration agents calculated from these data according to Eq. (10.4) are given in Tab. 10.6.

In conclusion, the results show that the amorphous regions of the polymer are not completely accessible to the diffusant and that the diffusion rate decreases with increasing molecular size of the penetrant. When structurally isomeric deuteration agents are compared, linear species exhibit higher sorption rates than spherical ones due to the possibility of reptational motion.

10.5.2
Rheo-optical FT-NIR Spectroscopy of Poly(Dimethylsiloxane)/Polycarbonate Block Copolymers

The unique properties of polysiloxanes such as chemical inertness, insensitivity to temperature extremes, surface properties and physiological indifference can be

Tab. 10.6 The diffusion coefficients of the different alcohols (OD) and D_2O for PA11 (50 °C)

Deuteration agent	***Diffusion coefficient (cm^2/s)***
methanol(OD)	1.74×10^{-7}
ethanol(OD)	3.8×10^{-8}
n-propanol(OD)	2.01×10^{-8}
n-butanol(OD)	1.53×10^{-8}
n-pentanol(OD)	2.41×10^{-9}
t-butanol(OD)	9.22×10^{-10}
3-ethyl-3-pentanol(OD)	2.96×10^{-10}
D_2O	9.91×10^{-9}

traced back to their molecular structure and have made these polymers extremely important materials for a broad variety of applications. Poly(dimethylsiloxane) (PDMS) is the most important of the silicone elastomers and has several applications (e.g. in gaskets and seals, wire and cable insulation and surgical and prosthetic devices) where high environmental stress may occur [70–72]. Alternatively, PDMS is often used in various copolymer systems, for example with thermoplastics such as polycarbonate (PC) to impart thermal stability, chemical inertness, easy mould release and chain flexibility [36, 73]. Copolymers of PDMS with thermoplastics are segmented block-copolymers composed of alternating soft PDMS and hard thermoplastic blocks. In these materials, physical crosslinking and reinforcement are provided by the hard thermoplastic segments whereas the elastomeric properties are contributed by the soft PDMS segments. These copolymers can be processed thermoplastically and have interesting automotive applications, due to their low-temperature resistance. The mechanical properties such as elasticity, modulus, toughness and stretchability of these materials depend on the internal structures and are generally attributed to the order of the physical cross-links that result from the phase separation of the soft and hard segments. During applications, these materials undergo various types of deformation which influence their microscopic structure. Thus, the study of segmental mobility as a function of the applied mechanical forces in these materials is of fundamental importance for the understanding of the dynamics of their molecular deformation mechanism and for the design of new materials for specific technical applications.

Rheo-optical vibrational spectroscopy is a powerful tool for the characterisation of transient structural changes *during* deformation and stress relaxation of polymers. The experimental principle is based on the simultaneous acquisition of polarisation spectra and stress-strain diagrams during elongation, recovery or stress relaxation of the polymer-film sample under investigation (Fig. 10.18). The instrumentation and the theoretical evaluation of the polarisation spectra have been described in numerous reviews and book chapters and the interested reader is referred to the relevant publications [74–76].

Generally, the absorption anisotropy is characterised by the dichroic ratio R

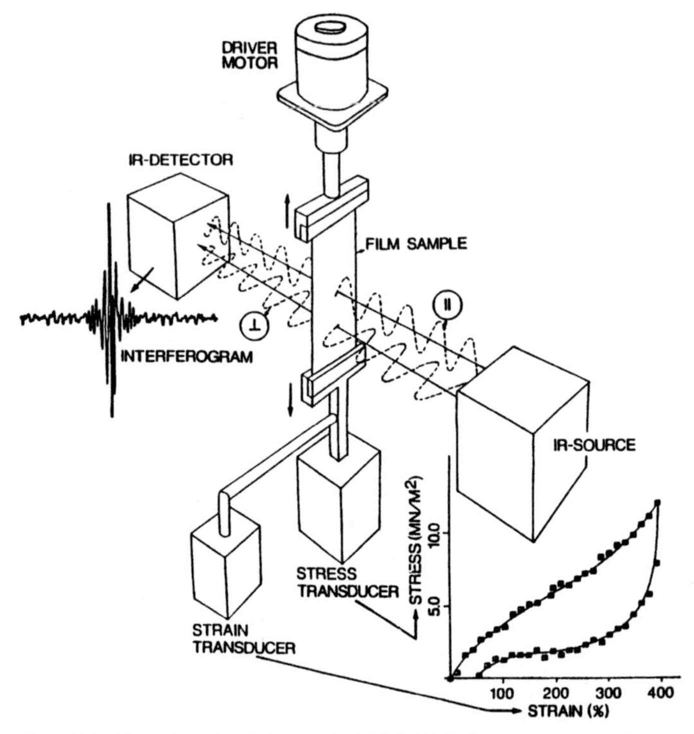

Fig. 10.18 The principle of rheo-optical FT-IR/FT-NIR spectroscopy of polymer films

$$R = \frac{A_{\parallel}}{A_{\perp}} \tag{10.5}$$

where $A_{\parallel}$ and $A_{\perp}$ are the absorbances measured with radiation polarised parallel and perpendicular to the draw direction, respectively [74–76]. If a uniaxial orientation model with an inclination angle θ of the polymer chains to the reference direction (e.g. the drawing direction) is assumed, an orientation function f can be defined, which is related to the dichroic ratio by the expression:

$$f = \frac{(R-1)(R_0+2)}{(R_0-1)(R+2)} = \frac{3 < \cos^2\theta > -1}{2} \tag{10.6}$$

where $R_0 = 2\cot^2\psi$ is the dichroic ratio for perfect uniaxial orientation ($\theta = 0$) with an angle ψ between the transition moment of the vibrating oscillator and the local polymer chain axis and R is the experimental dichroic ratio [74–76]. For an absorption band that has its transition moment parallel ($\psi = 0°$) or perpendicular ($\psi = 90°$) to the chain axis, f reads

$$f_{\parallel} = \frac{R-1}{R+2} \tag{10.7}$$

and

$$f_{\perp} = -2\frac{R-1}{R+2} \tag{10.8}$$

In previous studies with a miniaturised stretching machine that fits into the sample compartment of the spectrometer, it has been demonstrated that rheo-optical FT-MIR spectroscopy of polymer films with a thickness in the range of 10–50 μm is an extremely powerful tool for studying orientation phenomena in segmented polymers [74–76]. However, due to the very intense absorption bands of the fundamental vibrations of PDMS and PC, see for example Fig. 10.1(a), MIR spectroscopy cannot be applied to their copolymers because they had to be prepared in such small film thicknesses that they are far beyond a routine handling procedure for rheo-optical measurements. With the recent, more efficient exploitation of the NIR region, rheo-optical FT-NIR spectroscopy has become less restrictive with respect to sample geometry. Thus polymer test samples with a thickness even in the range of a millimetre are now accessible to this technique. Additionally, the mechanical measurements can be performed in a conventional tensile tester that is separated by polarisation-retaining quartz light fibres from the spectrometer. Finally, in future applications, on-line monitoring of orientation in polymer films under processing conditions may be applied (see Chapter 11).

Although NIR spectra are frequently considered less informative than their MIR analogues due to the overlap of overtones and combination bands, it can be clearly seen that in the polarisation spectra of a 300% elongated PDMS/PC-copolymer (Fig. 10.19), the first overtone of the $\nu(CH_3)$ mode of PDMS at 5872 cm^{-1} and the $3\times\delta(CH_3)$ absorption of PC at 4075 cm^{-1} reflect dichroic properties. Thus, two

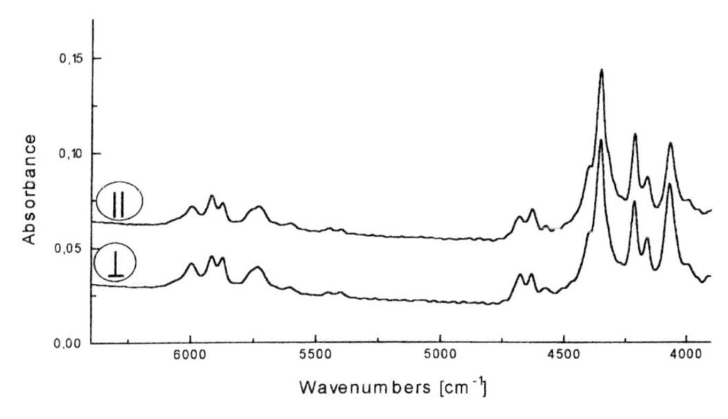

Fig. 10.19 FT-NIR polarisation spectra of a PDMS/PC 50:50 (% w/w) copolymer film (PDMS/PC block length 20:6) drawn to 300% strain at room temperature (‖/⊥: polarisation of the incident radiation relative to the drawing direction)

Tab. 10.7 Chemical structure and composition of the investigated PDMS/PC-copolymers

SOFT SEGEMENT	HARD SEGMENT
$\left[-O-Si(CH_3)_2- \right]_x$	$\left[-O-C_6H_4-C(CH_3)_2-C_6H_4-O-C(=O)- \right]_y$
Poly(dimethylsiloxane)	Polycarbonate on the basis of Bisphenol A

Sample	*%-Weight composition*		*Block length Pn*		*Molecular weight*	
	PDMS	*PC*	*PDMS*	*PC*	*PDMS*	*PC*
30Si6	30	70	6	4	440	1040
30Si12	30	70	12	8	890	2070
10Si20	10	90	20	52	1480	13320
40Si20	40	60	20	9	1480	2220
50Si20	50	50	20	6	1480	1480
30Si34	30	70	34	23	2520	5870
70Si30	70	30	30	4	2220	950
30Si75	30	70	75	51	5500	12950
40Si75	40	60	75	33	5550	8330
80Si75	80	20	75	5	5550	1390
50Si160	50	50	160	47	11840	11840
60Si160	60	40	160	31	11840	7900
60Si236	60	40	236	46	17460	11640
70Si236	70	30	236	30	17460	7480
80Si236	80	20	236	17	17460	4370
30Si350	30	70	350	238	25900	60430

specific absorption bands with transition moments perpendicular to the chain axis are available for the characterisation of the hard- and soft-segment orientation functions in the investigated copolymers according to Eq. 10.8.

A series of PDMS/PC-copolymers (Tab. 10.7) with varying compositions and block lengths was investigated by simultaneous mechanical and FT-NIR spectroscopic measurements to study the segmental orientation during elongation/recovery-procedures as a function of composition. The films, which were cast from dichloromethane, had a thickness of approximately 250 μm and were either elongated up to fracture or subjected to elongation/recovery cycles at 300 K. The data collected during the mechanical measurements were transformed into stress-strain diagrams by taking into account the original cross-section of the film sample [76]. For the spectroscopic data collection 20-scan interferograms were taken in 3 second intervals with a spectral resolution of 5 cm^{-1}.

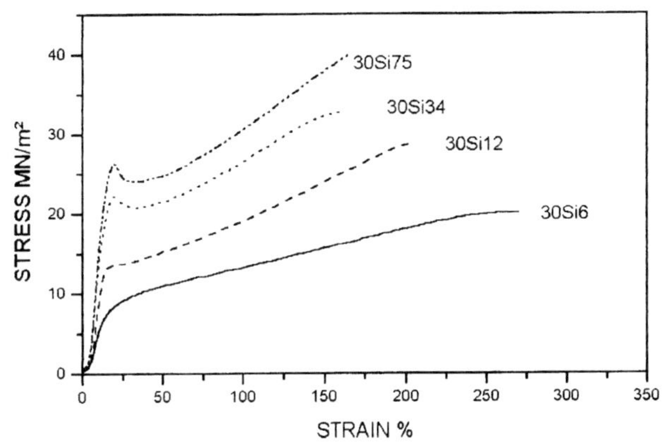

Fig. 10.20 Stress-strain diagrams taken during elongation-to-break of PDMS/PC 30:70 (% w/w) copolymer films with PDMS/PC block lengths of 6:4, 12:8, 34:23 and 75:51 (from bottom to top)

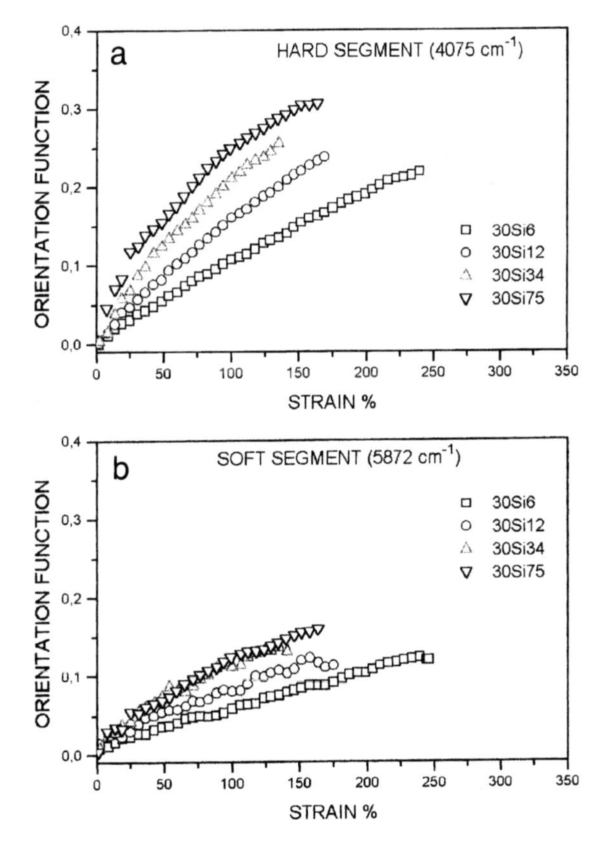

Fig. 10.21 Orientation function/strain diagrams for the (a) hard and (b) soft segments corresponding to the mechanical data shown in Fig. 10.20

Fig. 10.20 shows the effect of block length on the mechanical properties of the PDMS/PC (30:70% w/w) copolymers with the block length of PDMS and PC varying between 6–75 and 4–51, respectively (see also Tab. 10.7). Apart from the substantial increase in stress level with increasing block length of the soft and hard segments, the most obvious feature is the appearance of the transition to a typical thermoplastic polymer. The increase in stress level is accompanied by a decrease of the elongation-to-break and the development of a yield point below 25% strain. The mechanical behaviour of the 30:70 copolymer is dominated by the larger geometrical extension of the hard-segment PC domains, which aggregate/cluster increasingly as a function of block length. This is also well reflected in the observed melting endotherm for the PC domains in DSC thermograms of PC-rich PDMS/PC copolymers [77]. The orientation effects for the PDMS and PC segments in this copolymer as a function of strain are illustrated in Fig. 10.21. It is observed that the hard segments of the copolymers always orient better than the soft segments and that they reflect a more significant chain alignment with increasing block lengths. However, it is also

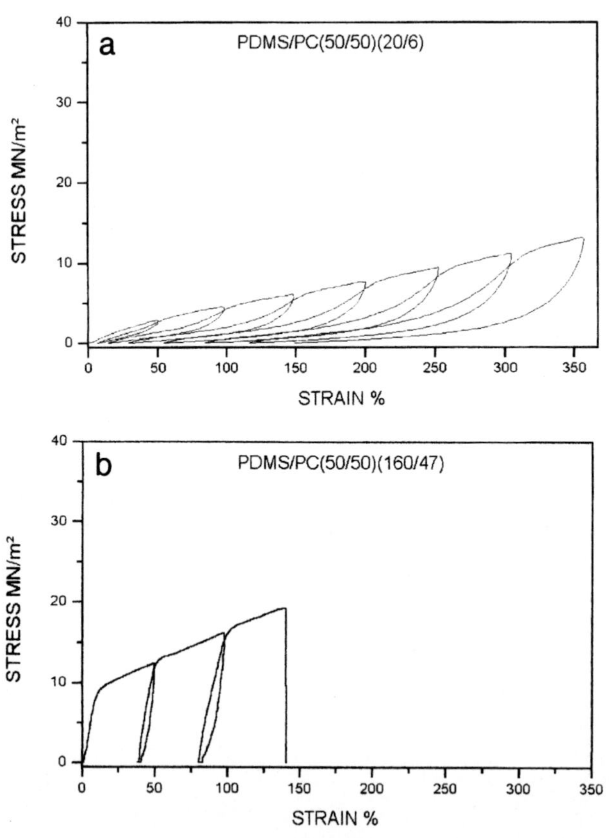

Fig. 10.22 Stress-strain diagrams taken during loading-unloading cycles of PDMS/PC 50:50 (% w/w) copolymer films with PDMS/PC block lengths of (a) 20:6 and (b) 160:47

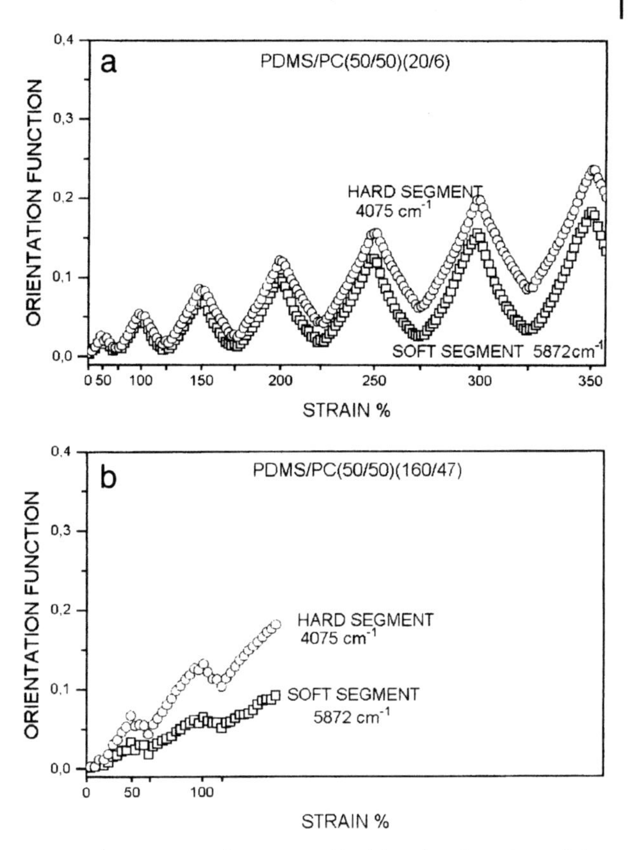

Fig. 10.23 Orientation function/strain-diagrams for the hard and soft segments of PDMS/PC 50:50 (% w/w) copolymer films with PDMS/PC block lengths of (a) 20:6 and (b) 160:47

observed that the block length exerts an appreciable influence on the reversibility of the strain-induced orientation. The stress-strain diagrams for successive loading/unloading-cycles for a 50:50 (% w/w) copolymer composition but with different block lengths 20:60 and 160:47 are shown in Fig. 10.22(a) and (b). The observed trend reflects a drastic decrease of the reversibility of the mechanical deformation with increasing block lengths. The orientation function/strain-plots corresponding to these mechanical treatments are shown in Fig. 10.23(a) and (b). Thus, Fig. 10.23(a) demonstrates, that at low block lengths, the orientation of the hard and soft segments shows reversibility, whereas Fig. 10.23(b) is clearly indicative of the strain-induced destruction of the hard- and soft-segment structure at high block lengths with the accompanying effect that the hard and soft segments recover only to a small extent during unloading.

Generally, for the investigated copolymers, the deformational stress decreases with increasing PDMS content. Concomitantly, the degree of hard- and soft-segment orientation at equivalent elongation decreases. On the other hand, the rever-

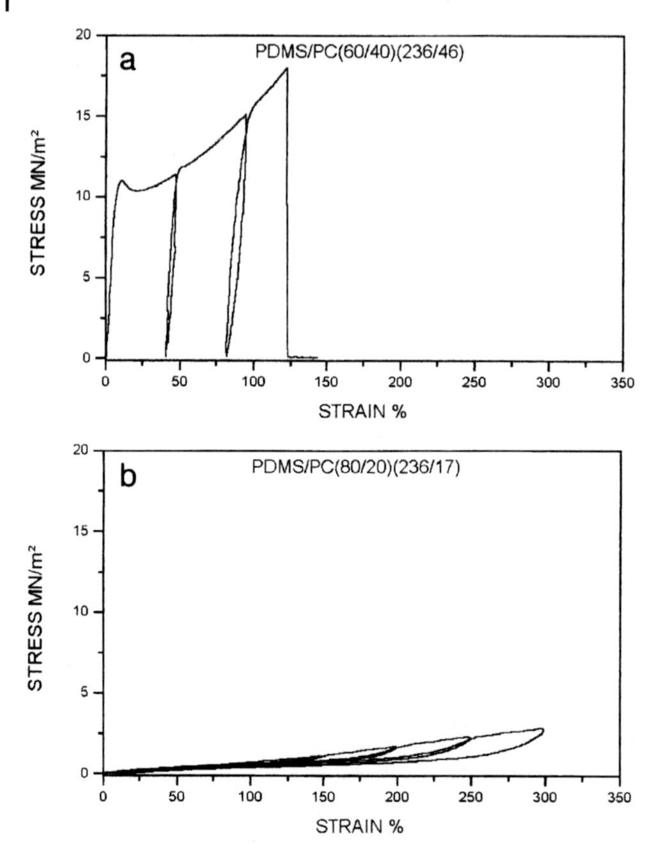

Fig. 10.24 Stress-strain diagrams of loading-unloading cycles of (a) a PDMS/PC 60:40 (% w/w) copolymer film and (b) a PDMS/PC 80:20 (% w/w) copolymer film with PDMS/PC block lengths of 236:46 and 236:17, respectively

sibility of the induced mechanical and orientational effects upon unloading increases. Up to a PDMS content of 60% w/w the hard segments of the copolymers always orient better than the soft segments. However, at PDMS contents ≥70% w/w an inversion of this trend is observed. In Fig. 10.24(a) and (b), the stress-strain diagrams of cyclic elongation-recovery procedures of PDMS/PC 60:40 and 80:20 (% w/w) with block lengths of 236:46 and 236:17, respectively, are shown. It is observed that starting at the threshold composition of PDMS/PC 70:30 only the soft segments exhibit detectable and highly reversible chain orientation whereas the hard segments do not orient any more (Fig. 10.25(a) and (b)). With increasing PDMS content from 60% to 80%, a drastic decrease of the elongational force and the hysteresis in the stress-strain diagrams is also observed. In fact, with increasing PDMS content in these copolymers, the PC blocks become significantly smaller than the PDMS blocks (Tab. 10.7). As a consequence, the predominant PC hard-segment aggregation disappears and PC blocks become dispersed in a soft

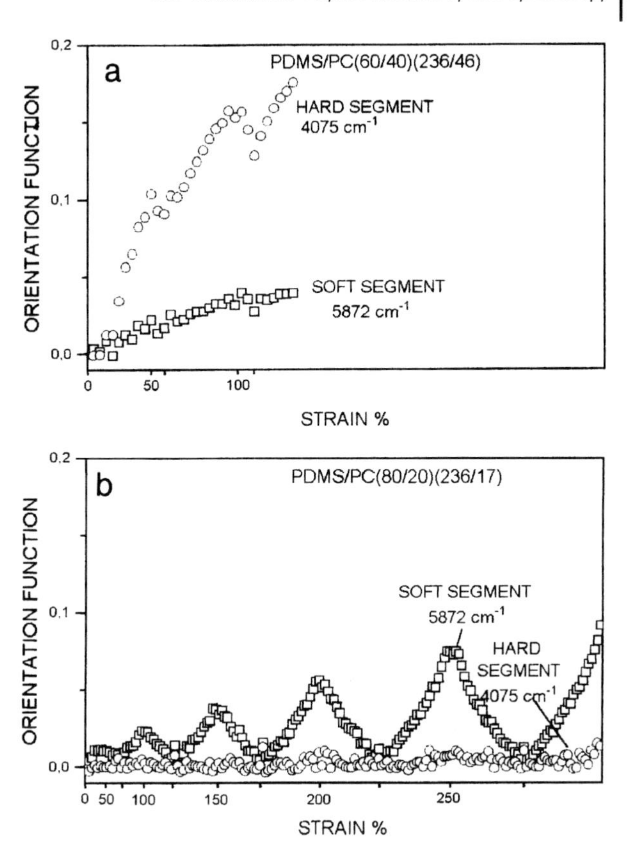

Fig. 10.25 Orientation function/strain diagrams for the hard and soft segments of (a) the PDMS/PC 60:40 (% w/w) copolymer film and (b) the PDMS/PC 80:20 (% w/w) copolymer film with PDMS/PC block lengths of 236:46 and 236:17, respectively

PDMS matrix. The mechanical properties of PDMS-rich copolymers are now dominated by the soft segments, which ultimately fail to induce any more hard-segment alignment due to decreasing extensional forces.

In the outlined example, NIR spectroscopy has been applied as a rheo-optical technique and has provided a potential tool for the detailed investigation of a phenomenon, otherwise not accessible by any other technique.

10.6 References

[1] E. W. Crandall, E. L. Johnson, C. H. Smith, *J. Appl. Polym. Sci.* **1975**, *19*, 897.

[2] E. W. Crandall, A. N. Jagtap, *J. Appl. Polym. Sci.* **1977**, *21*, 449.

[3] H. W. Siesler, K. Holland-Moritz, *Infrared and Raman Spectroscopy of Polymers* Marcel Dekker, New York, **1980**.

[4] L. G. Weyer, *Appl. Spectroscopy Revs.* **1985**, *21*, 1.

[5] V. P. Boiko, V. K. Grishchenko, *Acta Polymerica* **1985**, *36*, 459.

[6] C. Miller, *Appl. Spectroscopy Revs.* **1991**, *26*(4), 277-339.

[7] J. J. Workman, Jr., *Appl. Spectroscopy Revs.* **1996**, *31*(*3*), 251–320.

[8] D. L. Snavely, C. Angevine, *J. Polym. Sci. A* **1996**, *34*, 1669–1673.

[9] D. L. Snavely, J. Dubsky, *J. Polym. Sci. A* **1996**, *34*, 2575–2579.

[10] P. Wu, H. W. Siesler, *J. Near Infrared Spectrosc.* **1999**, *7*, 65–76.

[11] H. W. Siesler, *Makromol. Chem. Macromol. Symp.* **1991**, *52*, 113–129.

[12] H. W. Siesler, *NIR News* **1995**, *6*(*1*), 3–6.

[13] R. G. J. Miller, H. A. Willis, *J. Appl. Chem.* **1956**, *6*, 385.

[14] T. Takeuchi, S. Tsuge, Y. Sugimura, *J. Polym. Sci. A1* **1968**, *6*, 3415.

[15] A. J. Durbetaki, C. M. Miles, *Anal. Chem.* **1965**, *37*, 1231.

[16] R. F. Goddu, D. A. Delker, *Anal. Chem.* **1953**, *30*, 2013.

[17] E. A. Burns, A. Muraca, *Anal. Chem.* **1959**, *31*, 397.

[18] A. Elliott, W. E. Hanby, B. R. Malcolm, *Brit. J. Appl. Phys.* **1954**, *5*, 377.

[19] L. Glatt, J. W. Ellis, *J. Chem. Phys.* **1951**, *19*, 449.

[20] K. H. Bassett, C. Y. Liang, R. H. Marchessault, *J. Polym. Sci. A1* **1963**, 1687.

[21] P. Wu, H. W. Siesler, *Macromol. Symp.* **1999**, *143*, 323–336.

[22] T. Hirschfeld, E. Stark, in *Analysis of Food and Beverages* (Ed. G. Charambouлos), Academic Press, New York, **1984**, p. 505.

[23] F. W. Nees, M. Buback, *Ber. Bunsenges. Phys. Chem.* **1976**, *80*, 1017.

[24] M. Buback, *Z. Naturforsch.* **1984**, *39a*, 399.

[25] M. Buback, B. Huckestein, U. Leinhos, *Makromol. Chem. Rapid Commun.* **1987**, *8*, 473.

[26] M. Buback in *Infrared and Raman Spectroscopy* (Ed. B. Schrader), Wiley-VCH, Weinheim, **1995**.

[27] H. Martens, T. Naes, *Multivariate Calibration*, Wiley, Chichester, **1989**.

[28] H. Mark, J. Workman, *Statistics in Spectroscopy*, Academic Press, Harcourt Brace Jovanovich, London, **1991**.

[29] D. L. Wetzel, *Anal. Chem.* **1983**, *55*(*2*), 1165 A.

[30] E. Stark, K. Luchter, M. Magoshes, *Appl. Spectroscopy Revs.* **1986**, *22*, 335.

[31] P. Williams, K. Norris, *Near-Infrared Technology in the Agricultural and Food Industries*, AACC, St. Paul, Minnesota, **1987**.

[32] B. G. Osborne, T. Fearn, *Near Infrared Spectroscopy in Food Analysis*, Wiley, New York, **1986**.

[33] *Analytical Applications of Spectroscopy* (Eds. C. S. Creaser, A. M. C. Davies), Royal Society of Chemistry, London, **1988**.

[34] *Near Infrared Spectroscopy: The Future Waves* (Ed. A. M. C. Davies), Proceedings of the 7th International Conference on Near Infrared Spectroscopy, NIR-Publications, Chichester, UK, **1996**

[35] H. W. Siesler, *Kunststoffe* **1996**, *86*(*7*), 1007.

[36] A. Ameri, H. W. Siesler, *J. Appl. Polym. Sci.* **1998**, *70*, 1349–1357.

[37] P. Wu, H. W. Siesler, F. Dal Maso, N. Zanier, *Analysis Magazine* **1998**, *26*(4), 45–48.

[38] B. Kip (DSM, Geleen, Netherlands) and A. Landa (LT-Industries, Rockville, MD, USA), private communications

[39] H. W. Siesler in *Proceedings of the 2nd International Near-Infrared Spectroscopy Conference*, Tsukuba, Japan, May 29–June 2, 1989 (Eds. M. Iwamoto, S. Kawano), Korin, Japan, **1990**, p. 130.

[40] J. Birnie, *Spectroscopy World* **1991**, *3*(*3*), 12.

[41] S.N. VINOGRADOV, R. H. LINNELL, *Hydrogen Bonding*, Van Nostrand, New York, **1971**.
[42] *Intermolecular Forces* (Eds. P.L. HUYSKENS, W.A.P. LUCK, T. ZEEGERS-HUYSKENS), Springer, Berlin, **1991**, p. 157.
[43] W. GROH, *Makromol. Chem.* **1988**, *189*, 2861.
[44] W. GROH, D. LUPO, H. SIXL, *Angew. Chem. Adv. Mater.* **1989**, *101(11)*, 1580.
[45] T. KAINO, *J. Polym. Sci. Part A*, **1987**, *25*, 37.
[46] *Fibre Optics Handbook for Engineers and Scientists* (Ed. F. C. ALLARD), McGraw-Hill, New York, **1989**.
[47] M. D. DEGRANDPRE, L. W. BURGESS, *Appl. Spectroscopy* **1990**, *44(2)*, 273.
[48] J.-P. CONZEN, *PhD-Thesis*, University of Karlsruhe (Germany), **1992**.
[49] P. HEINRICH, R. WYZGOL, B. SCHRADER, A. HATZILAZARU, D. W. LÜBBERS, *Appl. Spectroscopy* **1990**, *44(10)*, 1641.
[50] V. RUDDY, S. MCCABE, *Appl. Spectroscopy* **1990**, *44(9)*, 1461.
[51] H. BATZER, U. KREIBICH, *Polymer Bull.* **1981**, *5*, 585.
[52] H. K. REIMSCHUESSEL, *J. Polym. Sci. Polym. Chem. Ed.* **1978**, *16*, 1220.
[53] M. KOHAN, *Nylon Plastics*, Wiley Interscience, New York, **1973**.
[54] H. B. HOFFBERG, V. STANNETT, G. M. FOLK, *Polym. Eng. Sci.* **1975**, *15*, 261.
[55] H. B. HOFFBERG, V. STANNETT, C. H. M. JACQUES, *J. Appl. Polym. Sci.* **1975**, *19*, 2439.
[56] G. T. FIELDSON, T. A. BARBARI, *Polymer* **1993**, *34*, 1146.
[57] A. KREITUSS, H. L. FRISCH, *J. Polym. Sci. Polym. Phys. Ed.* **1981**, *19*, 889.
[58] A. PETERLIN, *J. Macromol. Sci.-Phys.* **1975**, *B11(1)*, 57.
[59] M. H. COHEN, D. TURNBULL, *J. Chem. Phys.* **1975**, *31*, 1164.
[60] J. MANN, H. J. MARRINAN, *J. Trans. Faraday Soc.* **1956**, *52*, 481.
[61] N. CHAUPART, G. SERPE, *J. Near Infrared Spectrosc.* **1998**, *6*, 307.
[62] S. U. HONG, T. A. BARBARI, J. M. SLOAN, *J. Polym. Sci. Polym. Phys.* **1997**, *35*, 1261.
[63] E. G. CHATZI, M. W. URBAN, H. ISHIDA, J. L. KOENIG, *Polymer* **1986**, *27*, 1850.
[64] P. WU, *PhD Thesis*, University of Essen (Germany), **1998**.
[65] S. SINGH, *J. Mol. Structure* **1985**, *127*, 203.
[66] U. MUKAI, *Macromolecules* **1995**, *28*, 4899.
[67] M. DOI, S. F. EDWARDS, *J. Chem. Soc. Faraday Trans. II* **1978**, *74*, 1789.
[68] J. COMYN, *Polymer Permeability*, Elsevier Applied Science, New York, **1985**, p. 1.
[69] J. M. VANGUARD, *Liquid Transport Processes in Polymeric Materials: Modeling and Industrial Applications*, Prentice-Hall, Englewood Cliffs, **1991**.
[70] A.L. SMITH (Ed.), *The Analytical Chemistry of Silicones*, Wiley, Chichester, **1991**.
[71] R. G. ROCHOW, *Silicon and Silicones*, Springer, Berlin, Germany, **1987**.
[72] W. BÜCHNER, H. W. MORETTO, K. H. RUDOLPH, *Silikon-Chemie und Anwendung*, Bayer AG, Leverkusen, Germany, **1983**.
[73] M. SHIBAYAMA, M. INOUE, T. YAMAMOTO, S. NOMURA, *Polymer* **1990**, *31*, 349.
[74] H. W. SIESLER, *Adv. Polym. Sci.* **1984**, *65*, 1.
[75] H. W. SIESLER in *Advances in Applied FTIR Spectroscopy* (Ed. M. W. MACKENZIE), Wiley, Chichester, England, **1988**, p. 189.
[76] H. W. SIESLER in *Oriented Polymer Materials* (Ed. S. FAKIROV), Hüthig und Wepf, Heidelberg, **1996**, pp. 138–166.
[77] A. AMERI, *PhD Thesis*, University of Essen, Essen (Germany), **1998**.
[78] S. GOSH, J. RODGERS, in *Handbook of Near-Infrared Analysis* (Eds. D.A. BURNS, E.W. CIURCZAK), Marcel Dekker, New York, 2001, chapter 22, pp 573.
[79] C. KRADJEL, in *Handbook of Near-Infrared Analysis* (Eds. D.A. BURNS, E.W. CIURCZAK), Marcel Dekker, New York, **2001**, chapter 26, pp 659.
[80] K.A. BUNDING LEE, *Appl. Spectrosc. Revs.*, **1993**, *28(3)*, 231.
[81] J.-P. CONZEN, J. BÜRCK, H.-J. ACHE, *Appl. Spectrosc.*, **1993**, *47(6)*, 753.

11
Application to Industrial Process Control

H. W. Siesler

11.1
Introduction

Although NIR spectroscopy is now a well-established spectroscopic technique and despite numerous reports on applications and feasibility studies with reference to industrial on-line control [1–11, 56], its potential for this purpose is by far not exploited to its full extent. However, the pressure of increasing cost and legal requirements for quality control, processing conditions and environmental compatibility of chemical, pharmaceutical, petrochemical, food and agricultural products increasingly enforce the consideration of specific, non-destructive, rapid and flexible analytical techniques to meet these demands [12, 13]. The off-line methods applied so far (HPLC, NMR, MIR spectroscopy, mass spectrometry and wet chemical methods) and the current in-line and on-line control and measurement techniques of, for example, temperature, pressure, pH or dosing weight are in most cases either too time consuming, need solvents for their application which leads to problems of disposal, cannot be applied as remote control or are too non-specific to fulfil these requirements. In this context, NIR – and to a certain extent also Raman and MIR/ATR – spectroscopy have the potential to become, within the next years and in combination with new instrumental accessories and further optimised statistical evaluation procedures, indispensable analytical tools for industrial process monitoring in a wide field of applications ranging from chemistry to agriculture and from life sciences to environmental analysis [14–16]. This development will be further supported by the trend that for new chemical plants the implementation of NIR-spectroscopic on-line control methods at various points of operation can be planned in the very early stages of the construction phase and that experts in this technique can be integrated in the planning team. The difficulties encountered with a retrofit of an otherwise accepted NIR method into an old plant are often enough the crucial obstacle to a successful technology transfer.

One exception to the frequently encountered hesitation and reluctance of implementing NIR spectroscopy as an on-line tool is certainly the petrochemical industry. In this branch, the value of the NIR technique was recognised several years ago and its implementation has led to a drastic improvement in plant operation by improved process stability, better pollution control and more accurate blending procedures. Primarily the last factor has contributed to a significant increase of profit due to a drastic reduction of the so-called "give-away" [17, 18].

In a recent publication Hassell and Bowman [10] have given a very interesting overview of general process analytical aspects for spectroscopists. While a spectroscopist will generally focus primarily only on the analyser – the NIR spectrometer – Fig. 11.1 clearly shows that there are many interfaces involved in an industrial process analyser, which lead to a complex technical and human situation. Thus, departments or divisions with different expertise and diverse interests and people with different backgrounds and responsibilities are forced to co-operate on a common problem. Only people who are or who have been working in an industrial environment know which difficult situations can arise under such circumstances. Furthermore, regularly used expressions have been clearly defined by these and other authors [19]:

- off-line: manual sampling with transport to remotely located instrument (e.g. GC-MS)
- at-line: manual sampling with transport to analyser within the manufacturing area (e.g. simple titration)
- on-line: automated sampling and transport to an automated analyser (e.g. on-line GC)
- in-line: the sample interface is located in the process (e.g. pH-meter or insertion probe of an NIR spectrometer).

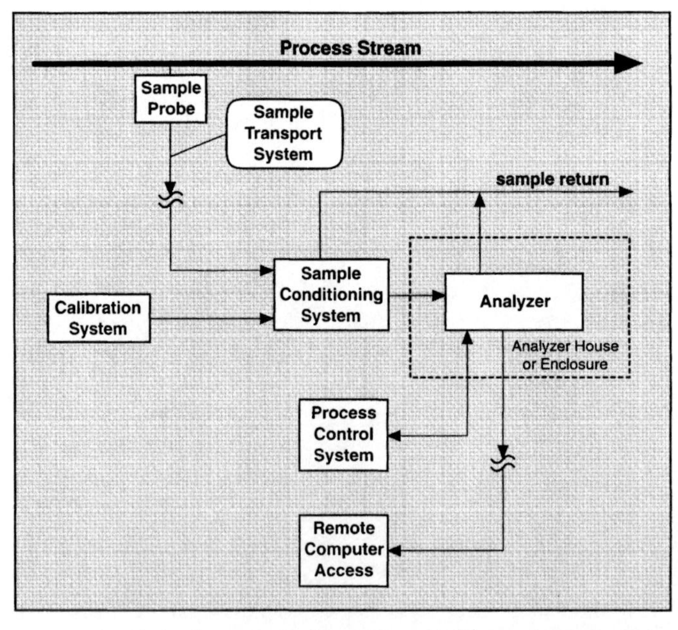

Fig. 11.1 Generalised process analyser schematic diagram (with permission from [10])

Finally, apart from the safety considerations, a thorough application evaluation with questions such as

- what chemical or physical properties are required to run the process?
- what is the required time-interval of control measurements?
- what are the extreme operating conditions of the process stream?
- how will the analytical information be used?

should precede the choice for a certain analytical technique in order to establish the priorities for example, in favour of precision (for a trend indicator) or accuracy (for the analysis of a quality parameter). Based on the evaluation of this analysis the appropriate technique will then be selected and installed.

The further discussion in this chapter will focus exclusively on the specific features of NIR spectroscopy as an industrial on-line/in-line tool. Finally, some selected examples of NIR spectroscopic industrial process monitoring cases as well as possible future applications will be discussed.

11.2 Advantages of NIR-Spectroscopic Process Analysers

Process analysers based on NIR spectrometers have a number of advantages over other techniques:

- they allow non-destructive – frequently contact-free – on-line analysis of multicomponent systems
- they can be applied to remote analysis (especially important for hazardous or toxic materials) by separating the analyser and the point of measurement by optical light fibres over distances of several hundred meters
- they can operate in hostile environments
- they have multiplexing capability
- the analytical wavelength region can be readily adjusted to convenient sample thickness requirements
- the analysis of aqueous systems is easy compared to MIR spectroscopy
- they can be applied not only to liquids, but also to any morphology of solid samples
- because no solvents are required, the technique is pollution free and energy saving
- the technique is highly cost-effective primarily due to savings of manpower; however, this should not be misinterpreted in terms of no maintenance requirements.

Despite these advantages, the acceptance of the technique for process monitoring is slow. The reasons for this development are manifold. Partly, it can be attributed to negative experiences of industrial users as a consequence of over-optimistic promises by instrument salesmen or to the neglect of proper calibration procedures or calibration maintenance by the responsible operators. Often enough, however, the systems did not perform satisfactorily due to hard- or software problems. Thus, changes in spectrometer performance due to a mechanically hostile environment, fouling of probe-heads or simply the lack of the possibility to establish a calibration by an independent reference method considerably hampered the implementation of NIR spectroscopy for process control purposes. Many of these problems occurring in the past have been minimised by rugged spectrometer designs, new concepts of probe maintenance without process interference and the development of chemometric algorithms for the evaluation of concentration profiles from on-line spectroscopic data without the need for prior calibration [20–22].

11.3 Instrumentation for NIR-Spectroscopic Process Analysers

The increasing need for very fast real-time process control measurements in a hostile plant environment has also led to the development of new monochromator/detector technologies. Fig. 11.2 presents an overview of monochromator/detection principles in current use for NIR spectroscopy [23–25]. Despite their simple design, systems using only selected frequencies (filter instruments and LED-based systems) should be applied as process monitoring tools only when a rather narrow information content is required. For the monitoring of several parameters in a complex process, these simple instruments will normally not be adequate and scanning instruments will have to be applied to provide the whole range of available wavelengths/wavenumbers. Of these scanning-type instruments the acousto-optic tuneable filter (AOTF) and diode-array systems have no mechanically moved parts and have also progressed furthest in terms of miniaturisation. Nevertheless, most of the other scanning systems have reached an extremely high standard of mechanical

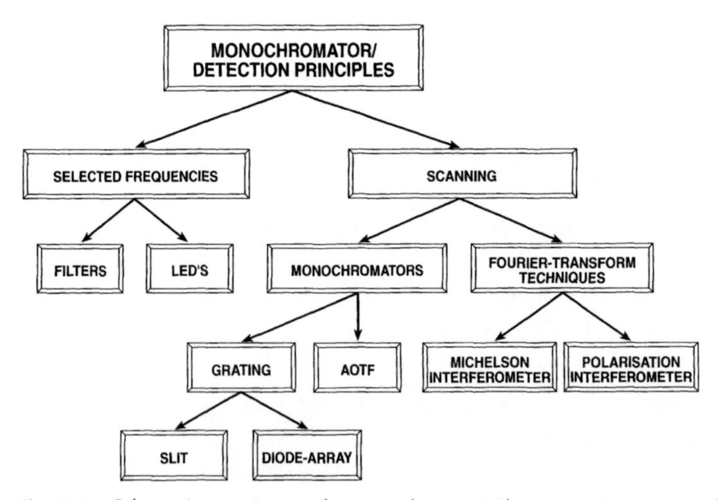

Fig. 11.2 Schematic overview on the monochromator/detector systems currently in use for NIR spectroscopy

stability and scanning speed, and are equally well adaptable to process-control conditions.

With the availability of low-loss, inexpensive silica fibres NIR spectroscopy has been launched into a new era of process control applications. Thus, specific fibre-optic probes (see below) allow the spectrometer to be separated from the location of sample measurement by several hundred meters and make the analysis of toxic and hazardous materials much safer [24, 26].

Despite the unique advantages of optical light fibres (low-loss, insensitivity to electrical interference) a word of caution is necessary. Light-fibres are sensitive to temperature and may break if handled inappropriately. Usually, the people in charge of the maintenance of cable runs are not familiar with this feature. Thus, in order to avoid fluctuations in calibrations due to temperature changes in the plant, care should be taken in terms of their thermal insulation and certainly also of their mechanical protection.

In most cases of NIR-spectroscopic process monitoring applications, the choice of the NIR fibre-optic probe configuration is the key issue to the success of the task. Examples of the most common probe configurations are shown in Fig. 11.3. While the probe types B and C are inserted in the product stream, probe A is applied contact-free. This type of probe based on the principle of diffuse reflection can not only be applied to turbid liquid products but can also be applied to solids (e.g. powders). Probe B is available as a transmission and as a "transflection" probe. Fig. 11.4 shows the design of such a true transmission in-line probe which has the additional advan-

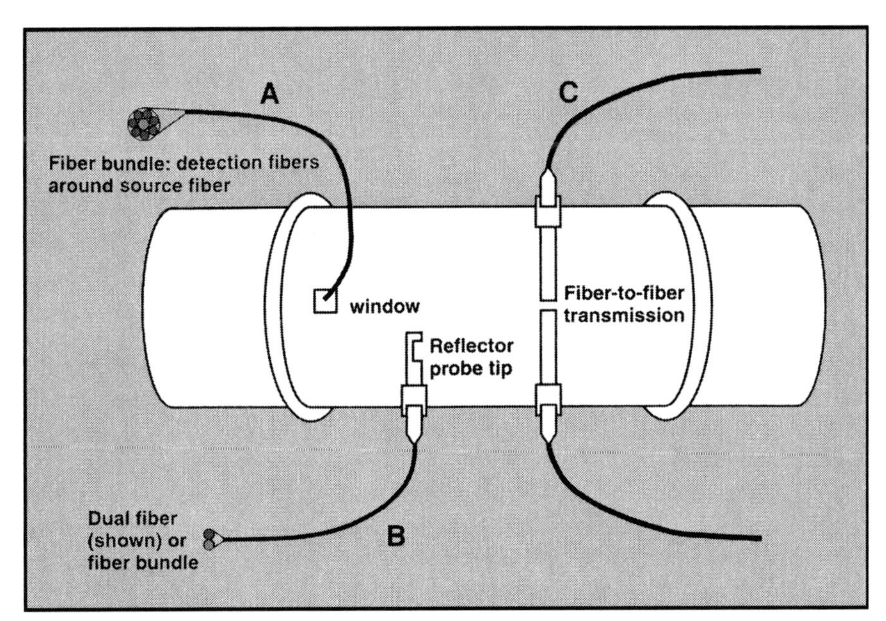

Fig. 11.3 Examples of NIR fibre-optic probe configurations. (A) Diffuse-reflection probe; (B) insertion probe based on transmission or transflection; (C) fibre-to-fibre transmission probe (with permission from [10])

tage that it is based on a ceramic body and sapphire windows to ensure resistance to chemical corrosion [27]. In the case of direct insertion probes, specific attention has to be paid to the possibility of removing the probe for cleaning or recalibration purposes without interrupting the process. Fig. 11.5 shows such a mechanism, which allows maintenance of the probe without shutting down the process stream.

NIR light-fibre spectroscopy can be further adjusted to plant control requirements by integrating an optical multiplexer into the system [24, 26]. Thus, radiation from a single instrument in a control room, for example, can be transmitted along one of several light-fibre lines to different plant locations, where the collimated beam interacts with the process stream through in-line or on-line probes designed to withstand process conditions. Such systems, which provide real-time results, will increasingly contribute to an improvement of loss control, product quality and throughput in the chemical industry.

11.4 Applications

The process media which are most easily accessible to the implementation of an NIR-spectroscopic monitoring technique are transparent organic solutions. They can be readily controlled with a light-fibre-coupled transmission or transflection probe with an adjustable sample pathlength or by a fibre-to-fibre transmission probe for a bypass stream. Such a bypass can also be thermostatted to the required temperature to perfectly simulate the reactor conditions. The determination of OH-number and water content in solvents in chemical plants and blending parameters in the petrochemical industry are current examples of where a successful lab-to-plant transfer of the NIR technology has been accomplished. Generally, current examples of NIR-spectroscopic process control applications are available in the proceedings of the International Conferences on Near Infrared Spectroscopy, for example [28, 29].

The first example discussed here in some detail refers to a feasibility study of the solution polymerisation of methylmethacrylate (MMA), shown in Fig. 11.6 [30], a polymer that has gained large technical importance due to its outstanding optical properties [31]. Although the described reaction was performed in a laboratory reactor, the necessary changes for a plant environment can be readily made. Generally, batch processes are typical cases in which the implementation of an efficient on-line monitoring technique can lead to significant savings in process time and energy.

The polymerisation was performed under temperature control (80 °C) in toluene at a starting concentration of 10% (w/w) methylmethacrylate monomer with azobisisobutyronitrile (AIBN) as radical initiator. To monitor the polymerisation progress FT-NIR spectra were recorded with a macro-routine in 5 min time intervals by accumulating 196 scans (scan-time 60 s) with a spectral resolution of 4 cm^{-1} over a period of about 7 hours. The measurements were performed in a temperature-controlled quartz-tube bypass (internal diameter 4 mm) with a light-fibre coupled fibre-to-fibre transmission probe (see Fig. 11.3, C). Alternatively, a probe

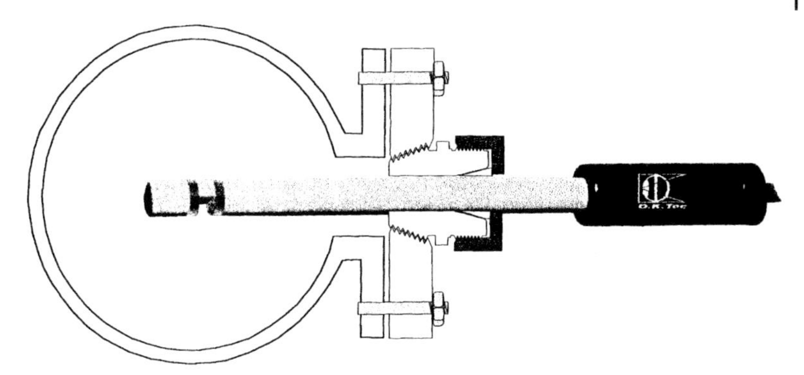

Fig. 11.4 Design of an insertion transmission probe based on a ceramic body and sapphire windows with a Swagelok™-like connection to a reaction vessel (with kind permission by O.K.-Tec, Jena, Germany)

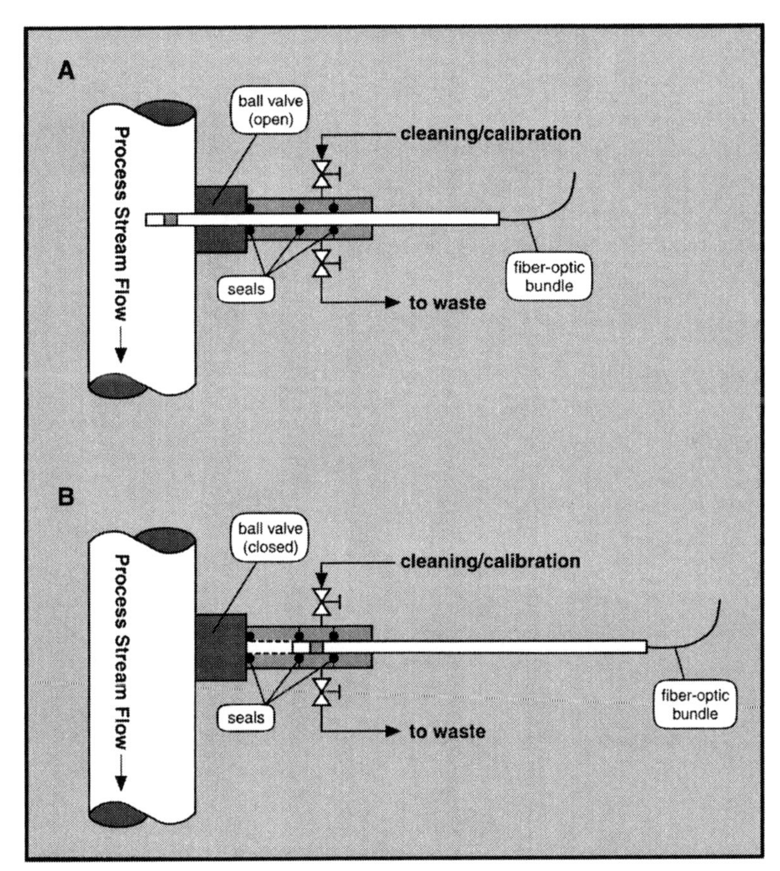

Fig. 11.5 Direct insertion sample probe assembly. (A) The probe is in the process stream for measurement; (B) the probe ha been retracted from the process stream for maintenance purposes (with permission from [10])

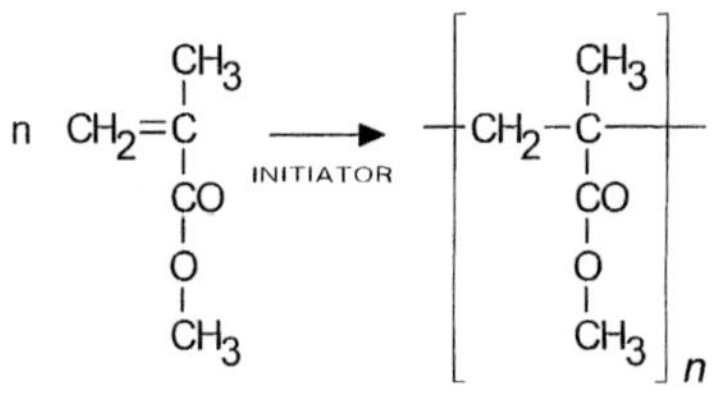

Fig.11.6 The polymerisation of methylmethacrylate (MMA)

(with a pathlength up to 10 mm) inserted directly into the reactor (see Fig. 11.4) could be used. Representative spectra recorded at the start and towards the end of the polymerisation reaction reflect only minor intensity changes (Fig. 11.7). Thus two monomer-specific bands can be identified in the FT-NIR spectra due to their intensity decrease: the band at 6163 cm^{-1} can be assigned to the first overtone of the $2 \times \nu(CH_2=)$ absorption, while the absorption at 4746 cm^{-1} is a combination band of the $\nu(CH_2=)$ and $\nu(C=C)$ modes.

Based on a preceding calibration procedure of 32 different toluene solutions with known composition of methylmethacrylate and poly(methylmethacrylate) without initiator (in order to suppress a polymerisation reaction and thereby a change of composition) whose NIR spectra were measured under identical experimental conditions with reference to the spectra recorded during the polymerisation reaction, a partial least squares (PLS) calibration model with 2 principal components was developed and used to predict the concentration profiles of polymerisation reactions. The actual/predicted plots of this PLS model for the monomer and polymer are shown in Fig. 11.8.

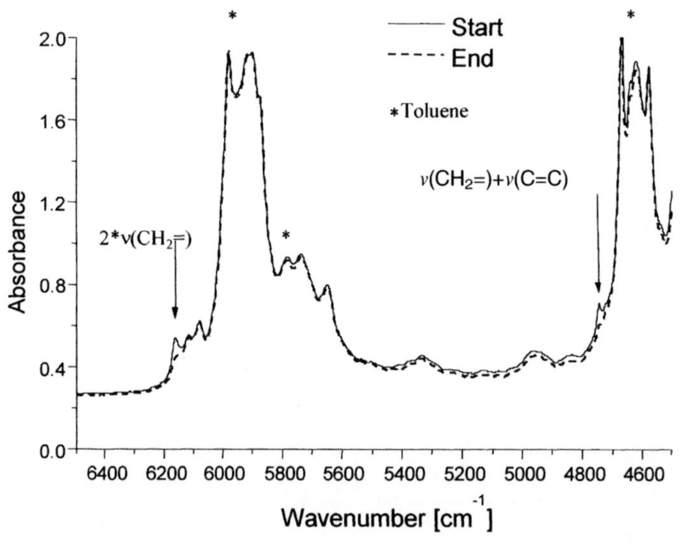

Fig. 11.7 FT-NIR spectra reflecting the changes during MMA polymerisation (* solvent absorptions)

A much clearer picture of the spectroscopic changes compared with Fig. 11.7, and especially their importance for the calibration, is obtained from the plot of the PLS regression coefficient versus wavenumber (Fig. 11.9). The peaks pointing upwards are characteristic for the monomer and disappear during the polymerisation reaction, whereas the peaks pointing downwards are specific for the evolving polymer. In this context Fig. 11.9 reflects that, despite their low intensity relative to the other absorptions, the two monomer-specific NIR absorption bands are very important for the calibration.

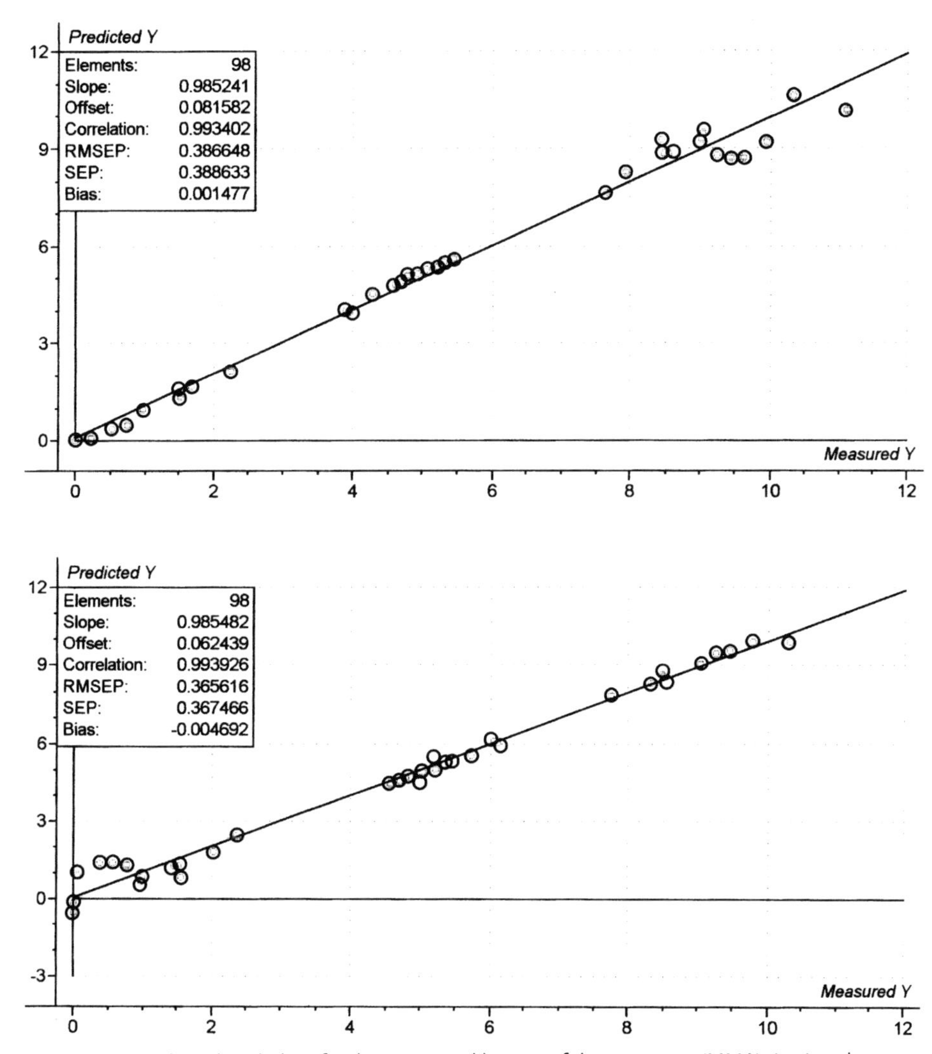

Fig. 11.8 Actual/predicted plots for the FT-NIR calibration of the monomer (MMA) (top) and the polymer (PMMA) (bottom)

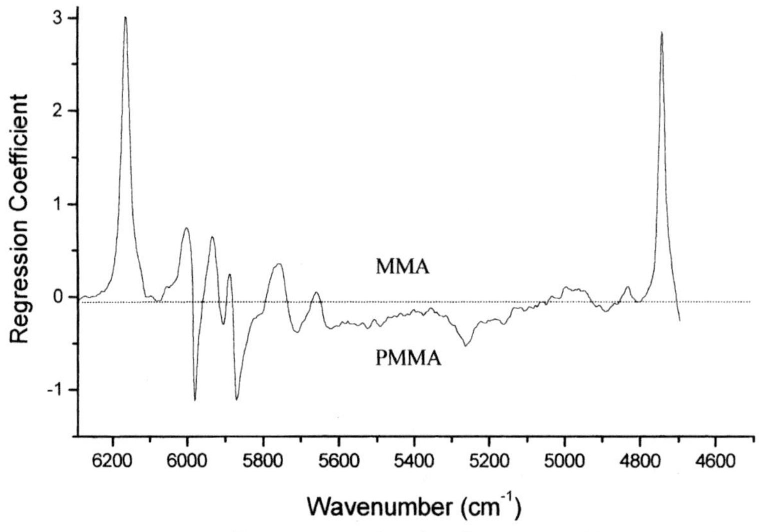

Fig. 11.9 Regression coefficient/wavenumber plots for the PLS calibration model

When the spectra recorded during the polymerisation reaction are predicted by the calibration model, the concentration profiles of the monomer and polymer shown in Fig. 11.10 are obtained. The discontinuities detected in these concentration profiles at the start of the reaction can be traced back to the addition of the initiator accompanied by a dilution of the monomer solution.

Certainly one of the most convincing examples for converting from conventional off-line laboratory analysis to a plant-located in-line process control by NIR spectroscopy is the monitoring of phosgene ($COCl_2$) concentration. Phosgene is a very important, but extremely toxic precursor material for the production of isocyanates (which are subsequently used to synthesise carbamate-pesticides and polyurethanes) and polycarbonate, and had to be analysed far off-line by a wet-chemical method (argentometric titration) under extreme safety precautions (gas mask, cooled-sample transport, etc.). In search for an analytical in-line technique to substitute this procedure, it was shown that NIR spectra obtained with a light-fibre (50 m) transmission probe (pathlength 10 mm) readily allowed the concentration level of phosgene in organic solvents to be monitored [24]. Fig. 11.11 shows the NIR spectra of three different solutions of phosgene in *o*-dichlorobenzene (55, 60 and 65% (w/w) phosgene) and demonstrates that the $3\times\nu(C{=}O)$ absorption at 1893 nm clearly reflects these concentration changes. Based on such measurements with samples covering the relevant concentration range under control, a calibration model can be developed for an in-line plant control.

Often material transported on a conveyor belt has to be subjected to a fast quality or sorting control or a chemical or physical parameter has to be determined without removing the material from the belt. This could be – to mention just two very different cases – the determination of water content in tobacco leaves or the sorting of polymer waste in a recycling process. The latter example will serve to

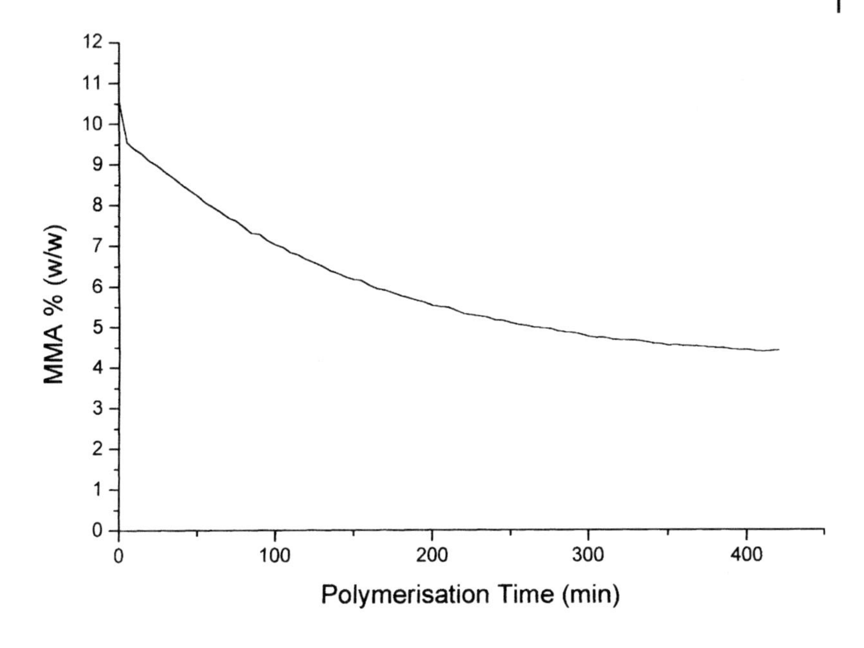

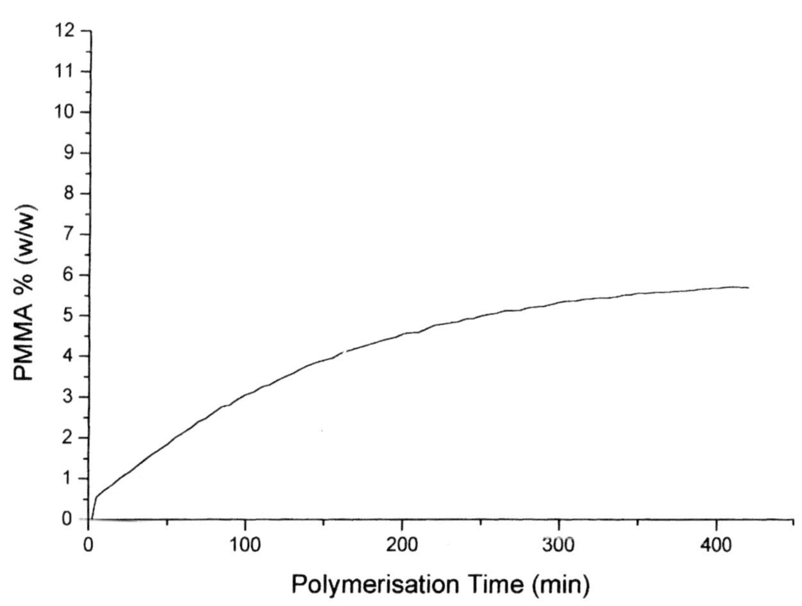

Fig. 11.10 Concentration (% w/w)/polymerisation time-plots for the test polymerisation (top: consumption of the monomer (MMA), bottom: evolution of the polymer (PMMA))

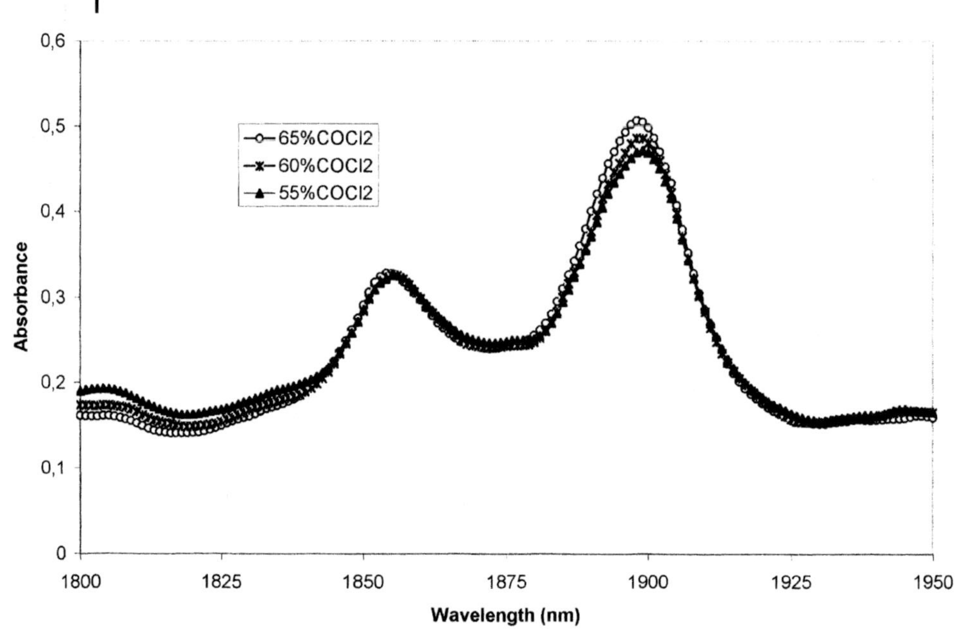

Fig. 11.11 NIR spectra of three different solutions of phosgene in o-dichlorobenzene (55, 60 and 65% (w/w) phosgene) measured with a transmission insertion probe (sample thickness: 10 mm, light-fibre length: 50 m)

demonstrate the potential of NIR spectroscopy for the solution of such process-analytical needs. Thus, although NIR spectroscopy has only low structural interpretative potential, it has gained large importance for identity control applications. In combination with discriminant analytical evaluation techniques it has become a widely used quality control tool for raw materials in the chemical, petrochemical, polymer and food industries.

In this aspect it has also been discovered for the rapidly growing needs of polymer recycling. Due to the increasing consumption of plastic products, the amount of plastic waste from electronic entertainment and household mass products and packaging material is expected to increase considerably in the near future. Additionally, the presently available techniques of waste disposal such as land fills and incineration plants are under permanent public discussion due to the limited available space in some countries and the steadily sharpened technical requirements to reduce the emissions. Therefore, consideration of the economic requirement of polymer recovery will become more important over the next years. In several publications it has been shown that, based on a factor analysis of their diffuse reflection spectra, the most frequently occurring polymers in waste – poly(ethylene), poly(propylene), poly(vinyl chloride), poly(styrene) and poly(ethyleneterephthalate) – can be readily differentiated [32–34]. Different approaches for the fast identification of polymers for sorting purposes have been proposed. Eisenreich et al. [34] for example, designed a funnel-like separation system in which the polymer waste falling through the bottle-neck of the funnel is identified by an integrated AOTF-detector and successively blown out with air jets into separate containers for the different polymers. Similarly, Bühler GmbH, Switzerland, demonstrated some time ago a polymer waste sorting system by mounting an extremely fast scanning NIR sensor (up to about 2000 scans per second) based on a so-called factor filter technology [11, 35] above a conveyor belt which moved with a speed of about 2.5 m s^{-1} (Fig. 11.12). With the exception of carbon-black filled polymers "real world" polymer waste could be very efficiently sorted with this instrument independently of shape, impurities and labels with a sorting accuracy of about 98–99%. Unfortunately, mostly due to political reasons, this system has had no break-through so far. Nevertheless, it demonstrated that even such complicated sample species like polymer waste could be identified on-the-flow on a conveyor belt thereby encouraging future applications of a qualitative and quantitative nature.

In Chapter 10 we demonstrated the application of rheo-optical FNIR spectroscopy for the characterisation of hard- and soft-segment orientation during uniaxial elongation of poly(dimethylsiloxane)/polycarbonate-block copolymer films with a miniaturised stretching machine on a laboratory scale. Generally, polymeric films are used extensively in the packaging, magnetic media and graphics industries and the benefits these films exhibit rely heavily on their physical properties. Thus, the required mechanical strength of poly(ethyleneterephthalate) – PET – films, for example, is imparted in a bidirectional drawing procedure. In this film-forming process (Fig. 11.13) an amorphous sheet (thickness approximately 600–1000 μm) is cast from the melt onto a cooled drum and drawn in the machine direction for 400% by rolls rotating with increasing speed at about 80 °C (close to the glass-transition temperature). In a subsequent sideway drawing operation by special stenter technology, the film is drawn in the transverse direction at 110 °C and finally subjected to

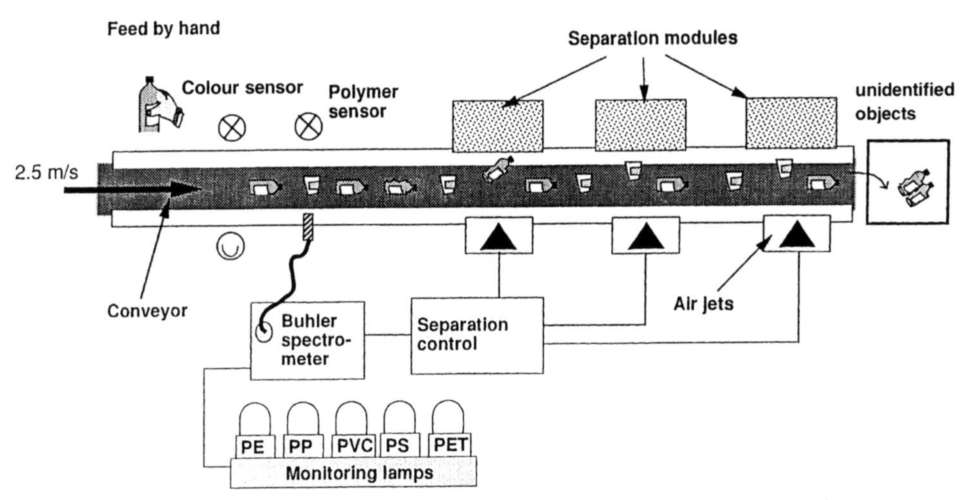

Fig. 11.12 Technical set-up of polymer waste sorting with a rapid-scanning NIR sensor (with kind permission of Bühler GmbH, Uzwil, Switzerland)

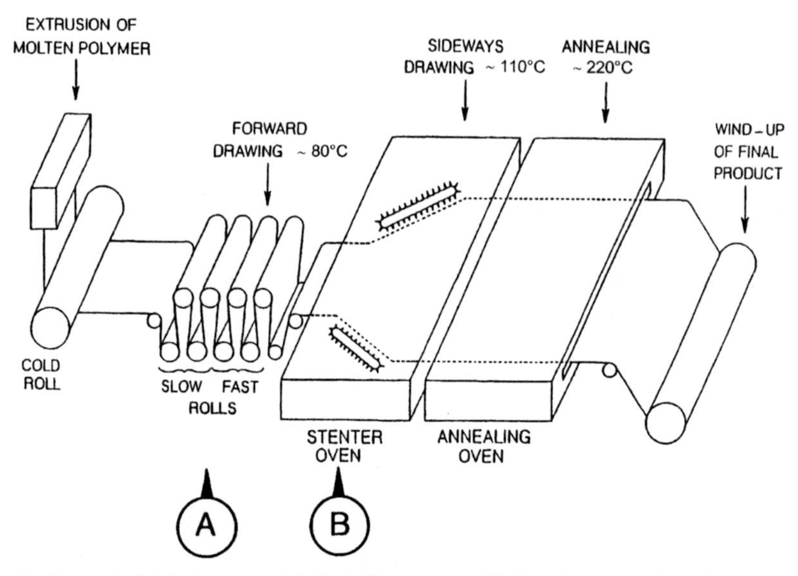

Fig. 11.13 Poly(ethylene terephthalate) film process. (A) Drawing in machine direction; (B) drawing in transverse direction

thermal annealing [36]. So far only on-line thickness measurements are performed and industrial companies are increasingly searching for techniques to monitor the orientational and crystallisation changes taking place during this process. Rheo-optical FT-NIR spectroscopy offers a means of shedding more light on the transient structural changes in this film-forming process. Although NIR spectra are generally less informative than their mid-infrared analogues due to the overlap of absorption bands in this wavelength region, it has been shown that the details of crystallinity, conformation and orientation are also contained in the overtone and combination vibration region [36–41]. In a detailed study of the film-forming process of PET [36] the changes in these physical parameters have been simulated in a miniaturised, temperature-controlled stretching machine with film samples withdrawn from the production process. The forward drawing process was investigated with a sample corresponding to the process conditions between the cold roll and the slow rolls (Fig. 11.13), whereas, for the sideways drawing experiment, a specimen was sampled just after the forward drawing process. The development of dichroic effects in the polarisation spectra measured after uniaxial elongation of the originally isotropic plain cast film of 600 μm thickness at 80 °C to 400% strain demonstrates the orientational anisotropy induced in this process. Thus, the polarisation spectra – measured with radiation polarised parallel and perpendicular to the drawing direction – do not show intensity differences before elongation (Fig. 11.14(a)) whereas significant parallel and perpendicular dichroism can be observed upon elongation (Fig. 11.14(b)). During the elongation procedure, polarisation spectra were recorded alternately with radiation polarised parallel and perpendicular to the drawing direction [11] in about 10 s intervals. Subsequently, the orientation functions of the crystalline, amorphous and total polymer were evaluated from these on-line spectra. For this purpose the absorption bands at 4463 cm^{-1} (CH_2 combination band, trans ethylene glycol segments), 4403 cm^{-1} (CH_2 combination band, gauche ethylene glycol segments) and the $3\times\nu(C{=}O)$ overtone (5123 cm^{-1}) were used, respectively (Fig. 11.16(a)). For the determination of the change in film thickness during the mechanical treatment, the first overtone of the $\nu(CH_2)$ absorptions at about 6000 cm^{-1} could be used. The same experimental and evaluation procedure was applied for the transverse drawing operation and it was shown that the orientation effects induced along the machine direction are reversed after about 200% strain in the transverse

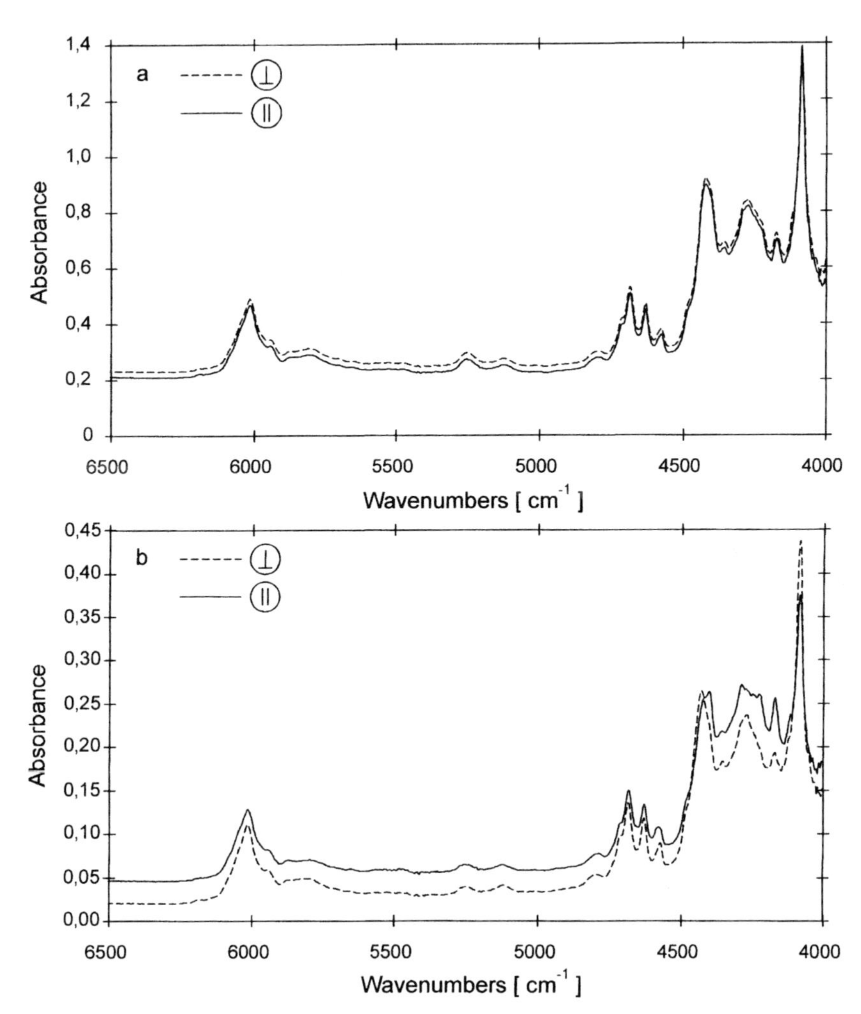

Fig. 11.14 NIR polarisation spectra of a PET film (d=600 μm). (a) before elongation. (b) after elongation for 400% strain at 80 °C (polariser II/⊥ to the drawing direction)

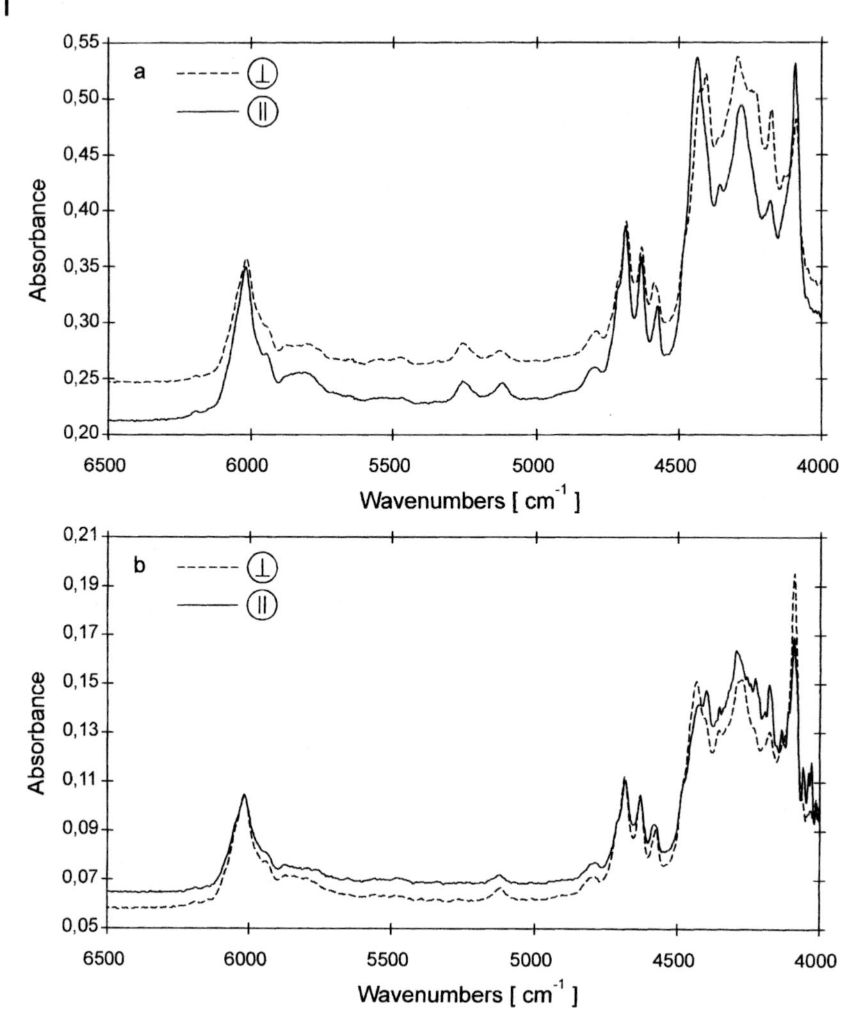

Fig. 11.15 NIR polarisation spectra of a uniaxially elongated (400%) PET film (d=250 μm, T=110 °C). (a) before elongation in the transverse direction. (b) after elongation in the transverse direction for 400% strain at 110 °C

direction (Fig. 11.16(b)). Therefore, the left part of Fig. 11.16(b) represents the decrease of chain orientation relative to the machine direction up to about 200% strain and then, beyond the inversion point, the subsequent orientation increase in the transverse direction is shown (in order to achieve positive orientation values after the inversion point, the direction of the reference axis had to be rotated by 90°). It is interesting to note that, although the films are different process samples for the experiments underlying Figures 11.16(a) and (b), the final orientation obtained on the miniaturised stretching machine in the laboratory simulation experiment (see Fig. 11.16(a) at 400% strain) is in reasonable agreement with the initial orientation

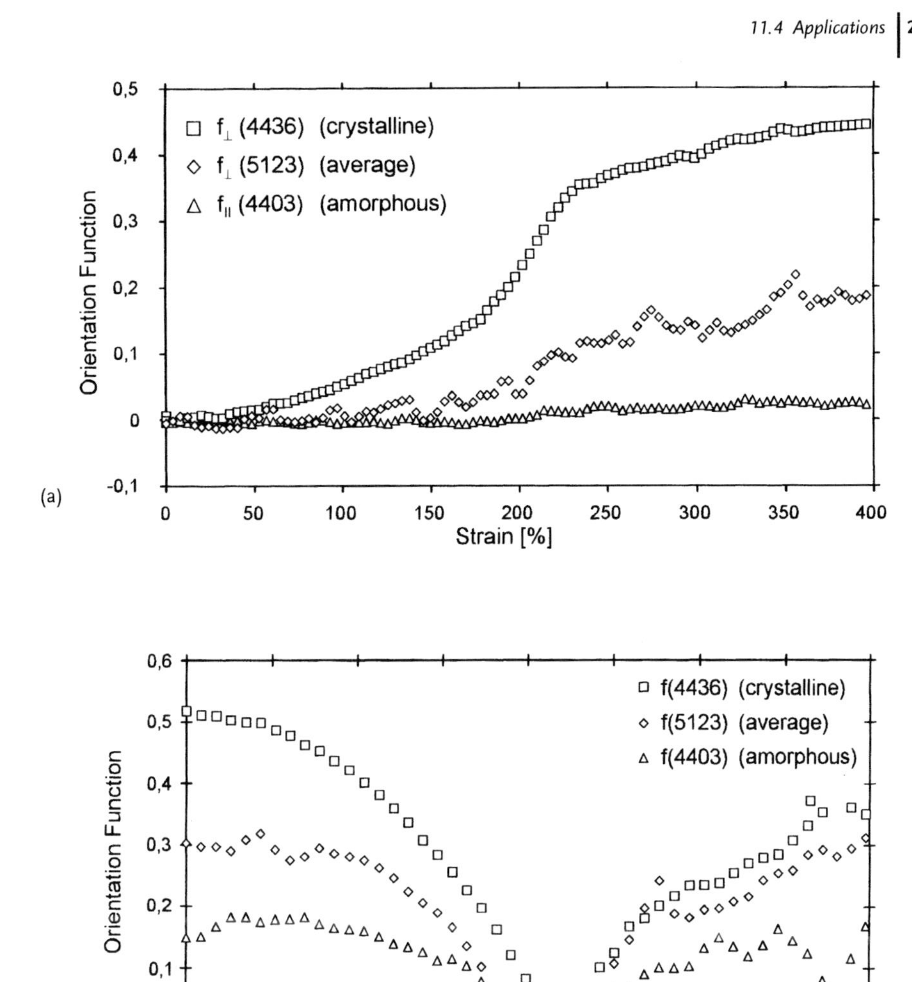

Fig. 11.16 Orientation function/strain plots: (a) for the uniaxial elongation of a plain cast PET film (d=600 μm, T=80 °C, 100% strain/min). (b) for the transverse elongation of a pre-drawn (400% strain) PET film (d=250 μm, T=110 °C, 100% strain/min)

values (see Fig. 11.16(b) at 0% strain) of the process sample used for the simulation of the sideways-drawing experiment. This proves that the operation performed on the laboratory stretcher is a realistic simulation of the industrial film-forming process.

Similar NIR-spectroscopic rheo-optical investigations on polypropylene films support this potential as a process monitoring technique [42]. In the future, these results will certainly contribute to the development of NIR-spectroscopic control systems with polarisation-retaining light-fibres and transmission sensors for the moving web at process positions where the polymer film has a thickness of about 100–1000 μm.

Another important technique in polymer processing is extrusion (Fig. 11.17) [43, 44]. This operation is used for the blending of different polymers or the compounding of a polymer with additives. Alternatively, reactive extrusion may be applied to substitute the corresponding polymer-analogous reaction in solution. The conveyance and blending procedure of the polymer melt involved in this operation takes place under extreme conditions of pressure (up to about 250 bars) and temperature (up to 350 °C). The resulting polymer melt is an intermediate product, which is subsequently fed into another processing step (moulding, film forming etc.). The material throughputs can range from a few kilograms to several tons per hour; this demonstrates the wide variety of extruder capacities and types. Nevertheless, it is important to control this operation in real time in order to obtain rapid information on the quality of this intermediate polymer melt. NIR spectroscopy has so far proved the only technique which combines the required robustness for plant applications with an adequate information content to perform in-line control on such polymer melts. For this purpose a suitable probe is usually mounted at the extruder exit or in a special adapter at the extruder exit. Reports on the implementation of NIR spectroscopy as an analytical tool in laboratory extrusion processes have been available since the early nineties [45–52]. Initially, the prediction of the composition of co-polymers and polymer blends was the main target of the analysis. Later on, researchers also focussed on the quantitative determination of additives and filler materials. Recently, rheological properties and physical parameters have also been investigated. In all cases, however, the data referred to a special material case under defined experimental conditions and little information has been produced on system distortions and the long term stability and transfer capabilities of the resulting calibrations. Very recently, however, a detailed study on the application of NIR spectroscopy to polymer extrusion processes became available [53] which addresses these important practical aspects for a successful development of an industrial NIR spectroscopic extrusion control system.

Based on a systematic analysis of the available literature data and with the aim of developing an integrated control system consisting of a light-fibre coupled sen-

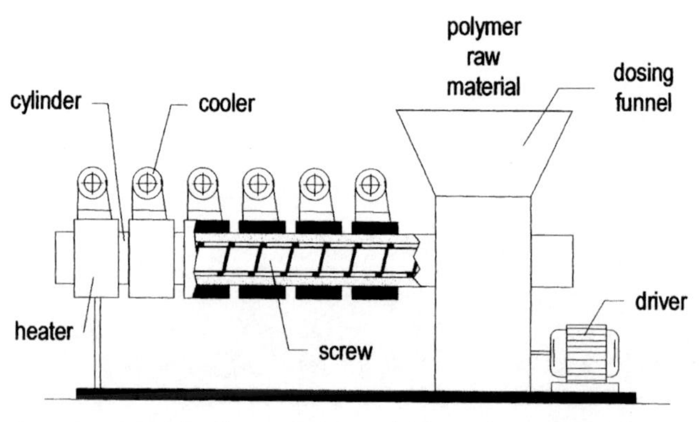

Fig. 11.17 Principle scheme of an extruder (according to [44] with permission from [53])

sor, an NIR spectrometer and the necessary evaluation software, the author identifies two indispensable criteria for the solution of this problem:

- construction of a light-fibre coupled sensor which can be mounted in a suitable extruder position and which fulfils the optical and mechanical property requirements to provide long-term stability with reference to the temperature and pressure conditions of the extruder
- development of a software routine for a simple calibration transfer from a laboratory to production plant extruders.

In this study three different sensor concepts, based on transmission, reflection and transflection measurements, are discussed in detail with reference to the monitoring of polyethylene/polypropylene (PE/PP) extrusion-blending procedures in the 40:60 to 60:40 (w/w) composition range. The NIR spectra were recorded on a diode-array spectrometer in the 950–1700 nm range. Typical spectra obtained in a long-term stability test with the transmission sensor (pathlength 8 mm) at a temperature of 220 °C from a series of different blend compositions are shown in Fig. 11.18. The two prominent absorption bands can be assigned to the second overtones of the $\nu(CH_2)$ and $\nu(CH_3)$ vibrations (1200 nm) and to combination bands of the $\nu(CH_2)/\nu(CH_3)$ and $\delta(CH_2)/\delta(CH_3)$ vibrations (1400 nm) of the two polymer components. For the transmission and reflection sensor systems, random mean-square errors of prediction (RMSEP) < 1% (w/w), based on PLS calibration models, were obtained. The capability for in-line measurements is demonstrated in Fig. 11.19 where the left-hand graph compares the PE-content (%; w/w) of the dosing weights with the corresponding NIR predictions measured at the extruder exit. The horizontal shift of the two lines reflects the residence time of the material in the extruder whereas the vertical shift corresponds to a constant bias of

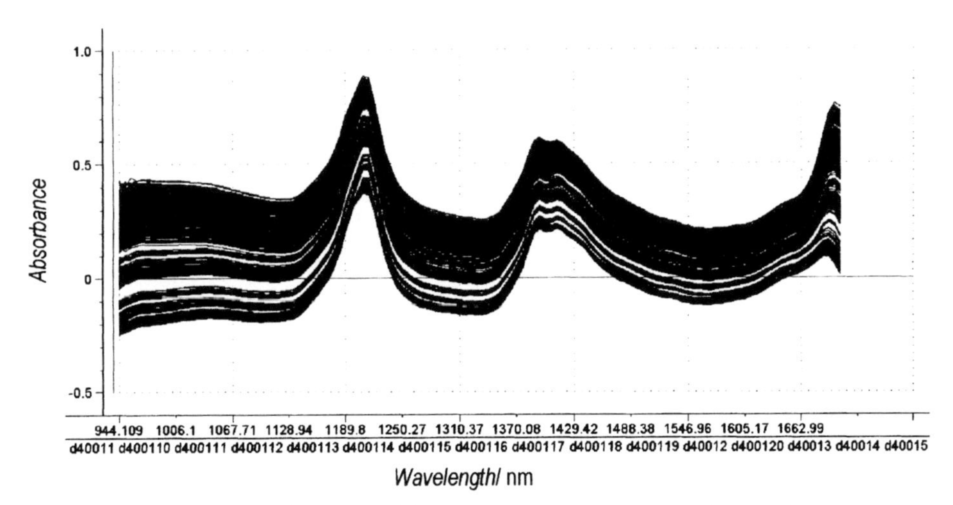

Fig. 11.18 NIR spectra of different PE/PP blends measured during extrusion (with kind permission from [53])

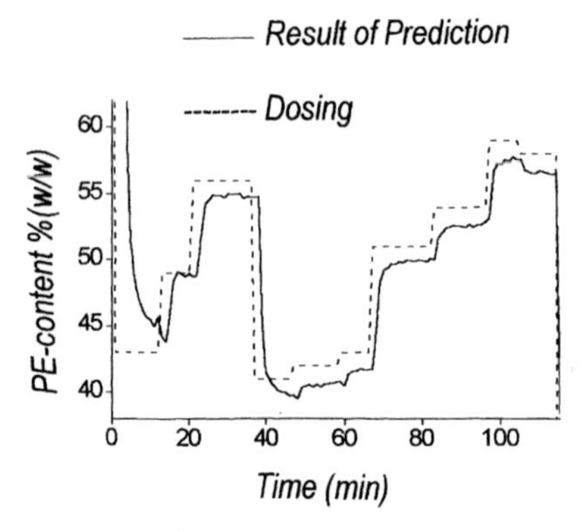

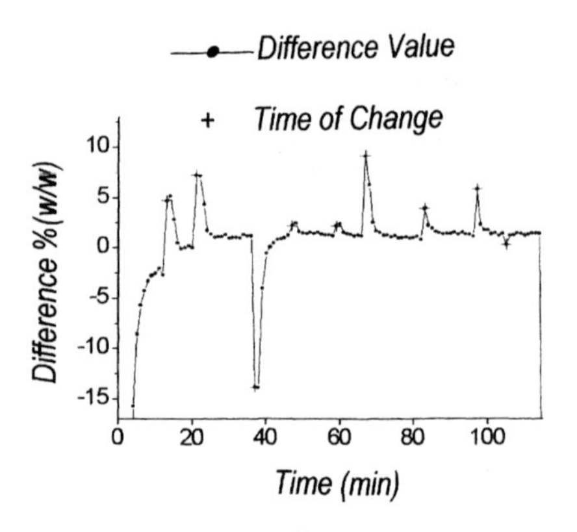

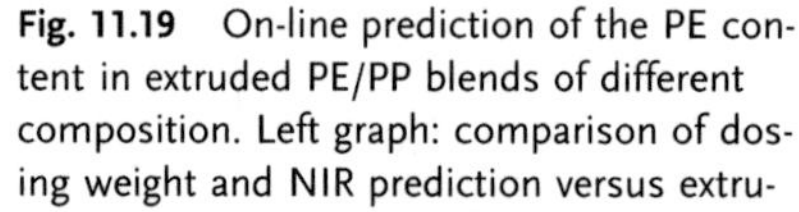

Fig. 11.19 On-line prediction of the PE content in extruded PE/PP blends of different composition. Left graph: comparison of dosing weight and NIR prediction versus extrusion time. Right graph: difference between dosing weight and NIR prediction as a function of extrusion time (with kind permission from [53])

about 1% due to inadequate optical compensation in this experiment. In the graph on the right the difference values of dosing weight and NIR prediction clearly mark the intervals of compositional changes by the spikes. Thus, following the trace from minute 35 to minute 60 proves that even very small compositional changes can be readily detected next to very large changes in dosing weight.

Within the same project an algorithm has been developed that automatically calculates the best spectral pretreatment for a calibration and validation set. Thus, the operator will become completely independent of subjective influences in the estimation of the quality of his calibration model. Finally, it is demonstrated that a calibration transfer for in-line NIR-spectroscopic extrusion data can be successfully performed either on the basis of this automated spectral pretreatment or by standardisation procedures such as Piecewise Direct Standardisation (PDS) [54].

Although most of the examples described here have been chosen from polymer synthesis and processing, numerous studies in other fields of materials are available, where an NIR technique has been applied as an on-line/in-line technique or has been successfully transferred from the laboratory to the plant [55–57]. Most of the examples presented here were chosen because of their inherent complexity with reference to the experimental conditions such as a rapidly moving analyte or hostile conditions of temperature and pressure.

11.5 References

[1] E. H. Baughman, D. Mayes, *Am. Lab.* **1989**, *21(10)*, 54-55, 57–58.

[2] G. K. Brown, *Proc. SPIE-Int. Soc. Opt. Eng.* **1992**, *1681*, 304–306.

[3] W. W. Daily, *Proc. Cont. Qual.* **1992**, *3*, 99–106.

[4] J. Workman, Jr., *J. Near Infrared Spectrosc.* **1993**, *1(4)*, 221–245.

[5] A. Kari, H. Hannu, J. Petri, *Proc. Cont. Qual.* **1994**, *6*, 125–131.

[6] B. D. Mindel, *Proc. Cont. Qual.* **1997**, *9(4)*, 173–178.

[7] M. G. Hansen, A. Khettry, *Proc. Annu. ISA Anal. Div. Symp.* **1995**, *28*, 107–13.

[8] M. M. Dumoulin, R. Gendron, K. C. Cole, *TRIP*, **1996**, *4(4)*, 109–114.

[9] K.G. Carr-Brion, J. R. P. Clark, *Sampling Systems for Process Analyser*, Butterworth-Heinemann, London, **1996**.

[10] D. C. Hassell, E. M. Bowman, *Appl. Spectrosc.* **1998**, *52(1)*, 18A–29A.

[11] H. W. Siesler, *NIR News* **1995**, *6(1)*, 3.

[12] K. J. Clevett, *Process Analyzer Technology*, Wiley, New York, **1986**.

[13] G. D. Nichols, *On-Line Process Analysers*, Wiley, New York, **1988**.

[14] *Spectroscopy in Process Analysis* (Ed. J. M. Chalmers), CRC Press, Boca Raton, USA, **2000**.

[15] H. W. Siesler "Quality-Control and Process-Monitoring by Mid-Infrared, Near-Infrared and Raman Spectroscopy" in *Near Infrared Spectroscopy: Proceedings of the 9th International NIR-Conference*, (Eds. A.M.C. Davies, R. Giangiacomo) NIR Publications, Chichester, UK, **2000**, pp. 331–337.

[16] *Analytical Applications of Raman Spectroscopy* (Ed. M. J. Pelletier), Blackwell Science Ltd., London, UK, **1999**.

[17] A. Espinosa, D. Lambert, M. Valleur, *Hydrocarbon Processing*, **1995**, *74*, 86-89.

[18] D. Lambert, P. Descales, S. Bages, S. Bellet, J. R. Llinas, M. Loublier, J. P. Maury, A. Martens, *Hydrocarbon Processing* **1995**, *74*, 103–106.

[19] J. B. Callis, D. L. Illman, B. R. Kowalski, *Anal. Chem.* **1987**, *59*, 624A.

[20] S. D. Brown, S. T. Sum, F. Despagne, B. K. Levine, *Anal. Chem.* **1996**, *68*, 21R.

[21] W. Blaser, R. Bredeweg, R. Harner, M. LaPack, A. Leugers, D. Martin, R. Pell, Workman, L. J. Wright, *Anal. Chem.* **1995**, *67*, 47R.

[22] K. de Brackeler, D. L. Massart, *Chemom. Intell. Lab. Syst.* **1997**, *39*, 127.

[23] H. W. Siesler, A. M. C. Davies, *NIR News* **1991**, *2(1)*, 8.

[24] H. W. Siesler, *Makromol. Chem. Macromol. Symp.* **1991**, *52*, 113–129.

[25] J. Coates, *Appl. Spectrosc. Revs.* **1998**, *33(4)*, 267.

[26] F. C. Allard, *Fiber Optics Handbook*, McGraw-Hill, New York, **1990**.

[27] O. K. Tec, Wildenbruchstr. 15, 07745 Jena, Germany; O.K.Tec@T-online.de.

[28] *Proceedings of the 7th International NIR Conference* (Eds. A. M. C. Davies, P. Williams), NIR Publications, Chichester, UK, **1996**, p. 239.

[29] *Proceedings of the 9th International NIR-Conference*, (Eds. A. M. C. Davies, R. Giangiacomo), NIR Publications, Chichester, UK, **2000**, p. 331.

[30] A. Olinga, R. Winzen, H. Rehage, H. W. Siesler, *J. Near Infrared Spectrosc.* **2001**, *7*, 223.

[31] H. G. Elias, *An Introduction to Plastics*, VCH, Weinheim, **1993**.

[32] T. Huth-Fehre, R. Feldhoff, T. Kantimm, L. Quick, F. Winter, K. Cammann, W. van den Broek, D. Wienke, W. Melssen, L. Buydens, *J. Mol. Struct.* **1995**, *348*, 143.

[33] U. Eschenauer, O. Henck, M. Hühne, P. Wu, I. Zebger, H. W. Siesler in *Near-Infrared Spectroscopy* (Eds. K. I. Hildrum, T. Isaksson, T. Naes, A. Tandberg), Ellis Horwood, Chichester, England, **1992**, pp 11–18.

[34] N. Eisenreich, J. Herz, W. Mayer, T. Rohe "Fast On-Line Identification of Plastics by Near-Infrared Spectroscopy for Use in Recycling Processes" in *Proceedings of ANTEC '96*, Indianapolis, **1996**, 3131–3135.

[35] Bühler GmbH, *Company Brochure on Polymer Sorting by Factor-Filter Technology*, Uzwil, Switzerland, **1994**.

[36] U. Hoffmann, F. Pfeifer, S. Okretic, N. Völkl, M. Zahedi, H. W. Siesler, *Appl. Spectroscopy* **1993**, *47* (9), 1531.

[37] N. Völkl, *PhD-Thesis*, University of Essen, Germany (**1995**).

[38] A. Ameri, H. W. Siesler, *J. Appl. Polym. Sci.* **1998**, *70*, 1349–1357.

[39] P. Wu, H. W. Siesler, F. Dal Maso, N. Zanier, *Analysis Magazine* **1998**, *26*(4), 45–48.

[40] C. E. Miller, B. E. Eichinger, *Appl. Spectrosc.* **1990**, *44*(3), 496-504.

[41] H. W. Siesler "Characterization of Polymer Deformation by Vibrational Spectroscopy" in *Oriented Polymer Materials* (Ed. S. Fakirov), Hüthig & Wepf, Heidelberg, **1996**, 138–166.

[42] H. W. Siesler, unpublished results.

[43] F. Hensen, W. Knappe, H. Potente, *Handbuch der Kunststoff-Extrusionstechnik*, Carl Hanser, München, *Vol. 1/2*, **1989**.

[44] W. Michaeli, *Einführung in die Kunststoffverarbeitung*, Carl Hanser, München, **1992**.

[45] A. Khettry, M. G. Hansen, *Polymer Engineering and Science*, **1996**, *36*(9), 1232–1243.

[46] D. Fischer, T. Bayer, K.-J. Eichhorn, M. Otto, *Fresenius J. Anal. Chem.* **1997**, *359*, 74–77.

[47] H. L. McPeters, S. O. Williams, *Proc. Cont. Qual.* **1992**, *3*, 75–83.

[48] M. G. Hansen, A. Khettry, *Polym. Eng. Sci.* **1994**, *34*(23), 1758–1766.

[49] M. G. Hansen, S. Vedula, *J. Appl. Polym. Sci.* **1998**, *68*(6), 859–889.

[50] R. Reshadat, S. Desa, S. Joseph, M. Mehra, N. Stoev, S. T. Balke, *Appl. Spectrosc.* **1999**, *53*(*11*), 1412–1418.

[51] T. Nagata, M. Ohshima, M. Tanigaki, *Polym. Eng. Sci.* **2000**, *40*(*5*), 1107–1113.

[52] S. Šašic, Y. Kita, T. Furukawa, M. Watari, H. W. Siesler, Y. Ozaki, *Analyst* **2000**, *125*, 2315.

[53] T. Rohe, *PhD Thesis*, University of Stuttgart, Germany, **2001**.

[54] J. Shenk, M. Westerhaus, *NIR News* **1993**, *4*(*5*), 13–15.

[55] *Proceedings of the 9th International Conference on Near Infrared Spectroscopy* (Eds. A. M. C. Davies, R. Giangiacomo), NIR Publications, Chichester, UK, **2000**, p. 331.

[56] G. J. Kemeny "Process Analysis" in *Handbook of Near-Infrared Analysis* (Eds. D. A. Burns, E. W. Ciurczak), Marcel Dekker, New York, **2001**, pp. 729–782.

[57] J. Workman, Jr., *J. Near Infrared Spectrosc.* **1993**, *1*(4), 221–245.

12
Application to Agricultural Products and Foodstuffs

Sumio Kawano

12.1
Introduction

In the food industry, NIR spectroscopy is being widely used for quality control of raw materials, intermediate products and final products. In this case, many kinds of sample presentation (or sample placement) are needed because there are various types of samples such as liquid, slurry, powder and solid. It is also not easy to prepare a sample set for calibration that has a wide range of chemicals or properties of interest, because intermediate and final products in a processing line have a narrow range.

In order to use an NIR instrument in the analysis of different types of samples, many kinds of sample cells have been developed. They are commercially available at present. Though the reflection method is generally used in the case of ground samples, various NIR measurement methods such as transmission, transflection and interaction are also used.

12.2
Grains and Seeds

Crops and seeds have been the objects of NIR studies for a long time. Therefore many studies dealing with these products were performed, and NIR spectroscopy is now used as a standard method for their routine analyses. While, at the beginning, NIR spectra were measured by using a ground sample, a NIR measurement method that uses whole kernels has become progressively employed because grinding is laborious and time-consuming.

According to the NIRT Handbook, NIR transmission spectroscopy (NIRT) with whole grains was employed in 1996 in place of NIR reflection spectroscopy (NIRR) with ground samples. NIRT is used as the official standard method for determining protein and oil content in soybeans, and protein content in wheat [1].

Williams et al. [2] studied the accuracy and precision of NIRT for determining protein and moisture content in wheat and barley. The standard error of prediction (SEP) for NIRT was slightly higher than that for NIRR. However, the elimination of grinding and cell-loading errors, and the large sample size, with a consequent reduction in sampling error, tend to compensate for the difference in accuracy. A comparison of commercial near-infrared transmission and reflection instruments for the analysis of whole grains and seeds was performed. Both approaches were comparable in accuracy and reproducibility [3, 4].

When grains and seeds are utilised as raw materials for the food industry, there are processing characteristics as well as constituents which should be evaluated. In the case of wheat, its kernel texture (degree of hardness or softness) is the most important factor affecting wheat-flour functionality. It exerts a significant influence on the yield of flour and on flour-damaged starch incurred during milling. Methods, including NIR spectroscopy, that define hardness by measuring a property of the wheat after it is ground are usually measuring some aspect of the resulting particle size distribution because harder wheats have a larger mean particle size than softer wheats. In the case of NIR reflectance spectroscopy with ground samples, the phenomenon that the spectral baseline is shifted by particle size variations between samples is used for measuring hardness.

Norris et al. [5] developed the Hardness Score that defines the hardness of wheat by using the phenomenon mentioned above. The NIR hardness score is scaled so that the soft wheats average 25 units and hard wheats average 75 units. The equation for Hardness Score (HS) is as follows:

$$\mathrm{HS} = a + b_1 L_{1680} + b_2 L_{2230}$$

where a and b are the standardisation constants and the L_S are the $\log(1/R)$ values at 1680 nm and 2230 nm. 1680 nm is the so-called reference wavelength for base-line correction in quantitative analysis. At this wavelength, $\log(1/R)$ is not affected by major constituents, such as protein, starch and oil, but only by particle size.

Williams [6] tried to measure the hardness or softness of wheat by NIR transmission (NIRT) spectroscopy using whole kernels. The investigation has shown that an NIRT method (whole grains) is capable of predicting wheat hardness with precision equal to that of the reference method (grinding-sieving method) and that it is slightly superior to the NIR reflection method (ground samples).

The fundamental composition of ash in wheat cannot be analysed by NIR spectroscopy because it does not absorb in the NIR region. However, many papers have indicated that it is possible to detect it with NIR spectroscopy, say, by using $\log(1/R)$ at 2345 nm, which seems to be due to cellulose [7]. Fig. 12.1 shows second derivative NIR spectra of non-defatted and defatted samples of 60% wheat-flour extraction and short bran [8]. The absorption at 2345 nm, which is stronger in short bran than in flour, clearly disappears in the spectrum of solvent-extracted or defatted flour; this indicates that the absorber at 2345 nm is not cellulose but a bran-related material that is also solvent extractable. Iwamoto et al. concluded that the absorber might be oil in the bran included in flour. Therefore, the NIR determination of the ash content of flour seems to be based on the fact that there is a high positive correlation between the ash and oil contents in wheat.

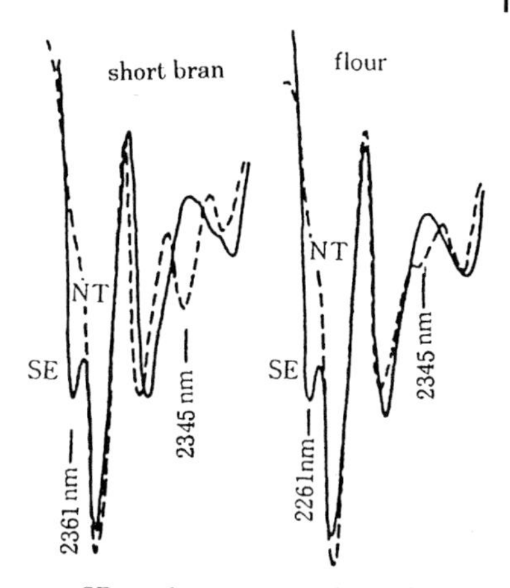

Fig. 12.1 Comparison of second-derivative NIR spectra of non-defatted and defatted samples of home-grown wheat flour and short bran [8]

One of the most interesting applications of NIR spectroscopy in the food industry is total quality evaluation, such as the quality of processing of raw materials. The quality of processing is defined by the balance of diverse chemical constituents and physicochemical properties. NIR spectra contain a lot of information about chemical constituents and physicochemical properties. By extracting the information by using chemometrics, such as principal component analysis (PCA) and discriminant analysis, the application area of NIR spectroscopy can be expanded.

In a rice milling plant, whiteness and taste are the most important factors for quality control. Whiteness, affected by the degree of milling, is related to the nutrition value of rice. It has been reported that whiteness can be measured by NIR spectroscopy [9]. However, whiteness does not seem to be measured directly by NIR spectroscopy but, rather, to be measured by determining the amount of bran remaining on the surface of the milled rice kernel. NIR determination of the surface lipid content (SLC) of milled rice has been investigated because there was a strong correlation between the degree of milling and SLC [10].

The taste of rice is generally assessed by sensory evaluation by using steamed rice. On the other hand, numerous studies on the relationship between rice taste and physicochemical properties have also been performed. It is well known that rice taste is a function of chemical constituents such as protein, moisture, amylose, fatty acid (fat acidity) and minerals.

The lipids in stored rice are hydrolysed and oxidised to free fatty acids or peroxides; this causes the acidity to increase and the taste to deteriorate significantly, as well as producing an "off-flavour". NIR determination of the fat acidity of whole-kernel and ground rough rice has been made, and a good calibration equation for the ground sample ($r^2=0.95$, SEP=0.73 mg of KOH/100 g of dry matter) has been obtained [11].

Amylose has an effect on the texture quality of cooked rice. NIR transmission spectroscopy (800–1050 nm) of unground brown rice or milled rice was shown to adequately screen for apparent amylose content (AAC) [12]. A method for deter-

Tab. 12.1 Commercially available rice taste analysers based on NIR spectroscopy

Company	*Factors*	*Treatment*	*Sample*	*Measurement*
Kett	Protein, amylosc, fatty acid 1), moisture	Whole	Milled rice Brown rice	Transmission
Kubota	Protein, amylose, moisture	Whole	Milled rice Brown rice	Transmission
Nireco	Mg, K, N	Whole	Milled rice Brown rice	Transmission
Satake	Protein, amylose, fatty acid, moisture	Whole	Milled rice Brown rice	Transmission
		Ground	Milled rice	Reflection
Shimazu	Protein, amylose, fatty acid 1), moisture	Whole	Brown rice Milled rice	Transmission
Shizuoka Seiki	Protein, iodine blue value, fatty acid, moisture	Whole	Brown rice Milled rice	Transmission
Yamamoto	Protein, iodine blue value, fatty acid, moisture	Whole	Brown rice Milled rice	Transmission

1) in the case of brown rice

mining AAC based on the NIR reflection spectrum (1100–2498 nm) of ground milled rice was also investigated [13]. The SEPs were 0.79% and 1.04%, respectively. In addition, prediction of cooked rice texture quality by using NIR reflection analysis of whole-grain milled samples has been performed [14].

Taste-related formulae that derive a taste score from the constituents have been developed. However, it is impractical for quality control to use the results of time-consuming chemical analyses. In order to overcome this problem, rice taste analysers based on NIR spectroscopy have been developed. There are currently seven different types of analysers, as shown in Tab. 12.1, which are commercially available.

The first rice-taste analyser, developed by the Satake Engineering Co. Ltd., contains an NIR instrument provided by Bran+Luebbe GmbH. The analyser is based on the experimentally proved result that rice taste is fixed by the balance of moisture, protein, amylose and fatty acid. In a practical procedure, milled rice is ground, and the ground sample is kept in a constant-temperature oven for more than 1 hour. NIR spectroscopy measurements are then performed to determine each constituent. From the constituents, a taste score can be calculated by using a taste-related formula that relates the constituents to the taste score. A taste score can be generated in only a few minutes. The rice taste analyser includes software that calculates the blending ratio necessary to give the same taste at the lowest price, or the best taste at the same price [15].

12.3 Fruits and Vegetables

Fruits and vegetables used not to be suited to NIR measurements because they contain a lot of moisture and have a relatively big size. However, after the development of high-performance NIR instruments and fibre optics in transmission or interaction mode, which can be used to measure intact fruits and vegetables, this application has become popular.

The first NIR application to intact fruits or vegetables was performed in the USA by using so-called "Biospect" (biological spectroscopy) designed for conveniently changing the source-sample-detector geometry; it deals with the determination of dry matter in onion [16], followed by soluble solid determination in cantaloupe melons [17]. After that, many studies on NIR applications to peaches [18, 19], apples [20], satsumas [21] and other fruits as shown in Tab. 12.2 were performed.

Kawano et al. used an NIR instrument with fibre optics as shown in Fig. 12.2; a commercially available "Interaction Probe" with a concentric outer ring of illuminator and an inner portion of receptor was used as the fibre optic probe. A cushion made of urethane foam was pasted onto the end of the probe to hold a sample. The NIR measurement was made by placing a sample on the end of the probe. A Teflon sphere 8 cm in diameter was used as a reference material because its light-scattering characteristics are similar to those of the sample. An absorbance was calculated by comparing the NIR energy reflected from the sample with that from the Teflon sphere. Multiple linear regression (MLR) based on $d^2\log(1/R)$ and Brix value gave good results ($r=0.97$, SEC=0.48°Brix, SEP=0.5 °Brix) [19] (SEC=standard error of calibration).

By using the interaction method mentioned above, it was hard to determine the composition of fruits such as satsumas, the peel of which is thick. Therefore, a NIR transmission method was used. A schematic for sample presentation is shown in Fig. 12.3. The top of the sample was illuminated by monochromatic light through fibre optics, and the intensity of the light transmitted through the sample was measured by a silicon detector located just below the sample. In transmission spectroscopy, the spectra are affected by the variation in diameter of the samples measured. In order to reduce the sample size effect, $d^2\log(1/T)$ at each wavelength was divided by $d^2\log(1/T)$ at 844 nm, which has a high correlation only to fruit size. The corrected spectra obtained here are called "normalised second derivative spectra" and are not affected by sample size. As a result of MLR based on normalised second derivative spectra and the Brix value, very good results ($r=0.989$, SEC=0.28°Brix, SEP=0.32°Brix) have been obtained [21].

The influence of various factors such as fruit variety, growing location, harvest season and producing year on the performance of the calibration equation for the Brix value of satsumas were also studied [22]. It was possible to obtain global and stable calibration equations when the calibration sample set had sufficient variation in all of the factors mentioned above. As a new approach of the use of NIR spectroscopy with fruits, a fundamental study for monitoring the growth stages of Japanese pear fruit was performed, based on sugar content [23].

Tab. 12.2 Reported NIR applications to intact fruits and vegetables

Authors	*Product*	*Constituents*	*Results*	*Methods*
Birth et al. (1985) [16]	Onion	D.M. [1] (%)	R=0.996, SEC=0.53, SEP=0.79	Biospect [2]
Dull et al. (1989) [17]	Melons	S.S. [3] (%)	R=–0.60, SEC=1.67, SEP=2.18	(Intact) Biospect
			R=–0.97, SEC=0.56, SEP=1.56	(Slice) Transmission
Dull et al. (1989) [26]	Potatoes	D.M. (%)	R=–0.92, SEC=1.04, SEP=1.52	(Intact) Biospec
			R=–0.95, SEC=1.25, SEP=1.69	(Slice) Transmission
Kawano et al. (1990) [18]	Peaches	Brix	R=0.96, SEC=0.46, SEP=0.50	Reflection InfraAlyzer 500
			R=0.95, SEC=0.52, SEP=0.56	Interaction Pacific Scientific 6250
Tenma et al. (1990) [20]	Apples	Brix	R=0.95, SEC=0.48, SEP=0.31	Interaction Pacific Scientific 6250
Horiuchi et al. (1991) [27]	Melons	Brix	R=0.94, SEC=1.38, SEP=1.09	Interaction NIRSystems 6500
Dull et al. (1991) [28]	Dates	Moisture (%)	R=0.977, SEC=0.89, SEP=1.50	Transmission
Kawano et al. (1992) [19]	Peaches	Brix	R=0.97, SEC=0.48, SEP=0.50	Interaction Pacific Scientific 6250
Fujiwara et al. (1992) [29]	Strawberry	Brix	R=0.90, SEC=0.61, SEP=0.65	Reflection InfraAlyzer 500
Dull et al. (1992) [30]	Melons	S.S. (%)	R=0.91, SEC=0.82, SEP=1.85	Transmission
Kawano et al. (1993) [21]	Satsuma Mandarin	Brix	R=0.989, SEC=0.28, SEP=0.32	Transmission Pacific Scientific 6250
Bellon et al. (1993) [31]	Peaches	Brix	R=0.81, SEC=1.25, SEP=1.04	Interaction CCD camera
Kojima et al. (1993) [32]	Japanese Pear	Brix	R=0.96, SEC=0.50, SEP=0.49	Reflection InfraAlyzer 500
Onda et al. (1994) [33]	Plum	Brix	R=0.92, SEC=0.41, SEP=0.61	Reflection InfraAlyzer 500
		Acidity (%)	R=0.78, SEC=0.11, SEP=0.14	
		Firmness (kg)	R=0.83, SEC=0.11, SEP=0.21	

Tab. 12.2 (continued)

Authors	*Product*	*Constituents*	*Results*	*Methods*
Kawano et al. (1995) [25]	Peaches	Brix	With temperature compensation R=0.96, SEC=0.41, SEP=0.42	Interaction Pacific Scientific 6250
Miyamoto et al. (1995) [22]	Satsuma Mandarin	Brix	R=0.94, SEC=0.50, SEP=0.48	Transmission Pacific Scientific 6250
Onda et al. (1995) [34]	Mume	Firmness (kg)	R=0.91, SEC=0.32, SEP=0.35	Reflection InfraAlyzer 500
Slaughter et al. (1996) [35]	Tomatoes	Brix	R=0.92, SEC=0.27, SEP=0.33	Interaction NIRSystems 6500
Guthrie et al. (1997) [36]	Pineapple	Brix	R=0.95, SEC=0.69, SEP=1.20	Interaction NIRSystems 6500
	Mango	D.M. (%)	R=0.95, SEC=0.80, SEP=0.79	
McGlone et al. (1998) [37]	Kiwifruit	Brix	R=0.95, RMSEP=0.39	Interaction NIRSystems 6500
		D.M. (%)	R=0.95, RMSEP=0.42	
		Firmness (kg)	R=0.81, RMSEP=0.80	
Osborne et al. (1999) [38]	Kiwifruit	Brix	RMSECV=0.32	MMS1-NIR, Zeiss
		D.M. (%)	RMSECV=0.32	

1) D.M.: Dry matter,
2) Biospect: Biological spectrophotometer
3) S.S.: Soluble solid

In high moisture products, such as fruits and vegetables, there is a high negative correlation between Brix value and moisture content. Therefore, optical data (say, log $(1/R)$ or $d^2\log(1/R)$) at a moisture band is often selected by mistake as the first variable term of a calibration equation for Brix determination because absorption at a moisture band is stronger than that at a sugar band. The calibration equation developed by using the internal, or secondary, correlation is not stable and gives a big bias in prediction. Fig. 12.4 shows the simple correlation coefficients, so called "correlation plots", between $d^2\log(1/R)$ and the Brix value of intact peaches when the first optical term is selected. Positive and negative high correlation coefficients can be observed at specific wavelengths. For example, a positive high correlation coefficient of 0.83 is obtained at 950 nm, due to reverse correlation between the Brix value and moisture content, while the negative high correlation coefficient obtained at 906 nm is a sugar band. Tab. 12.3 shows the predicted

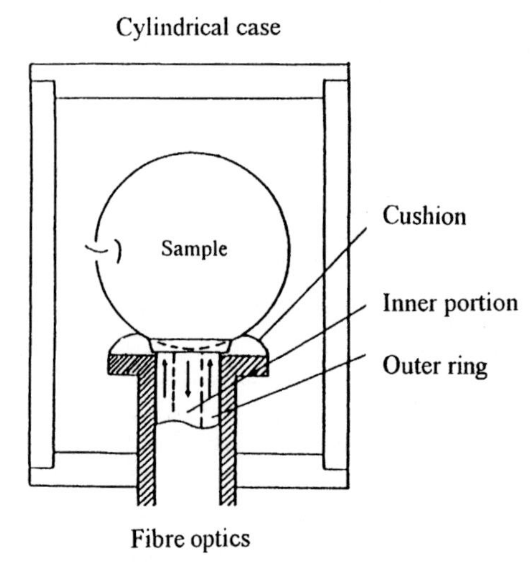

Fig. 12.2 Schematic of the geometry used to make interaction measurements on intact peaches with fibre optics [19]

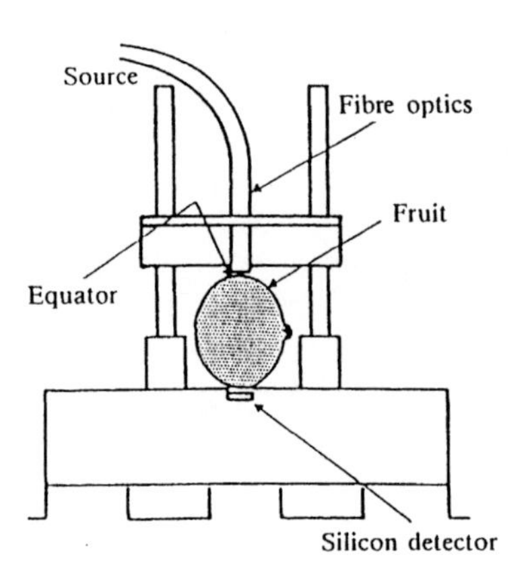

Fig. 12.3 Schematic of the geometry used to make transmission measurements on intact satsuma with fibre optics [21]

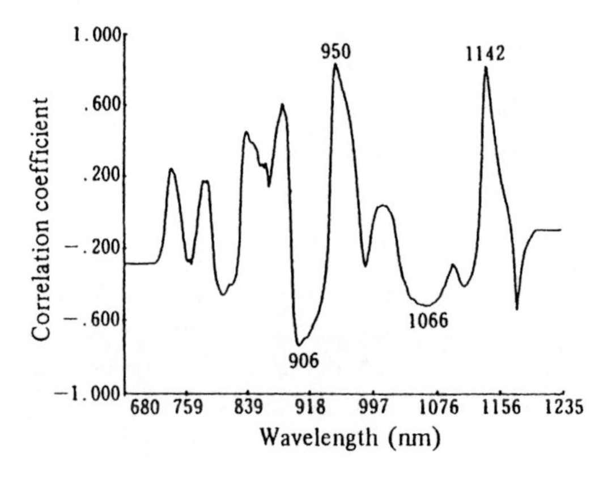

Fig. 12.4 Correlation plots between $d^2\log(1/R)$ and the Brix value of intact peaches when the first optical term is selected [19]

Tab. 12.3 Prediction results calculated by using each of four optical-terms-calibration equations having $d^2\log(1/R)$ at 906 nm or 950 nm as the first optical term

Selected wavelength (nm)				*SEP* °*Brix*	*Bias* °*Brix*
λ_1	λ_2	λ_3	λ_4		
906	878	870	889	0.50	0.01
950	877	1146	884	0.53	–3.45

results calculated by using each of four optical-terms-calibration equation having $d^2\log(1/R)$ at 906 nm or 950 nm as the first optical term. A calibration equation based on a sugar band is more stable than that based on internal or secondary correlation.

A study on the measurement of apple firmness by NIR spectroscopy has been performed and good results have been obtained. The first wavelength selected in the calibration for apple firmness was due to the second overtone of the O-H stretch, which was assigned to pectic material closely related to fruit firmness [24].

The influence of sample temperature on the performance of a NIR calibration equation was evaluated, and the development of a calibration equation with temperature compensation was examined. It was reported that a calibration equation with temperature compensation could be developed by using a calibration sample set which covered a variation in temperature of future samples [25]. This finding is important in developing a practical sweetness tester or sorting machine because it is not easy to maintain samples at a constant temperature in a field or a packing house.

In 1989, the Mitsui Mining and Smelting Co. Ltd. developed and introduced the first peach-sweetness sorting machine by which fruits can be sorted nondestructively depending on the Brix value. When fruits on a line-up conveyor are illuminated by two focused tungsten halogen lamps, the scattered reflected radiation is measured by a sensor unit which consists of a grating and a diode array. A lens causes the reflected radiation to converge and be projected onto a grating to extract the required wavelength and intensity data. The intensity of the radiation at any one wavelength is measured by the diode array (Fig. 12.5). The Brix value of a peach is calculated from the measured reflection intensity of NIR radiation by using a previously developed calibration equation. The sorting rate is 3 fruits/sec/lane.

If the sweetness sorting machines are introduced to packing houses, then taste grading will become possible. If the quality data from the sweetness sorting machine are correlated with data corresponding to the producing conditions such as soil condition and climate condition, guidance for producing high-quality products will also come about.

Different types of sweetness sorting machines for other products, say, oranges, melons or water melons, have become possible at the time of writing.

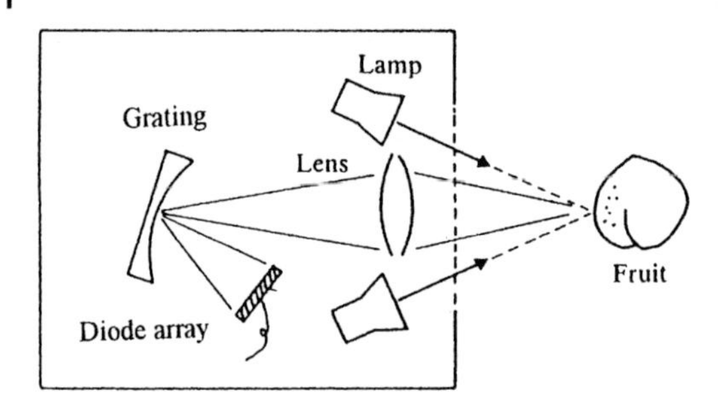

Fig. 12.5 Sweetness sorting machine for peaches based on NIR spectroscopy

12.4 Livestock Products

For the determination of protein, fat, lactose and total solids (TS) in milk, infrared spectroscopy (IR method) was employed as a standard method by the AOAC (Association of Official Analytical Chemists, Inc.) [39]. Studies on NIR application to milk analysis have been performed for a long time. When compared with an IR method, the NIR method had a higher accuracy in measuring TS as well as equivalent accuracy in measurements of fat, protein and lactose contents. With the NIR method it is also possible to measure casein in milk directly [40]. However, the NIR analysis of milk is strongly affected by fat globules as they are scattering particles which produce spectral deformations. Therefore, homogenisation of the sample is a fundamental need prior to the NIR analysis.

The MIR analysis of milk is based on absorption of IR energy at specific wavelengths of 6.465 μm (peptide linkages between amino acids of protein molecules), 5.723 μm (carbonyl groups in ester linkages of fat molecules) and 9.610 μm (OH groups in lactose molecules). On the other hand, as a result of principal components analyses and correspondence factorial analyses of NIR spectra of milk, it has been found that the specific wavelengths for NIR analyses of milk are 2050 nm and 2180 nm for protein, 1724 nm, 1752 nm, 2308 nm and 2344 nm for fat, and 2094 nm for lactose [41].

In a unique NIR application, identification of a particular cow's feed has been made by using PCA of raw milk spectra [42], and detection of foreign fat adulteration of milk fat has been performed [43].

NIR analyses of milk products other than raw milk, such as milk powder [44], nonfat dry milk [45], cheese [46–49], whey [50, 51] and other products have been performed. With analysis of cheese, very good calibration equations for fat (SEP=0.34%), protein (SECV=0.55%) and moisture (SEP=0.33%) could be obtained.

Many NIR application studies on meat or meat products have also been done. In 1968 for example, a study on NIR determinations of moisture and fat in meat products was reported [52]. In this study, NIR spectra of samples of meat emulsion 2 mm thick were measured by the transmission method. Data analytical methods that are in use today, such as MLR, were not used for calibration. Simple regression analysis was performed by using moisture content and ΔOD (1800–1725 nm) for moisture determination, and by using fat content and ΔOD (1725–1650 nm) for fat determination, where ΔOD is the difference of absorbance, which used to be called "Optical Density", at two arbitrary wavelengths. The ΔOD values predicted the fat content and moisture content to within a standard error of 2.1% and 1.4%, respectively.

Based on a similar idea, an infrared-emitter fat meter of ground beef has been developed. In this meter, NIR radiation of 937 nm and 943 nm illuminated the sample, and then the diffusely reflected light from the sample was measured by solar cells. The fat content was calculated by using the following formula:

$$\text{Fat} = A + B(R_{937} - R_{943})/(R_{937} + R_{943})$$

where R_{937} and R_{943} are reflectance values at 937 nm and 943 nm, respectively, and A and B are constants. The correlation coefficient between fat contents determined by the fat meter and by chemical analysis was 0.8 with a standard error of 1.98%. If the fat meter is used, NIR fat measurement becomes possible even for a film-wrapped sample [53] and NIR determination of protein, fat and water of homogenised meat has also been performed on film-wrapped samples [54].

Besides these constituents, it was reported that NIR spectroscopy had a high performance for the determination of sodium chloride in hams [55] and sausage [56], soybean flour in ground beef [57], starch content in gravy [58], calories in raw pork and beef [59], pH in ham [60] and the physical and chemical characteristics of beef cuts [61].

With the analysis of salt content in ham by NIR reflection, the correlation coefficients between $d^2\log(1/R)$ at 1806 nm and salt content were high at 0.960–0.997. Here the wavelength of 1806 nm is not a salt band but one of the water bands. The reason why salt, which does not absorb in the NIR region, can be measured by NIR spectroscopy is due to the shift in the water spectrum caused by salt-induced changes in the amount of hydrogen bonding as shown in Fig. 12.6 [55].

High accuracy for NIR analyses of meat and meat products is found with ground samples. However, it is possible to improve the accuracy by emulsifying the sample after grinding as pretreatment. It will also be necessary in the future to study the effect of the degree of emulsification, surface condition of the sample in the cell, storage, freezing, types of protein and fat, pH and sample temperature on NIR performance.

As a unique NIR application, the differentiation of frozen and unfrozen beef has also been investigated [62].

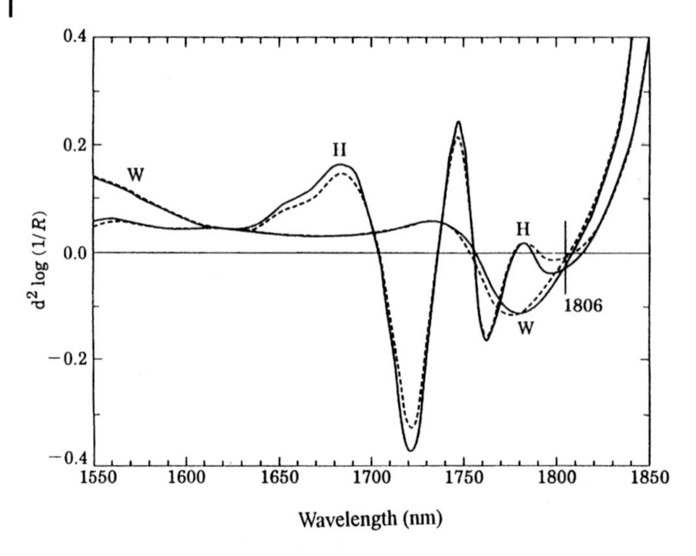

Fig. 12.6 Second derivative spectra of fresh ham (H) with 0.0% and 4.8% added salt and water (W) with 0.0% and 5.0% added salt. The spectra of samples without added salt are dashed [55]

12.5 Marine Products

NIR spectroscopy has been applied to the determination of moisture, protein, oil and salt content of Surimi or ground fish meat. Reflection spectroscopy in the NIR region (900–1800 nm) was used to predict moisture, crude lipid and protein nitrogen of either fresh or frozen rainbow trout muscle [63]. In the case of fish-meat analysis, not enough accuracy was provided in the reflection mode because of the high moisture content. Sample homogenisation was also needed as pre-treatment. On the other hand, transmission spectroscopy in the NIR region (850–1050 nm) was used for prediction of the fat content of fresh salmon fillets [64, 65]. Even simple homogenisation of the sample prior to NIR measurement gave good results.

Nondestructive NIR measurements of fish have been made [66–68]. Yamaguchi and Sawada used an interaction probe to measure NIR spectra in the short NIR wavelength region of 700–1100 nm in the same manner as for peaches [19], while Isaksson et al. used a remote reflection probe for the 400–2500 nm region. Fairly good results were obtained in both cases. For the assessment of fish quality, NIR rapid analysis of the free fat acid in fish oil was also investigated [69].

The state of the water in the fish flesh during the preparation process of fish gel, or Kamaboko, was investigated by using NIR spectroscopy. Absorptions at 1428 nm and 1942 nm were assigned to free water and bonded water, respectively. It was reported that the ratio of absorption intensities at these wavelengths showed the state of the water, and that free water decreased during gel formation and, conversely, bonded water increased [70].

The assignments of about 40 absorption bands observed in the spectrum (1100–2500 nm) of dried seaweed ("Nori") were clarified. At wavelengths of 2345 nm, 2305 nm, 2168 nm, 2056 nm, 1278 nm and 1188 nm, positive correlation was obtained between $\log(1/R)$ and the quality grade scored by an inspector, while at wavelengths of 2409 nm, 2365 nm, 2205 nm and 2110 nm, a negative one was obtained. The absorption bands at 2345 nm and 2305 nm were assigned to an organic solvent-extractable substance, and the bands at 2205 nm, 2110 nm and 2168 nm were assigned to a water-extractable one. The strong absorption bands at 2168 nm and 2056 nm were assigned to protein, and the bands at 2205 nm and 2110 nm were assigned to starch [71].

12.6 Beverages

In the analysis of beverages by NIR spectroscopy, either the transmission or the transflection method has been used. One major application in beer and wine has been to determine alcohol content [72, 73]. In the case of quantitative analysis of the alcohol content of beer, sample colour does not affect the analysis if the calibration sample set covers a wide range [72]. In the brewing process, NIR spectroscopy has been used to determine β-glucan [74] and malt extract values [75] related to the malting quality of the barley, as well as the α-acid of the hops. NIR determinations of acidity, amino acid and total sugar as well as alcohol have also been done for "sake", or rice wine [77]. Kojima et al. and Aramaki et al. independently determined the mycelial weight in rice koji by NIR spectroscopy. Rice koji is a steamed rice inoculated with *Aspergillus oryzae*, the amylase that converts the steamed rice into sugar. The former group dried rice koji samples at 100 °C for 60 minutes and ground them; NIR measurements of ground samples were then made over 1100–2500 nm. A good calibration was obtained (R: 0.98, SEC: 0.56 mg/g, SEP: 0.31 mg/g, bias: 0.16 mg/g) by MLR analysis based on the NIR spectra and the mycelial weight as measured by enzymatic hydrolysis. The absorbances at 1730 nm, 1738 nm, 2348 nm and 2360 nm have been included in the calibration [78]. The calibration equation obtained by the latter group also included the absorbance at 2348 nm. An additional examination made it clear that the absorption at 2348 nm should be assigned to fat in the mycelia of rice koji [79].

In the NIR analysis of juice, major constituents such as total sugar could be determined either by the transmission or the transflection method [80]. However, in order to analyse minor constituents such as glucose, fructose and sucrose, a special method, the so called Dry-Extract System for Infrared or "DESIR" had to be used [81]. In this method, a precise aliquot of the sample is deposited on a glass fibre disk, and a NIR reflection measurement of the disk is made after fast drying. The relation between $\log(1/R)$ and chemical concentration C can be modelled mathematically by the following formula:

$$\log(1/R) = h + s(1 - \exp(-kC))$$

where h, s and k are constants [82]. The relation is not linear. This method has also been applied to the analyses of milk, wine, beer, coffee and tea. The details have been reviewed by Thyoll and Isaksson [83]. In a unique application, the detection of the adulteration [84] or authenticity [85] of orange juice has been performed.

12.7 Other Processed Food

For wheat products in processed foods, the NIR method is widely used for the quality control of raw materials, intermediate products and final products. In the case of wheat flour, the NIR reflectance method has been successfully used to determine the degree of starch damage of commercially milled flour [86–88], this is important to the baking properties of the flour because the degree of starch damage has a strong influence on water absorption. If the starch has been much damaged during flour milling, it becomes hydrated, and then the water absorption properties of the flour are affected. It has been reported that the degree of starch damage can be measured in reflection at wavelengths corresponding to overtones and combinations of hydrogen-bonded O-H vibrations.

In relation to the quality of biscuits and biscuit doughs, the possible use of NIR reflection methods to monitor their fat, sucrose, dry flour and water has been reported [89], and the simultaneous analysis of ascorbic acid, azodicarbonamide and L-cysteine in bread improver mix has been achieved [90]. As for bread, NIR analyses of sliced white bread and air-dried bread have been done, and very good results have been obtained [91].

In the analyses of wheat products, NIR spectroscopy is widely used for, say, determination of fat and sucrose in dry cake mixes [92] and fibre in processed cereal foods [93–95], although insufficient accuracy has been obtained in NIR determination of the α-amylase activity of wheat [96], and loaf volume at bread-baking [97].

In determining the fat content in chocolate, it was reported that temperature did not influence NIR accuracy [98]. Generally, NIR determination of moisture or protein is easily influenced by temperature, but the temperature effect on fat is small. In order to estimate the degree of deterioration of fats and oils, NIR analyses on the degree of saturation of fatty acid moieties [99], iodine value [100, 101] and acid value have been done. As the degree of *cis*-unsaturation increased, the maximal peak around 1725 nm shifted to the lower wavelength, 1717 nm, and further to 1712 nm. Significant changes in absorption were observed at 1720 nm ($-CH_2-$) and 2140 nm ($-CH{=}CH-$) as the iodine value changed. The wavelength region from 1600 to 2200 nm was also used to successfully classify vegetable oils (soybean, corn, cottonseed, olive, rice bran, peanut, rapeseed, sesame and coconut oil) [102] and to detect the adulteration of olive oil [103] and of sesame oil [104].

During the shelf-life of coleslaw or a mayonnaise-based salad, major changes take place in the quality of the mayonnaise and the cabbage tissue. In order to make the quality changes clear, the oil in the mayonnaise and the cabbage tissue were determined by NIR spectroscopy [105]. It was found that cabbage tissue becomes translucent within 6 hours after being mixed with mayonnaise to form coleslaw and that this was caused by the migration of only a small quantity of oil from the mayonnaise into the tissue. Unlike cabbage, carrot and onion tissue did not absorb oil.

The Howard Mould Count (HMC), which is a microscopic method, has been used to measure the amount of mould in tomato puree. However, the HMC requires a high degree of operator skill, and therefore a more objective analysis based on chemical differences between mould and tomato was required. A preliminary experiment on the NIR determination of the amount of mould added to fresh tomato homogenate was made, and the possibility was suggested [106]. The wavelengths selected for the calibration were 2346 nm, 2224 nm, 1716 nm and 1692 nm, and it was concluded that the wavelengths could be associated with absorption bands of chin, the major carbohydrate in the cell walls of mould, because each wavelength was assigned to the C-H functional group.

As for NIR sugar analysis, it is possible to determine Spindle Brix, polarisation (pol), purity and reducing sugar at the sugar refining and manufacturing factory [107, 108]. NIR on-line analysis of Brix and pol is put to practical use in a beet sugar factory in Italy [109]. A sugar-cane-juice automatic analyser based on the NIR method has also been developed, and sugar-cane quality inspection by using this analyser has been implemented in each sugar-cane processing factory in Japan starting in 1994 [110].

In soy sauce manufacturing, quality control of intermediate and final products by many components such as total nitrogen, sodium chloride, alcohol, reducing sugar, lactic acid, glutamic acid and glucose is demanded. Wet chemistry analytical methods need much labour and time for these analyses.

In order to compensate for this problem, the Kikkoman company, a soy sauce manufacturer, developed an automatic chemical-composition analyser for soy sauce [111]. The analyser consists of an NIR spectrometer (InfraAlyzer 400 or InfraAlyzer 500), temperature controller, automatic sampler and pumps as shown in Fig. 12.7. A certain amount of soy sauce, collected by the automatic sampler, is sent to the NIR spectrometer at a constant flow rate controlled by the pump through the temperature controller, which is maintained at 20 °C. The NIR measurement is made automatically. After the measurement, the sample cell and tube are washed with cleansing liquid. It takes about 3 minutes to analyse one sample, including the washing process. Raw soy sauce in each fermentation vessel had a different chemical composition; therefore, to maintain the preferred quality of soy sauce, certain amounts of sauce from different fermentation vessels are blended in the bottling process. Components of each lot of soy sauce, such as sodium chloride (SEP: 0.09%), total nitrogen (0.01%), alcohol (0.12%), lactic acid (0.12%) and glutamic acid (0.05%), are analysed automatically by NIR spectroscopy. In 1993, the NIR method was adopted as the JAS (Japanese Agricultural Standard) Quality Inspection method in Japan for screening soy sauce samples. In the assay

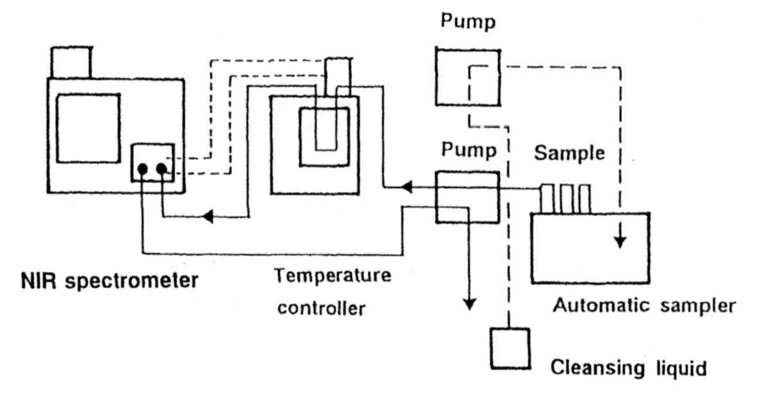

Fig. 12.7 Automatic chemical composition analyser of soy sauce

of soy sauce, high accuracy could be provided in determining sodium chloride in the same manner as in the NIR analysis of ham. The first wavelength of 1445 nm in the calibration for sodium chloride was assigned to the water band.

12.8 References

[1] USDA Near-Infrared Transmission (NIRT) Handbook, Federal Grain Inspection Service, USDA, **1996**.

[2] P. C. Williams, K. H. Norris, D. C. Sobering, *J. Agric. Food Chem.* **1985**, *33*, 239–244.

[3] P. C. Williams, D. C. Sobering, *J. Near Infrared Spectrosc.* **1993** *1*, 25–32.

[4] S. R. Delwiche, R. O. Pierce, O. K. Chung, B. F. W. Seabourn, *J. AOAC Int.* **1998**, *81*, 587–603.

[5] K. H. Norris, W. R. Hruschka, M. M. Bean, D. C. Slaughter, *Cereal Foods World* **1989**, *34*, 696–705.

[6] P. C. Williams, *Cereal Chem.* **1991**, *68*, 112–114.

[7] P. C. Williams, B. N. Thompson, D. Wetzel, G. W. McLay, D. Loewen, *Cereal Foods World* **1981**, *26*, 234–237.

[8] M. Iwamoto, N. Kongseree, J. Uozumi, T. Suzuki, *Nippon Shokuhin Kogyo Gakkaishi (J. Jpn. Soc. Food Sci. Tech.)* **1986**, *33*, 842–847.

[9] S. R. Delwiche, K. S. McKenzie, B. D. Webb, *Cereal Chem.* **1996**, *73*, 257–263.

[10] H. Chen, B. P. Marks, T. J. Siebenmorgen, *Cereal Chem.* **1997**, *74*, 826–831.

[11] W. S. Li, J. T. Shaw, *Cereal Chem.* **1997**, *74*, 556–560.

[12] C. P. Villareal, N. M. D. Cruz, B. O. Juliano, *Cereal Chem.* **1994**, *71*, 292–296.

[13] S. R. Delwiche, M. M. Bean, R. E. Miller, B. D. Webb, P. C. Williams, *Cereal Chem.* **1995**, *72*, 182–187.

[14] W. R. Windham, B. G. Lyon, E. T. Champagne, F. E. Barton II, B. D. Webb, A. M. McClung, K. A. Moldenhauer, S. Linscombe, K. S. McKenzie, *Cereal Chem.* **1997**, *74*, 626–632.

[15] Y. Hosaka, *Proceedings of the International Symposium on Agricultural Mechanization and International Cooperation in High Technology era*, Tokyo, Japan, **1987**, pp. 357–360.

[16] G. S. Birth, G. G. Dull, W. T. Renfroe, S. J. Kays, *J. Am. Soc. Hort. Sci.* **1985**, *110*, 297–303.

[17] G. G. Dull, G. S. Birth, D. A. Smittle, R. G. Leffler, *J. Food Sci.* **1989**, *54*, 393–395.

[18] S. Kawano, H. Watanabe, M. Iwamoto, *Proceedings. of the 2nd International NIRS Conference* (Ed. M. Iwamoto, S. Kawano), Korin, Tokyo, Japan, **1990**, pp. 343–352.
[19] S. Kawano, H. Watanabe, M. Iwamoto, *J. Jpn. Soc. Hort. Sci.* **1992**, *61*, 445–451.
[20] T. Temma, E. Ueda, K. Matsue, F. Shinoki, T. Tsushima, *Proceedings of the 6th NIR Tsukuba Meeting*, Japanese Society for Food Science and Technology, **1990**, pp. 98–102 (in Japanese).
[21] S. Kawano, T. Fujiwara, M. Iwamoto, *J. Jpn. Soc. Hort. Sci.* **1993**, *62*, 465–470.
[22] K. Miyamoto, Y. Kitano, *J. Near Infrared Spectrosc.* **1995**, *3*, 227–237.
[23] M. Tanaka, T. Kojima, *J. Agric. Food Chem.* **1996**, *44*, 2272–2277.
[24] R. K. Chen, **1998**, private communication.
[25] S. Kawano, H. Abe, M. Iwamoto, *J. Near Infrared Spectrosc.* **1995**, *3*, 211–218.
[26] G. G. Dull, G. S. Birth, R. G. Leffler, *Am. Potato J.* **1989**, *66*, 215–225.
[27] M. Horiuchi, K. Sato, N. Sato, Y. Suzuki, *Bull. Shizuoka Agr. Exp. Stn.* **1991**, *36*, 47–55 (in Japanese).
[28] G. G. Dull, R. G. Leffler, G. S. Birth, A. Zaltzman, Z. Schmilovitch, *Hort.-Sci.* **1991**, *26*, 1303–1305.
[29] T. Fujiwara, T. Honjyou, *Norugiken Kaihou* **1992**, *154*, 14–16, (in Japanese).
[30] G. G. Dull, R. G. Leffler, G. S. Birth, D. A. Smittle, *Transactions of the ASAE* **1992**, *35*, 735–737.
[31] V. Bellon, J. L. Vigneau, M. Leclercq, *Appl. Spectrosc.* **1993**, *47*, 1079–1083.
[32] T. Kojima, Y. Inoue, M. Tanaka, *Bull. Fac. Agr. Saga Univ.* **1994**, *77*, 1–10 (in Japanese).
[33] T. Onda, M. Tsuji, Y. Komiyama, *Nippon Shokuhin Kogyo Gakkaishi (J. Jpn. Soc. Food Sci. Tech.)* **1994**, *41*, 908–912.
[34] T. Onda, S. Iino, C. Otoguro, *Nippon Shokuhin Teion Hozou Gakkaishi (J. Jpn. Soc. Cold Preserv. Food)* **1995**, *21*, 139–142 (in Japanese).
[35] D. C. Slaughter, D. Barrett, M. Boersig, *J. Food Sci.* **1996**, *61*, 695–697.
[36] J. Guthrie, K. Walsh, *Aust. J. Exp. Agric.* **1997**, *37*, 253–263.
[37] V. A. McGlone, S. Kawano, *Postharvest Biol. Technol.* **1998**, *13*, 131–141.
[38] S. D. Osborne, R. Kunnemeyer, R. B. Jordan, *J. Near Infrared Spectrosc.* **1999**, *7*, 9–15.
[39] Official Methods of Analysis, (Ed. K. Helrich), AOAC, Inc. USA, **1990**, pp. 816–817.
[40] T. Sato, M. Yoshino, S. Furukawa, Y. Someya, N. Yano, J. Uozumi, M. Iwamoto, *Jpn. J. Zootech. Sci.* **1987**, *58*, 698–706.
[41] P. Robert, D. Bertrand, M. F. Devaux, R. Grappin, *Anal. Chem.* **1987**, *59*, 2187–2191.
[42] P. Robert, D. Bertrand, C. Demarquilly, *Anim. Feed Sci. Technol.* **1986**, *16*, 215–224.
[43] T. Sato, S. Kawano, M. Iwamoto, *J. Dairy Sci.* **1990**, *73*, 3408–3413.
[44] G. Vuataz, *Proceedings of the International NIR/NIT Conference* (Eds. J. Hollo, K. Kaffka, J. L. Gonczy), Akademiai Kiado, Budapest, Hungary, **1987**, pp. 303–309.
[45] R. J. Baer, J. F. Frank, M. Loewenstein, *J. Assoc. off. Anal. Chem.* **1983**, 66, 858–863.
[46] R. L. Wehling, M. M. Pierce, *J. Assoc. off. Anal. Chem.* **1989**, *72*, 56–58.
[47] M. M. Pierce, R. L. Wehling, *J. Agric. Food Chem.* **1994**, *42*, 2830–2835.
[48] J. L. Rodriguez-Otero, M. Hermida, A. Cepeda, *J. AOAC Int.* **1995**, *78*, 802–806.
[49] S. J. Lee, I. J. Jeon, L. H. Harbers, *J. Food Sci.* **1997**, *62*, 53–56.
[50] R. J. Baer, J. F. Frank, M. Loewenstein, G. S. Birth, *J. Food Sci.* **1983**, *48*, 959–989.
[51] M. Pouliot, P. Paquin, R. Martel, S. F. Gauthier, Y. Pouliot, *J. Food Sci.* **1997**, *62*, 475–479.
[52] I. Ben-Gera, K. H. Norris, *J. Food Sci.* **1968**, *33*, 64–67.
[53] D. R. Massie, *Quality Detection in Foods*, American Society of Agricultural Engineers, **1976**, pp. 24–26.
[54] T. Isaksson, C. E. Miller, T. Naes, *Appl. Spectrosc.* **1992**, *46*, 1685–1694.
[55] T. H. Begley, E. Lanza, K. H. Norris, W. R. Hruschka, *J. Agric. Food Chem.* **1984**, *32*, 984–987.
[56] M.R. Ellekjaer, K. I. Hildrum, T. Naes, T. Isaksson, *J. Near Infrared Spectrosc.* **1993**, *1*, 65–75.

[57] L. T. Black, A. C. Eldridge, M. E. Hockridge, W. F. Kwolek, *J. Agric. Food Chem.* **1985**, *33*, 823–826.

[58] W. Zen, H. Zhang, T. Lee, *J. Agric. Food Chem.* **1996**, *44*, 1460–1463.

[59] E. Lanza, *J. Food Sci.* **1983**, *48*, 471–474.

[60] H. J. Swatland, *J. Anim. Sci.* **1983**, *56*, 1329–1333.

[61] M. Misumoto, S. Maeda, T. Misuhashi, S. Ozawa, *J. Food Sci.* **1991**, *56*, 1493–1496.

[62] K. Thyholt, T. Isaksson, *J. Sci. Food Agric.* **1997**, *73*, 525–523.

[63] B. A. Rasco, C. E. Miller, T. L. King, *J. Agric. Food Chem.* **1991**, *39*, 67–72.

[64] H. Sollid, C. Solberg, *J. Food Sci.* **1992**, *57*, 792–793.

[65] J. P. Wold, T. Jakobsen, L. Krane, *J. Food Sci.* **1996**, *61*, 74–77.

[66] M. H. Lee, A. G. Cavinato, D. M. Mayes, B. A. Rasco, *J. Agric. Food Chem.* **1992**, *40*, 2176–2181.

[67] T. Isaksson, G. Togersen, A. Iversen, K. I. Hildrum, *J. Sci. Food Agric.* **1995**, *69*, 95–100.

[68] S. Yamaguchi, T. Sawada, *Proceedings of the 13th NIR Tsukuba Meeting*, Japanese Society for Food Science and Technology, **1997**, pp. 33–38 (in Japanese).

[69] H. Zhang, T. Lee, *J. Agric. Food Chem.* **1997**, *45*, 3515–3521.

[70] E. Niwa, T. Nakayama, I. Hamada, *Bull. Japan. Soc. Sci. Fish.* **1980**, *46*, 1147–1150.

[71] M. Iwamoto, T. Hirata, T. Suzuki, J. Uozumi, T. Ishitani, *Nippon Shokuhin Kogyo Gakkaishi (J. Jpn. Soc. Food Sci. Tech.)* **1983**, *30*, 397–403.

[72] A. G. Coventry, M. J. Hunston, *Cereal Food World* **1984**, *29*, 715–718.

[73] K. Kaffka, Z. Jeszenszky, *Hung. Sci. Instrum.* **1984**, *58*, 69–74.

[74] J. Szczodrak, Z. Czuchajowska, Y. Pomeranz, *Cereal Chem.* **1992**, *69*, 419–423.

[75] C. F. McGuire, *Cereal Chem.* **1982**, *59*, 510–511.

[76] Y. Wakai, Y. Inoue, Y. Nishikawa, J. Murata, T. Miura, *J. Brew. Soc. Jpn.* **1984**, *79*, 445–446 (in Japanese).

[77] Y. Wakai, *Jyokyo* **1992**, *87*, 492–496 (in Japanese).

[78] Y. Kojima, Y. Asai, Y. Hata, E. Ichikawa, A. Kawato, S. Imayasu, *Nippon Nogeikagaku kaishi* **1994**, *68*, 801–807, (in Japanese).

[79] I. Aramaki, K. Fukuda, T. Hashimoto, T. Ishikawa, Y. Kizaki, N. Okazaki, *Seibutsu kougaku kaishi* **1995**, *73*, 33–36, (in Japanese).

[80] E. Lanza, B. W. Li, *J. Food Sci.* **1984**, *49*, 995–998.

[81] W. Li, P. Goovaerts, M. Meurens, *J. Agric. Food Chem.* **1996**, *44*, 2252–2259.

[82] G. Alfaro, M. Meurens, G. S. Birth, *Appl. Spectrosc.* **1990**, *44*, 979–986.

[83] K. Thyholt, T. Isaksson, *J. Near Infrared Spectrosc.* **1997**, *5*, 179–193.

[84] M. Twomey, G. Downey, P. B. McNulty, *J. Sci. Food Agric.* **1995**, *67*, 77–84.

[85] D. G. Evans, C. N. G. Scotter, L. Z. Day, M. N. Hall, *J. Near Infrared Spectrosc.* **1993**, *1*, 33–44.

[86] B. G. Osborne, S. Douglas, *J. Sci. Food Agric.* **1981**, *32*, 328–332.

[87] B. G. Osborne, S. Douglas, T. Fearn, *J. Food Technol.* **1982**, *17*, 355–363.

[88] J. E. Morgan, P. C. Williams, *Cereal Chem.* **1995**, *72*, 209–212.

[89] B. G. Osborne, T. Fearn, A. R. Miller, S. Douglas, *J. Sci. Food Agric.* **1984**, *35*, 99–105.

[90] B. G. Osborne, *J. Sci. Food Agric.* **1983**, *34*, 1297–1301.

[91] B. G. Osborne, G. M. Barrett, S. P. Cauvain, T. Fearn, *J. Sci. Food Agric.* **1984**, *35*, 940–945.

[92] B. G. Osborne, T. Fearn, P. G. Randall, *J. Food Technol.* **1983**, *18*, 651–656.

[93] D. Baker, *Cereal Chem.* **1983**, *60*, 217–219.

[94] S. E. Kays, W. R. Windham, F. E. Barton II, *J. Agric. Food Chem.* **1996**, *44*, 2266–2271.

[95] S. E. Kays, F. E. Barton II, W. R. Windham, D. S. Himmelsbach, *J. Agric. Food Chem.* **1997**, *45*, 3944–3951.

[96] B. G. Osborne, *J. Sci. Food Agric.* **1984**, *35*, 106–110.

[97] C. Starr, D. B. Smith, J. A. Blackman, A. A. Gill, *Anal. Proc.* **1983**, *20*, 72–74.

[98] C. M. Gardam, *Proceedings of the International Symposium on Near Infrared Reflectance Spectroscopy* (Eds. D. Miskelly, D. P. Law, T. Clucas), Royal Australian Chemical Institute, Melbourne, **1984**, pp. 119–125.

[99] T. Sato, S. Kawano, M. Iwamoto, *J. Am. Oil Chem. Soc.* **1991**, *68*, 837–833.

[100] H. Watanabe, *Proceedings of the 11th NIR Tsukuba Meeting*, Japanese Society for Food Science and Technology, **1995**, pp. 57–61, (in Japanese).

[101] Anon. *Food Industries and Sensor* (Ed.: On-line Sensor R & D Association), Korin Publishers, Tokyo, **1991**, pp. 307–322 (in Japanese).

[102] T. Sato, *J. Am. Oil Chem. Soc.* **1994**, *71*, 293–298.

[103] I. J. Wesley, R. J. Barnes, A. E. J. McGill, *J. Am. Oil Chem. Soc.* **1995**, *72*, 289–292.

[104] R. K. Cho, *Proceedings of the 6th NIR Tsukuba Meeting*, Japanese Society for Food Science and Technology, **1990**, pp. 95–97 (in Japanese).

[105] A. M. C. Davies, T. F. Brocklehurst, *J. Sci. Food Agric.* **1986**, *37*, 310–316.

[106] A. M. C. Davies, C. Dennis, A. Grant, *J. Sci. Food Agric.* **1987**, *39*, 349–355.

[107] G. Vaccari, G. Mantovani, G. Sgualdino, *Sugar J.* **1988**, *51*, 4–8.

[108] J. Nomura, H. Makino, H. Nakajima, K. Horiuchi, *Seito Gijutsu Kenkyo Kaishi* **1990**, *38*, 13–18 (in Japanese).

[109] G. Marchetti, *Int. Sugar J.* **1990**, *92*, 210–215.

[110] R. Sekiguchi, K. Fuchigami, S. Hara, C. Tutumi, *Near Infrared Spectroscopy: The Future Waves* (Eds. A.M.C. Davies, P. Williams), NIR publication, Chichester, UK, **1996**, pp. 632–637.

[111] K. Kobayashi, K. Iizuka, T. Okada, H. Hashimoto, *Proceedings of the 2nd International NIRS Conference* (Eds. M. Iwamoto, S. Kawano), Korin, Tokyo, Japan, **1990**, pp. 178–189.

13
Applications of Near-Infrared Spectroscopy in Medical Sciences

H.M. Heise

13.1
Introduction

Near-infrared spectroscopy has found many applications, which are described in part in other chapters of this book. For chemical quality and process control, near-infrared spectroscopy has received much attention because of its speed and attribute of requiring little or no sample preparation. Fast measurement equipment and techniques have been established within the pharmaceutical industry for the identification of raw materials on receipt, as well as for verifying the composition of pharmaceutical formulations before the final products leave the premises. Within the life sciences arena, several standard methods for food analysis have been accepted by a number of different associations, even as early as in the 1980s, one example being the determination of protein and moisture in cereals. Other applications are concerned with the analysis of meat and fish, for example, for the determination of fat, protein etc., whereas convenient analyses of beer and other alcoholic beverages are carried out by transmission spectroscopy. The authentication of food and food ingredients is another field in which near-infrared spectroscopy has been successfully applied.

The relatively late emergence of near-infrared spectroscopy into the medical field is mainly due to the complexity of the samples under investigation, analytes of low concentrations and problems associated with the intense absorptions of water, usually the major constituent of biomedical samples and, in particular, of biofluids. Therefore, it was of great importance to have good reproducibility for the recording of spectra and a high signal-to-noise ratio, which could be accomplished by using Fourier-transform spectrometers. Clinical chemistry has profited from recent improvements in instrumentation, and the successful near-infrared spectroscopic analysis of human blood plasma for a number of important analytes has marked the beginning of a new era [1].

The rationale behind medical diagnosis by spectroscopy was substantiated by Mantsch and co-workers through the fact that diseases are accompanied by changes in the biochemistry of the cells and tissues that make up the different organs in our body. Since the infrared spectroscopic technique can provide information on biological molecules like proteins and peptides, chromophores and chromophoric proteins, nucleic acids, carbohydrates, lipids and others, it can, in principle, be used as a tool for medical diagnostics [2]. After having worked out the fundamental aspects, the molecular basis of the disease process can be approached by molecular spectroscopy. Although the quantification of individual compounds is often the first step in information gathering for medical diagnosis, a straightforward classification of a sample to a disease state is possible by chemometrics.

A noticeable momentum has gathered since then because fast and reagentless multicomponent assays, designed by spectroscopists and clinical chemists, are expected to be available soon. The analysis of tissues or smears of cells has also created large expectations in the field of pathology. Despite this, the first results have not yet been translated into immediate technological or commercial success. Other research activities concentrated on the development of noninvasive methodologies. For pulse oximetry, the red and infrared intensity fluctuations arising from cardiac-induced rhythmic changes in the arterial blood volume are evaluated for deoxyhaemoglobin and oxyhaemoglobin concentrations to give arterial oxygen saturation. This is an established measurement technique, but a few problems still await their solutions: for example, avoiding motion artefacts.

Another noninvasive technology uses the transmission or diffuse reflection of near-infrared radiation for the measurement of intact tissue of centimetre thickness in order to monitor, for example, cerebral haemodynamics, as well as the intracellular redox level of cytochromes. This technology resulted from the extension of noninvasive optical measurement from the visible to the near-infrared spectral range (780–1300 nm), also called the short-wave near-infrared. In 1977, the first application of in vivo monitoring of the redox behaviour of cytochrome c oxidase (or cytochrome aa_3) was published by Jöbsis [3]. A rapid development has taken place since then, because it soon became apparent that brain and muscle tissues could also be reached by near-infrared spectroscopy. The early development in this field has recently been reviewed by the same author [4].

In the following sections, near-infrared spectroscopy applied to clinical chemistry, the diagnosis of disease states by correlating tissue pathology and spectroscopy data, and the development of noninvasive techniques will be described. This review on the main streams within the various medical spectroscopy areas and the selection of papers is certainly subjective and reflects the interests of the author in the past.

13.2 Applications in Clinical Chemistry

13.2.1 Measurement Techniques and Chemometrics

Biofluids such as blood and urine play an important role in clinical chemistry. Conventionally, blood samples can be obtained by punctation or by using syringes, other samples can be taken by biopsies. For example, for self-monitoring of their blood-glucose concentration, many diabetic patients puncture their finger tips regularly to draw a drop of capillary blood, although some minimally invasive techniques have been developed recently for harvesting interstitial fluid from the dermis tissue. Microporation and suction techniques have also been proposed, which require sensitive methods for the analysis of biofluids of submicrolitre volumes.

Water is nature's biosolvent and its near-infrared spectrum must be discussed – see Fig. 13.1(a). The hydrogen bonding network of the water molecules in the liquid phase is not only sensitive to temperature changes, but also to changes in concentration of electrolytes and biopolymers, such as proteins or carbohydrates, or changes in pH. As a consequence, combination and overtone bands in the near-infrared, including the OH-stretching vibrations, are rather broad and show a large variance when samples with a wide variation in composition are studied. For such reasons, near-infrared in vitro measurements of biofluids are predominantly carried out under strict thermostating conditions. Dry film measurements

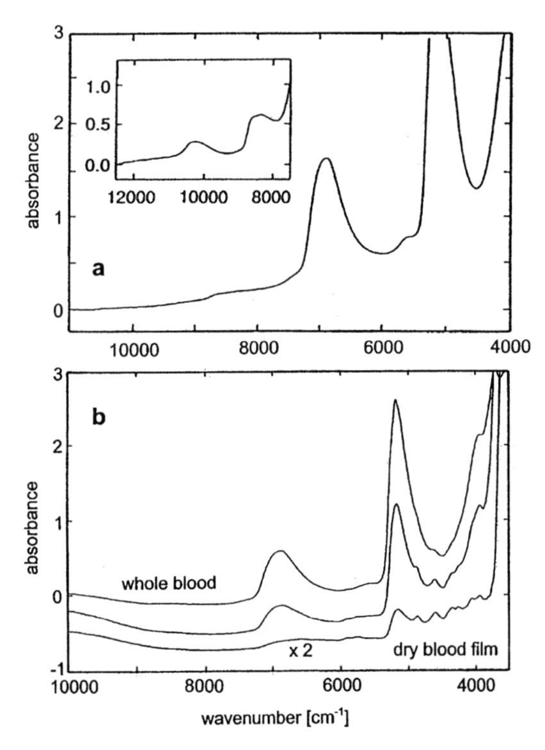

Fig. 13.1 (a) Near-infrared absorbance spectra of water measured by transmission at a temperature of 25 °C and a cell pathlength of 1 mm (inset: cell pathlength 10 mm); (b) absorbance spectra of whole blood as measured by diffuse transflection in a cell backed by a gold-coated diffuse-reflection substrate (upper trace), mid-spectrum after part water evaporation, and dry blood film spectrum (bottom trace; the last two spectra were offset for clarity and enlarged by a factor of 2)

have repeatedly been proposed to avoid complications from the water matrix and allow in addition an enrichment of the dissolved biofluid constituents – see Fig. 13.1(b).

NIR spectra of biological molecules usually comprise bands arising from overlapping absorptions, which mainly correspond to overtones and combinations of vibrational modes involving C–H, O–H and N–H molecular bonds. In principle, the concentrations of constituents such as protein, lipids, carbohydrates and others can be determined by using classical absorption spectroscopy. In the conventional spectroscopy experiment, transmission measurements are performed to analyse the radiation absorption by the sample. Nowadays, one powerful measurement technique is diffuse-reflection spectroscopy, which can be applied to scattering samples such as tissues, dried blood samples and other disperse systems. It often requires less sample preparation than is needed for a spectrum recorded in transmission. After sample penetration, the radiation is diffusely scattered and partially absorbed before a proportion emerges back to the surface, whence it is detected by using various optics and detectors. For best performance and high precision, instrumentation and sample preparation must often be adapted to meet the demanding applications in the medical field. In Fig. 13.2, the diffuse-reflection spectra of several substances of interest to the clinical chemist are shown.

Owing to the component complexity of the integrally probed biosamples, multivariate spectroscopic measurement strategies are necessary to realise clinical assays, and efficient chemometric tools are indispensable for reaching such goals. For these assays, it is nearly impossible to have quantitative information on all the components contributing to the spectrum. Fig. 13.3 illustrates some problems with the measurement of aqueous solutions, in which additional effects of the interactions of the solute with the solvent molecules are evident from the difference spectra shown (see the spectral derivative structures that arise from incomplete compensation for the water-absorption bands in the solution spectra due to changes in the hydrogen-bond network of the water matrices). Therefore, the classical least-squares approach of fitting all component spectra to that of the sample spectrum is seldom successful, although this is still the method of choice when only a limited number of chromophores contribute to the spectrum. This situation is usually found for non-invasive measurements that were performed in the short-wave near-infrared region to derive tissue oxygenation and blood parameters.

For the more complex near-infrared spectra of biosamples, the so-called soft-modelling approach with statistical calibrations works well. In such a case, however, the collection of calibration samples can be laborious and time consuming, since it is essential that the calibration data span the range of variations to be expected for all future samples to be analysed. Usually, factor methods such as partial least squares (PLS) are applied by using the information from broad spectral intervals, which may be selected by a priori knowledge about the position of absorption bands of the compound of interest or by avoiding spectral regions that, for example, show large variance due to strong water absorptions.

An example of successful PLS calibrations in clinical chemistry is the multicomponent assay on human blood-plasma samples from a population of more than

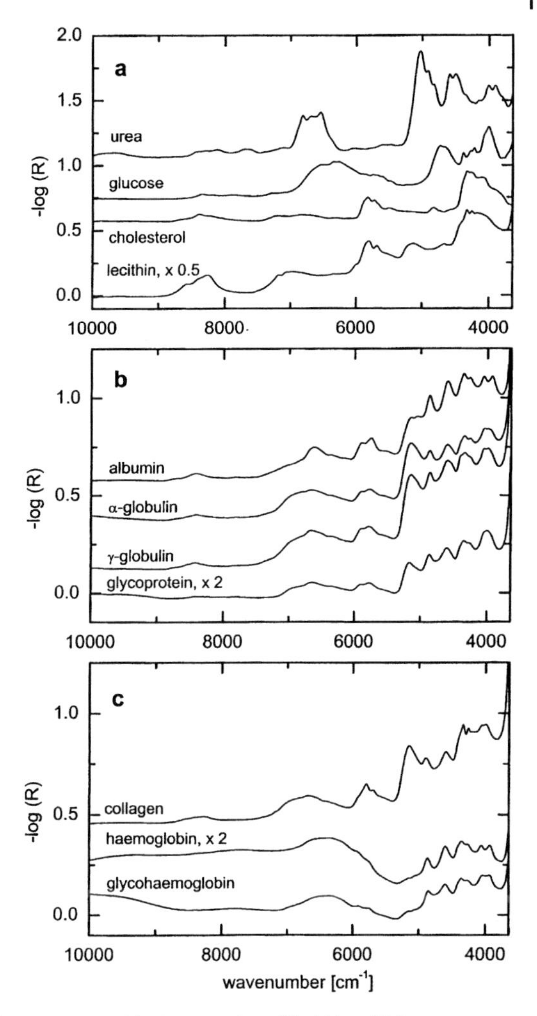

Fig. 13.2 Near-infrared diffuse-reflection spectra of several compounds important in clinical chemistry; measured from powders with the exception of lecithin, which was prepared as a smear

100 different patients by using near-infrared transmission spectroscopy [5–7]. In Fig. 13.4, results for glucose are compared, which were obtained by the conventional enzymatic hexokinase/glucose-6-phosphate dehydrogenase (G6P-DH) method, as well as by spectroscopic assays, which employed the ATR technique for mid-infrared measurements and near-infrared transmission experiments on blood plasma. One finding from our studies was that logarithmised single-beam Fourier-transform (FT) spectra could be successfully employed as calibration input data. An example near-infrared spectrum is given in Fig. 13.5(a), in which the curved baseline is evident, compared with an absorbance spectrum. The latter requires,

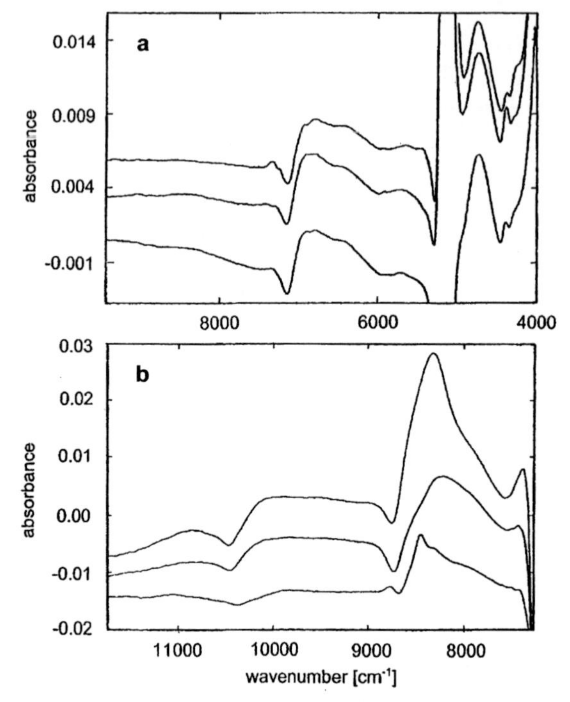

Fig. 13.3 (a) Near-infrared absorbance spectra of monosaccharides in aqueous solution (concentration of 1.5%, cell pathlength of 0.5 mm) with water absorbance compensation: glucose (upper trace), fructose (mid trace) and galactose (bottom trace); (b) absorbance spectra of an aqueous hydroxyethylstarch solution (10%, upper trace, cell pathlength of 10 mm), glucose (5%, mid trace) and ethanol (1.5%, bottom trace)

by definition, a background spectrum for its calculation (cf. Fig. 13.1). Photometric artefacts for the maximum of the water band with ordinate values above two at around 5000 cm^{-1} are evident. Absorbance spectra of two blood plasma samples, as measured against a water-filled cell, are presented in Fig. 13.5(b) and clearly show the absorption bands of dissolved proteins. For each analyte, an individual calibration with optimum spectral ranges was calculated and validated by 'leave-one-out' cross-validation; this led to a standard error of prediction (SEP) of the assay studied (for details, see Tab. 13.1).

Robust calibration models can be set up after the selection of suitable spectral variables; this has been practised in the past with multiple linear regression (MLR) methods. Identifying the spectral variables that bear useful information for use in generating efficient calibrations is still a subject of hot debate. A number of papers considering search strategies such as genetic algorithms, simulated annealing or artificial neural network approaches have been cited in papers by McShane et al. [8, 9]. Their approaches were applied to glucose and lactate calibrations and intended for monitoring cell culture media.

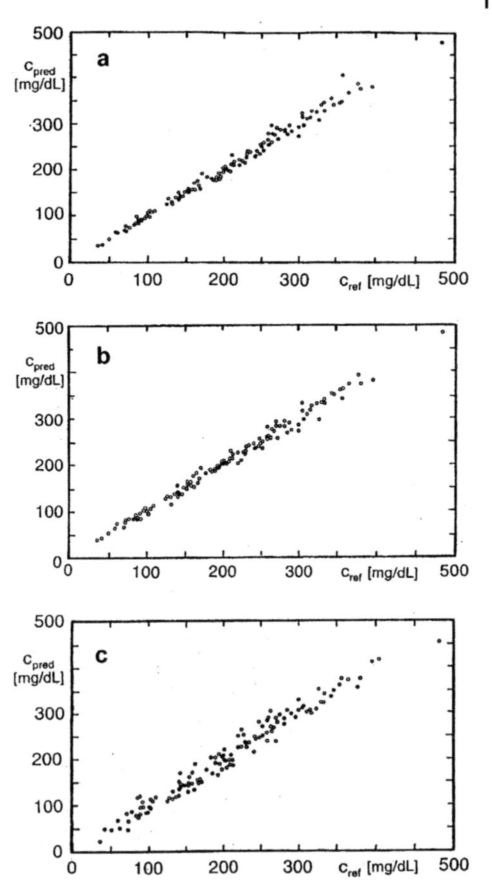

Fig. 13.4 Results from different glucose assays based on 126 blood plasma samples from different patients plotted as ordinate values [68] (all abscissa values were obtained from triplicate determinations with a hexokinase/glucose-6-phosphate dehydrogenase (G6P-DH) method): (a) Concentration values obtained by repeat G6P-DH clinical analyser measurement; (b) concentration values predicted from a PLS calibration model by using mid-infrared spectral data obtained by the attenuated total-reflection technique (1200–950 cm^{-1}); (c) concentration values obtained from a near-infrared PLS calibration model (cell pathlength of 1 mm, –log (single beam intensities), spectral data from the intervals 6800–5460 cm^{-1} and 4750–4200 cm^{-1})

A fast and efficient algorithm had been proposed and tested by us, which utilises spectral variables from pairs of minima and maxima of the optimum PLS-calibration regression vector that was calculated from broad spectral interval data. Results of this calibration strategy for several blood plasma substrates were published recently [10]. Figs 13.6 and 13.7 provide detailed results for total protein and glucose, respectively. For total protein, four wavelength variables were found to be sufficient to produce the same calibration performance (SEP = 0.98 g L^{-1}) as obtained with whole spectral data between 6000 and 5500 cm^{-1}. On the other

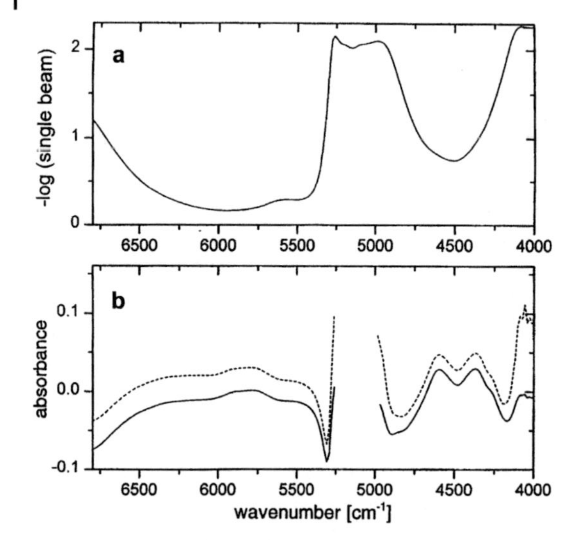

Fig. 13.5 (a) Logarithmised single-beam spectrum of blood plasma showing the windows containing useful analytical information, and (b) plasma absorbance spectra as measured against a water-filled cell for water absorbance compensation (the dashed spectrum trace was from a turbid lipaemic plasma sample) [10]

Tab. 13.1 Comparison of standard errors of prediction (SEP) obtained by leave-one-out cross-validation for different infrared spectrometric assays based on logarithmised single beam spectra of a blood plasma sample population from 126 patients

Compound	*Standard errors of prediction*	
	SW-NIR a)	*NIR b)*
Total protein	1.08 g L^{-1}	1.07 g L^{-1}
Glucose	47.3 mg dL^{-1}	16.2 mg dL^{-1}
Total cholesterol	15.4 mg dL^{-1}	7.7 mg dL^{-1}
Triglycerides	23.4 mg dL^{-1}	12.1 mg dL^{-1}
Urea		4.7 mg dL^{-1}

a) Spectral calibration interval for all substrates 11015–7621 cm^{-1}, measurements with a 10 mm transmission quartz cell [6, 7].

b) Spectral range for protein 6001–5508 cm^{-1}, for cholesterol and triglycerides 6001–5508 and 4520–4212 cm^{-1}, for glucose 6788–5461 and 4736–4212 cm^{-1}, measurements with a 1 mm quartz cell [5].

hand, 26 spectral variables were necessary to reach an optimum prediction (SEP = 16.3 mg dL^{-1}) for glucose. Using single beam spectra, which result from near-infrared FT-spectrometers after transformation of the primary sample interferogram data, was advantageous, as noise from the recording and processing of a background spectrum can be omitted.

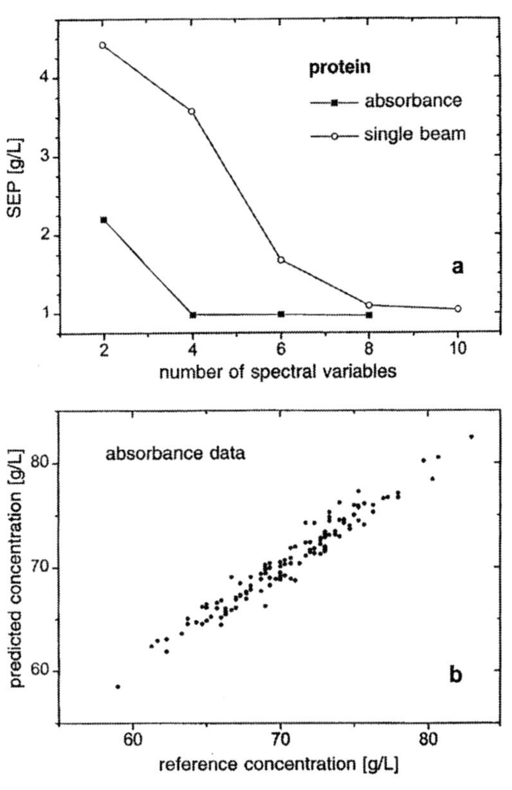

Fig. 13.6 Calibration results for total protein in human EDTA (ethylene diamine tetraacetate) blood plasma: (a) SEP-statistics for cross-validated PLS-calibration models versus a different number of spectral variables; (b) predicted concentrations versus measured reference values based on a calibration model from a four specially selected wavenumber set [10]

Nonlinear calibration-model-building techniques have been applied in several investigations. A recent publication with an interesting clinical chemistry background can be given [11]. Improvements in prediction performance for glucose concentrations were observed by means of such a strategy for aqueous mixtures that contained additional bovine serum albumin and triacetin, by using spectral data from the interval between 4700 and 4250 cm^{-1}. Calibration was either based on quadratic PLS, or stepwise, quadratic PLS regression, as well as on PLS followed by artificial neural networks (PLS-ANN).

An alternative technique, known as 'kromoscopy' as the near-infrared analogue of human colour perception, has been discussed in the context of noninvasive in vivo experiments [12]. It is based on sample transmission detection by a few detector channels with complementary broad band-pass functions, which provide a low dimensionality of the analytical measurement data. The scheme of kromoscopic analysis works for samples of low complexity that contain only a small number of components, as it was recently critically evaluated for aqueous solutions of glu-

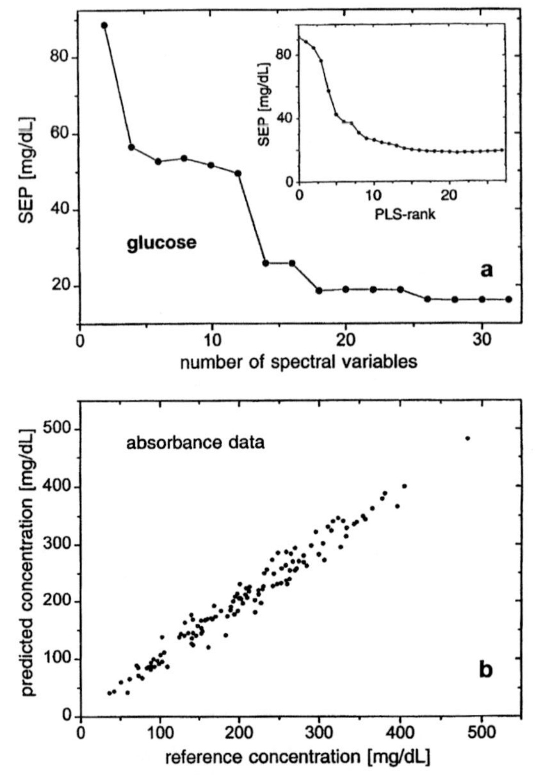

Fig. 13.7 Calibration results for glucose in human EDTA blood plasma: (a) SEP-statistics for calibration models of different numbers of spectral variables obtained from cross-validation (the inset diagram is the SEP in dependence of the number of PLS-factors for a full spectrum model); (b) predicted concentrations versus measured reference values from a calibration model based on a specially selected 26 wavenumber set [10]

cose and urea [13]. However, given the complexity of real biosamples, a more elaborate multivariate technique is necessary to fulfil the demands for appropriate selectivity.

13.2.2
Biofluid Assays

Besides blood and fluids derived from blood, other important biofluids such as urine, cerebrospinal fluid, synovial and interstitial fluid, aqueous humour and saliva etc. must be mentioned. Substrates and metabolites have been quantified in such biofluids. Total protein is often found at rather high concentrations (for example in blood serum at between 60 and 85 g L^{-1}, see also Fig. 13.6(b)), whereas other analytes such as glucose and urea are found at per mil concentrations and below.

Some of our results from multicomponent assays for blood plasma that were based on near-infrared transmission spectroscopy, have been presented above. Similar activities from the group of Arnold and Small are worth reporting. Their early investigations were often concerned with aqueous solutions of noncomplex composition and, as an extension of those, they recently reported the effect of sample complexity on the quantification of analytes in aqueous samples containing up to ten different components [14]. In a further publication, decimated time-domain filtered interferogram data, rather than spectral intensity data, were employed as calibration input [15].

Rigorous testing of the applicability of the near-infrared spectroscopic assays for the clinical laboratory must be carried out with natural specimens, which show a much greater complexity than artificially prepared solutions with a limited number of analytes. Such a study was carried out by Arnold et al. who reported on extensive calibrations for total protein, serum albumin, globulin protein, triglycerides, cholesterol and urea. In addition, concentrations of glucose and lactate were predicted by using about 240 undiluted human serum samples and their spectra between 5000 and 4000 cm^{-1} [16]. Raw and digitally filtered spectral data were used as PLS-calibration input. Except for lactate, the concentrations of which were below the detection limit, satisfactory near-infrared assays could be established (for glucose: SEP = 23.3 mg dL^{-1}). Calibration transfer to data recorded on a modified spectrometer after 19 months resulted in a slight, but constant, positive prediction bias (after off-set correction, the prediction errors for the glucose concentrations of the 50 new samples was still a factor of two larger than previously obtained: 45 mg dL^{-1}).

Hall and Pollard provided an extensive study of human sera by using the NIR approach for the clinical analysis of proteins, triglycerides and glucose, as has already been mentioned above [1]. A further study from these authors was devoted to total protein and protein fractions, for example albumin and globulins, as well as to urea [17]. Their previous NIR validation results for human sera were summarised in ref. [18]. In addition, spectra of whole blood as measured in transmission and reflection were presented. Rapid transmission measurements of whole blood were carried out by Norris and Kuenstner, who tested calibration models for several blood substrates such as total protein, albumin, cholesterol and urea [19]. The robustness of multivariate calibration models for six analytes, including glucose, based on absorbance spectra of human sera, was tested by Gatin et al. by employing two data sets recorded by two operators at different times [20]. The prediction errors obtained by cross-validation were similar to those calculated from calibration models derived from one data set and applied to the second set for validation.

An unsuccessful assay for lactate in blood serum was mentioned above. To increase the concentration range favourably within another study, plasma samples were taken from exercising test persons [21]. The normal human physiological range for lactate concentrations is given by the interval of 0.3–25 mmol L^{-1}, which was taken into account by the authors. Second derivative absorbance spectra between 2050 and 2400 nm, recorded with a thermostated transmission quartz cell, were used for PLS-calibration and a standard prediction error of 0.5 mmol L^{-1} was achieved.

A recent investigation was concerned with the determination of protein and β-lipoprotein in blood serum by using transflection measurements [22]. Further near-IR studies on lipoproteins were carried out by Demsey et al. [23], who planned the development of an in vivo assay for apolipoproteins that were immobilised in the walls of human arteries and part of the carotid plaque. Usually, the different particles are classified on the basis of their density by ultracentrifugation: low-density lipoproteins (LDL), high-density lipoproteins (HDL). Among the major lipid components, triglycerides and cholesterolesters are important markers for probing deviations from the normal metabolism of lipoproteins.

Human albumin and γ-globulin measurements have attracted clinical chemists, since their ratio (the A/G ratio) is usually employed as an indicator for various infections or nutritive conditions. Spectra of these two proteins in phosphate buffer solutions at physiological concentrations were recorded over the spectral range of 1300–1850 nm [24], and the corresponding SEP values were found to be 1.5 g L^{-1} and 1.2 g L^{-1} for albumin and γ-globulin, respectively. Another study was carried out by the same group, but with a control serum solution as matrix for both proteins with similar results [25].

A clinically important parameter is haemoglobin. In order to determine its concentration, visible and near-IR transmission spectra of unlysed blood samples were recorded in an open cell under a vertical path of the probing IR radiation. The haemoglobin could be quantified with a standard error of prediction of 0.43 g dL^{-1} by using spectral second-derivative data at 1740 and 1346 nm [26]. Since the noninvasive measurement of haemoglobin is certainly advantageous, spectral data of different haemoglobin species, including oxy-, deoxy-, carboxy- and methaemoglobin, were recorded within the visible and the near-IR regions from 620 to 2500 nm by Kuenster and Norris [27]. An extension of that work was carried out to test calibrations based only on three wavelengths chosen for a modified pulse oxymeter that were to be employed for future use in noninvasive haemoglobinometry [28]. An extensive study was also carried out by Yoon et al., who tested spectra of whole blood samples from 165 patients [29]. For calibration, whole spectra between 500–900 nm were evaluated, as well as special three wavelength schemes that showed a similar assay performance to that obtained with broad spectrum data (521 nm, 615 nm and 894 nm; SEP = 0.35 g dL^{-1}). The optical properties, including scattering characteristics, of circulating human blood were recently investigated for developing noninvasive methodology, and results within the spectral range between 400 and 2500 nm were reported [30].

A different near-infrared assay was developed to monitor changes in human blood during storage by reflectance spectroscopy through the original plastic bags as used for blood transfusion in the cases of severe blood losses in accidents or in surgery. Spectra were recorded for different blood groups, but global calibrations could be achieved for predicting storage time by using first and second derivative data between 1000 and 1600 nm [31].

Laboratory examination of biofluids other than blood is frequently carried out. To test for renal and metabolic diseases, urine is used, which is easily accessible, available in large quantities and often collected over 24 hours to get representative results. In addition, it can provide diagnostic clues on infections of the urinary tract. An early near-infrared investigation of glucose in urine exists [32], which is important because this analyte has been an index component for metabolic disorders for a long time, in particular for diabetes. For glucose, 19 different wavelengths between 1440 and 2350 nm were selected to establish a calibration model. However, the applicability of the model was limited, because the glucose concentrations were rather large, namely between 900 mg dL^{-1} and 6.3 g dL^{-1}.

In a different study, a multicomponent assay was presented for urea, creatinine and protein based on 123 different urine samples for calibration, and a further 50 samples for validation [33]. The average prediction errors were 0.10 g dL^{-1} (16.6 mmol L^{-1}) for urea, 8.9 mg dL^{-1} (0.79 mmol L^{-1}) for creatinine and 0.23 g L^{-1} for protein. An alternative calibration model was calculated by using urine samples with protein concentrations lower than 1 g L^{-1}, which are the most frequently found. The resulting standard error of prediction was reduced by a factor of two compared with the global calibration model. Additionally, precision estimates for near-IR predicted concentrations were presented for between-day and within-day measurements, and the effect on concentration prediction from spectroscopic noise was also estimated.

Dry biofluid films have already been mentioned. Such methodology is suited for automation, as it avoids the need for the cuvette always required for the measurement of fluids. Different carrier substrates, which can be considered as a reagent-free analogue to a test strip with dry chemistry reagents, have been proposed: namely transparent materials such as glass or gold coated, diffusely reflecting substrates (see also Fig. 13.1). An early interesting study was carried out with dried human serum samples on glass-fibre filters [34]. The performance of the urea assay based on diffuse reflectance was comparable to the analysis of unmodified serum samples. However, improvements in dry-film preparation should lead to even better assay performance than by using aqueous biofluids because of an improved signal-to-noise ratio for the water-free sample.

Multivariate studies on arthritis diagnosis have been carried out based on the near-IR spectra of 109 synovial-fluid samples. Although the spectra were available for the range between 400–2500 nm, optimal quantitative chemical information was extracted from the wavelength range between 2000 and 2400 nm; this allowed distinctions to be made between specimens from patients suffering from rheumatoid arthritis, osteoarthritis, or spondyloarthropathy. Classification was made by linear discriminant analysis (LDA) [35]. The accuracy of this approach was surprising, because conventional wisdom denies that synovial fluid composition is useful for arthritis diagnosis. The protein level change, as manifested between inflammatory and noninflammatory diseases, seems to play a role in the correct classification. There were interesting parallels between mid-IR studies of dry films by using the CH-stretching modes within the interval of 3000–2850 cm^{-1} only [36] and the range in the near-IR examined for the fluid spectra, where combination bands involve the same molecular moieties, which give rise to the absorption bands below 3000 cm^{-1}. The assumption was supported by the fact that a similar information content was available in both spectral sections. Further investigations relied upon 13 specific spectral subregions selected between 3500 and 2800 cm^{-1}, and classifications were in excellent agreement with clinical diagnosis (96.5% correct results). MIR spectroscopy was stated to be a suitable tool for an objective differential diagnosis of arthritis by using an adequately trained classifier. This methodology is especially attractive, since only fluid volumes of 20 μL needed to be aspirated, and even more importantly, a diagnostic result was available within minutes.

For ante-natal diagnostics, amniotic fluid is important. It is the liquid medium that bathes the foetus throughout its gestation. By guidance through ultrasound imaging, amniocentesis, that is the collection of this fluid, is now a wide-spread and safe obstetric procedure. Respiratory distress syndrome (RDS) is a common cause of death in preterm delivered babies and caused by insufficient surfactant production within the newborn's lungs, which is essential for proper functioning. To test foetal pulmonary maturity, three phospholipids (lecithin (L), sphingomyelin (S) and phosphatidylglycerol (P)) are routinely determined. The aim of the study by Lin et al. [37] was to correlate near-IR spectra of amniotic fluid with data obtained from thin layer chromatography (TLC); in particular the ratio of the first two compounds mentioned above (L/S) plays an important role as a diagnostic indicator. The spectral range between 2000 and 2500 nm of the second-derivative spectra of the centrifuged fluids was used for a PLS calibration. The results from TLC correlated satisfactorily with the L/S ratios obtained from the PLS calibration (correlation coefficient $r=0.91$).

In a further paper published by the same group, different biofluids (serum, amniotic and synovial fluids) were prepared as dry films on inexpensive glass substrates, which limited the accessible spectral range to wavenumbers above 2400 cm^{-1}. The most prominent absorption bands corresponded to the CH and NH groups of the constituents, which were clearly observable due to the drying process and well suited for calibration input. The results for triglycerides, by using just the spectral range of 2800 to 3000 cm^{-1} for a PLS calibration of undiluted serum films without compensation for variability in film thickness, were promising as a standard error of prediction of 0.22 mmol L^{-1} (19 mg dL^{-1}) was yielded [38]. The same strategy can be used for a near-infrared spectroscopic analysis.

13.3 Near-Infrared Spectroscopy of Tissues

13.3.1 General Pathology Studies

Near-infrared spectroscopy enables us to determine characteristic molecular changes in the composition and structure of tissues. This chemical probing can also be used for the study of diseased tissue. Classical pathology embraces several fields such as cytology, histology, haematology and microbiology. These are fields where investigations are currently carried out for testing novel spectroscopic tools.

Spectroscopy can be used to reveal malignant features in various tissues, whereas further projects look, for example, for neurological disorders that affect brain tissues in different ways. In most cases, samples come from biopsies or post-mortem investigations.

It is certainly advantageous that infrared spectroscopic analysis does not require any extrinsic labelling or staining of the biomaterial; however, interpretation of the spectra is sometimes complicated given just the assignments by conventional pathological judgement. In recent years, "infrared pathology" has been a primary domain of mid-infrared microscopy, based either on scanning devices or on focal-plane detectors for fast imaging. In such cases, a spectral range is exploited that contains the so-called fingerprint region, which has an enormous information content on molecular substructures. An overview on current activities on microscopy and imaging of tissues can be found, for example, in a recent review and book [39, 40].

It is evident from the type and number of vibrational bands that the information content within the near-infrared range with respect to component identification is reduced compared with the opportunities which are rendered by the mid-infrared. Despite this, several near-infrared spectroscopic applications for tissue characterisation and medical diagnosis have been published so far. Certainly, the possibility for exploitable, noninvasive opportunities or the ease of sample handling have been arguments for evaluating such methodology.

An early near-infrared spectroscopic investigation into the protein and lipid composition of whole gerbil brains was carried out to analyse ageing and stroke-related effects. In addition, noninvasively recorded diffuse-reflection spectra between 1.1 and 2.5 μm were evaluated to reveal age effects and the amount of brain-tissue oedema by the changes in protein and lipid composition. The aim was to support therapies for reducing ischaemic and post-ischaemic damage to the brain [41].

Other investigations were devoted to the development of a near-infrared imaging system, by using a fibre-optic probe to produce chemical maps, with a special focus on lipoprotein and apolipoprotein composition of the surfaces of living arteries [42]. Further developments on the in vivo analysis of human carotid plaque were reported by Lodder and co-workers, who obtained near-infrared images of blood vessels at specific wavelengths by employing an InSb focal plane detector array and tuneable radiation sources [23, 43].

In the context of coronary heart disease, diets with low saturated fat and cholesterol are important. A quantitative near-IR technique was developed by Windham and Morrison [44] for the prediction of fatty-acid content in bovine muscle tissue. Homogenised meat samples were scanned with a near-IR monochromator spectrometer, and spectral data between 2000 and 2470 nm were used for PLS calibration. Satisfactory calibration models were obtained for a few individual totally saturated and unsaturated fatty acids.

The chemical changes in the composition of cardiac rat tissue after myocardial infarction were followed by near- and mid-infrared spectroscopy [2, 45]. One conclusion was that the major spectral changes in infarcted ventricular tissue could be attributed to an increased deposition of collagen, which must be seen as part of the extracellular matrix secreted by the tissue cells and which had already become apparent a week after the ischaemic event in the heart. The near-infrared spectroscopic analysis was carried out by using second-derivative spectra in the combination region between 3900 and 5100 cm^{-1}. The results were consistent with the findings from the parallel mid-infrared studies.

Carcinomatous tissue and normal surrounding fibro-glandular tissue from several breast-cancer cases were studied by diffuse reflection spectroscopy by Wallon et al. [46]. Four specific wavelength intervals between 1200 and 2400 nm were found that provided the basis for tissue discrimination. The diagnostic predictions from the near-infrared method coincided with the histology diagnosis. A continuation of their work was presented recently [47].

A rapid method for the analysis of Pap smears, which are used for diagnosis of cervical cancer, was developed by Ge et al. [48]. Normal, atypical and malignant cells were measured by near-infrared spectroscopy between 10000 and 4000 cm^{-1}. Original and second-derivative spectra were subjected to principal-component analysis and discriminant analysis with Mahalanobis distances for the classification of the spectra. The results were consistent with those obtained from the established pathological methods.

A variety of chemometric tools was recently applied to species identification of raw homogenised meat from chicken, turkey, pork, beef and lamb on the basis of their reflectance spectra, as measured in the mid- and near-infrared and in the visible ranges, which could also be employed for medical diagnosis. The algorithms tested were factorial discriminant analysis (FDA), soft independent modelling of class analogy (SIMCA), K-nearest neighbour (KNN) analysis and discriminant PLS regression [49]. While the lean-meat composition of beef and lamb was relatively constant, a major source of variation could be related to the content of lipids. In Fig. 13.8, the diffuse-reflection spectra obtained for lean bovine muscle tissue and fat from fatty tissue are shown. Similar component spectra have been

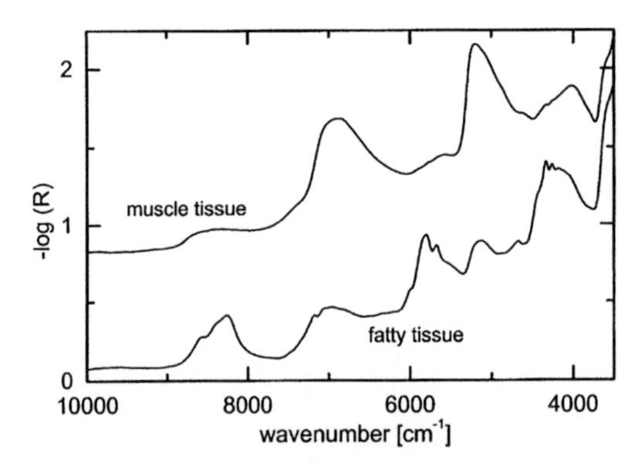

Fig. 13.8 Comparison of diffuse-reflection near-infrared spectra of bovine lean muscle and fatty tissue measured by using a diffuse-reflection accessory based on an on-axis ellipsoidal mirror [67]

observed in transcutaneous in vivo measurements of human tissues (see below). The spectral features of the fat material are similar to the spectrum of lecithin (phosphatidylcholine, see Fig. 13.2(a)), which is a phospholipid. Another lipid class is the triglycerides, for which the overtone and combination bands involving alkyl-chain vibrational modes of the acyl groups similarly dominate the near-infrared spectra.

13.3.2 Near-Infrared Spectroscopic Analyses of Skin

One body tissue which is accessible for various noninvasive spectroscopic analyses is skin. It is an extremely important organ, since it renders many vital services to the body. The most obvious one is to provide mechanical and chemical protection, which is possible through its layered structure and the chemical composition seen within these layers. Among other tasks is heat regulation through sweat production and an enormous variation in blood flow, which is evident from the fact that large blood volume changes can even be seen. The probing of blood has led to several approaches for the development of noninvasive assays.

One is seldom aware of the enormous heterogeneity of living skin, but a thorough review of its anatomy is far beyond the scope of this chapter. The uppermost layer, called the epidermis, possesses multilayered substructures. The stratum corneum is the skin material at the surface, consisting of keratin and several lipids. Its thickness can vary tremendously, depending on the body site, although for mucous tissue such as oral mucosa, a stratum corneum layer is missing. The next main layer, the dermis, is mainly connective tissue consisting of collagen, but also containing important elements such as sweat glands, blood vessels, lymphatics, nerves and others. The third layer, below the dermis, is the subcutis, mainly consisting of fatty tissue.

Apart from the structural variability of skin, its principal optical characteristics in the visible and short-wave near-infrared spectral ranges are dependent on factors such as water content, changes in vasodilation and blood volume, the concentration of haemoglobin in the blood, its oxygenation status and the extent of colouration from melanin and other pigments. Fig. 13.9(a) provides some absorptivity data on such chromophores, which is needed when assessing tissue haemodynamics, that is, dynamic changes in the concentration of oxy- and deoxyhaemoglobin or in the total blood volume. In Fig. 13.9(b), some examples of diffuse reflection spectra of skin from different body locations are shown, which were recorded with integrating sphere optics. The most pronounced effect is from differences in concentrations of total haemoglobin, which is found here, particularly in the dominating oxygenated form (the small spikes in the spectra were from measurement artefacts produced by faulty detector elements within the detector array).

Gentle diagnostics, for example for an in vivo differentiation between malignant and benign tissues, can be accomplished with diffuse-reflection spectroscopy by using fibre probes. Spectroscopy in the visible spectral range up to the short-wave near-infrared coupled with endoscopic techniques was employed to study different

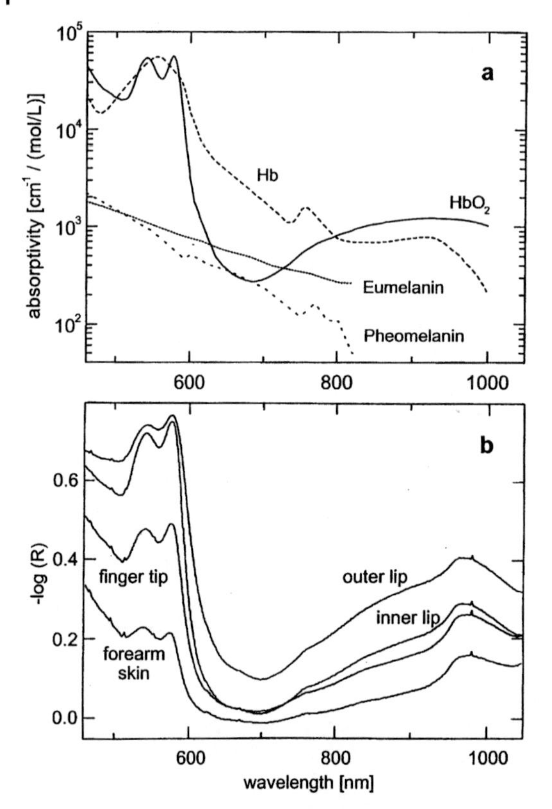

Fig. 13.9 (a) Visible/near-infrared spectral absorptivity data for haemoglobin, its oxygenated form and two melanin variants found as skin pigments (haemoglobin after Prahl and melanin after Jacques (Oregon Medical Laser Center; URL: http://omlc.ogi.edu); (b) diffuse-reflectance spectra of human skin measured by using an integrating sphere: outer lip, inner lip, finger tip and forearm skin); spectra were offset for clarity

colonic tissues [50]. Apart from non-neoplastic tissue samples consisting of normal mucosa and hyperplastic polyps, the neoplastic samples were identified as adenomatous polyps and adenocarcinomas. Several pattern recognition algorithms were used to discriminate between the two tissue categories, but the major spectral differences observed between the tissues originated from changes in blood volume, haemoglobin oxygenation status, blood vessel distribution and tissue scattering.

The near-infrared diffuse-reflection spectrum of normal skin – some example spectra [51] are displayed in Fig. 13.10 – is dominated to a very large part by water (cf. Fig. 13.1). This is also most often the main component in other soft body tissues. Scattering can also be observed: for example, at shorter wavelengths around $10\,000\ \mathrm{cm}^{-1}$, the slope of skin spectra is different from that of the water spectrum. The scattering is mainly due to discontinuities in the refractive index on the subcellular level. For further elucidation of the relevant mechanisms of light scatter-

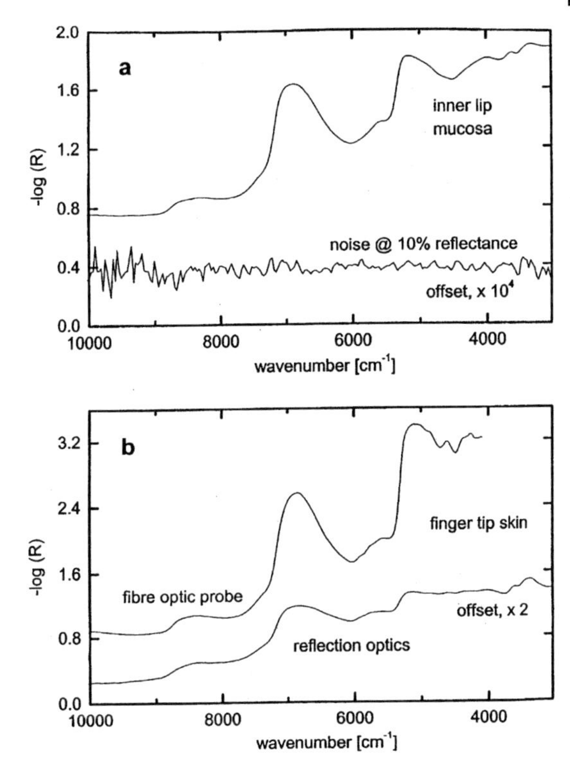

Fig. 13.10 (a) Diffuse-reflection spectrum of inner-lip mucosa as measured by a diffuse-reflection accessory based on ellipsoidal mirror optics versus white Spectralon, and –log (*R*) noise level estimated from two subsequent measurements by using a grey Spectralon standard of 10% reflectance. (b) Diffuse-reflection spectra of the finger tip skin as measured with a bifurcated fibre-optic probe and with the accessory as used for the spectra shown in the upper diagram [51]

ing from biological cells, investigations were recently presented by Mourant et al. [52, 53]. Furthermore, the so-called anisotropy parameter is necessary to describe the nonisotropic scattering in tissue, which is usually highly forward directed. The scattering is not only a requisite for reflection measurements, but also for optical coherence tomography (OCT), which was developed to provide an in vivo optical probe for structural analysis, see for example [54]. Applications within dermatology, with radiation of 1300 nm wavelength and a bandwidth of 50 nm, rendered a much improved spatial resolution (maximum scanning depth 1 mm, respective axial and lateral resolution about 10 μm) compared with high frequency sonography [55]. Besides probing tissue morphology, OCT can be used for the retrieval of important optical parameters, such as refractive indices and scattering coefficients [56].

Coming back to skin reflection spectroscopy; different accessory types are often involved in such measurements. Besides the integrating sphere mentioned above, reflection optics based on segmented ellipsoidal mirrors or fibre optics can also be used. For the latter probes, concentric arrangements are found with multiple fibres positioned for tissue illumination and central fibres used for the collection of back-scattered radiation. Other set-ups exploit fixed source-detector distances. For imaging purposes, even larger arrays of such multiple source-detector probes have been employed.

Our group has been engaged for some time in probing biochemical aspects of tissue. In particular, the development of a noninvasive blood-glucose assay by near-infrared diffuse-reflection spectroscopy of oral mucosa has been one of our goals. This needs extremely high spectral signal-to-noise ratios, see also Fig. 13.10(a) [57]. The two reflectance spectra of finger-tip skin, as shown in Fig. 13.10(b), were measured with a quartz-fibre- and a reflection-optics-based accessory. The different intensities clearly illustrate the influence of the geometry of the radiation-collecting optics, that is, the solid angle for the detected photons. There is photon scattering by the stratum corneum of the skin, which has a dramatic impact on spectral intensities when spectra are recorded with reflection optics (cf. the spectrum of oral mucosa in Fig. 13.10(a)). On the other hand, the inner lip was coupled optically in a perfect way to the immersion lens of the diffuse reflection accessory, by which a refractive index matching between the two media was also accomplished, thus avoiding Fresnel reflection from the skin surface.

Fibre-optic probes are the most convenient for in vivo skin measurements, due to their flexibility for remote sensing and ease of handling. For fibre-optical arrangements with a fixed source-detector probe distance, the influence of the separation gap and of the scattering coefficients of the matrix was studied in the spectral region of 800 to 1700 nm by Schmitt and co-workers [58]. The water content of a tissue sample influences its bulk refractive index, which in turn affects the wavelength-dependent tissue scattering coefficients. These investigations are still important for an improved quantitative interpretation and the processing of diffuse-reflection spectra of biological tissues such as skin. They allow us to answer the question of whether a modified Lambert-Beer law is applicable for quantitative analysis. In addition, the variable offsets and the baseline tilt of $\log(1/R)$ spectra of skin could be explained. A pathlength multiplication factor was also previously derived from diffusion theory for photon transport in a scattering medium [59].

In Fig. 13.11, several near-infrared measurements on different skin tissues are shown, for which a fibre-optic probe with a random arrangement of illuminating and detecting fibres was employed (numerical aperture of 0.22). Compared with a second probe based on an on-axis rotational ellipsoid, a larger average penetration depth for the photons subsequently detected from the tissue was realised by using the bifurcated quartz fibre-optic probe. This was confirmed by calculation of the transmission-equivalent water-band absorbencies, as well as by Monte Carlo simulations of the radiation transport within the tissue [60].

A few clues can be obtained from the spectral features below 6000 cm^{-1} in Fig. 13.11, which provide information on the probing depth and the layered skin structure. For the lip-mucosa spectrum a contribution from the subcutaneous fatty tissue is evident, which is missing in the recorded spectrum of the outer lip.

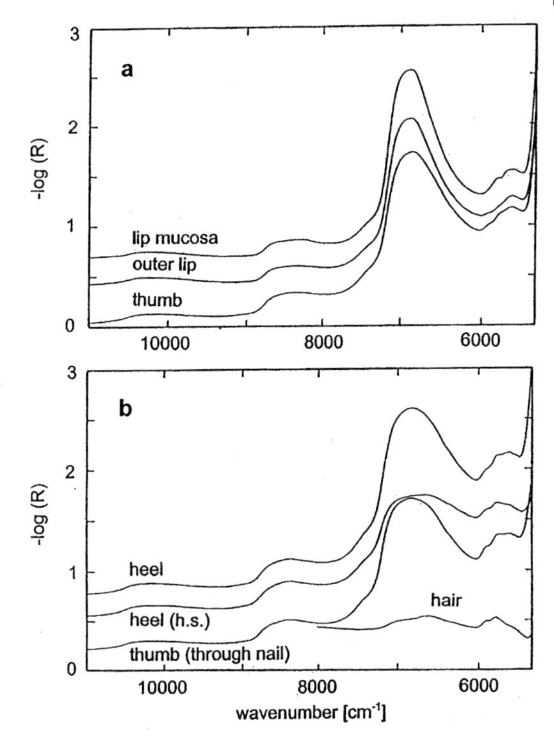

Fig. 13.11 Diffuse-reflection spectra of various human skin tissues and hair measured by a fibre-optic probe; the heel skin samples had thick layers of stratum corneum, and the lower spectrum was obtained from highly scattering (h.s.) material [60]

The other spectra show significant contributions from the stratum corneum, which are most noticeable below 6000 cm^{-1} (for comparison, the spectrum of a hair sample is also given in Fig. 13.11(b)). A topic which is of interest in cosmetics or dermatology is the water content within the stratum corneum. This was analysed by Martin for living skin samples by diffuse-reflection spectroscopy [61]. It was found that different hydrogen bonding environments existed for water associated with the lipid and keratin components of the outermost skin layer, apart from the bulk water found in deeper tissue.

The chemistry of nails and hair is similar, as both consist mainly of keratin. In this context, near- and mid-infrared spectroscopic studies by Mantsch and coworkers on the structure and chemistry of human nails must be mentioned [62]. Complementing their reflection measurements, the attenuated total-reflection technique and photo-acoustic spectroscopy were also employed for depth-dependent probing.

Diffuse-reflection spectroscopy can provide information useful for noninvasive skin diagnosis. A special field is the discrimination between pigmented lesions, melanoma and naevi (moles). Malignant melanoma is the most dangerous form of skin cancer, but it is often difficult to diagnose because of its similarities to a naevus, so biopsies taken by excision are still needed. Fredericks and co-workers recently compared results from mid-infrared ATR-fibre measurements with those from diffuse-reflection spectroscopy by using a probe head constructed from silica fibres [63]. Near-infrared spectra between 10000 and 4000 cm^{-1} were recorded from the normal skin and moles of 51 volunteer subjects. Discriminant analysis on all near-infrared spectra measured, based on PLS with mean-centred, second derivative data, showed only little separation of skin and mole spectra, when the analysis was based on data recorded over several days. A much better discrimination was obtained with spectra collected within a single working day, but further studies are needed.

Wallace and co-workers have also recently been involved in the study of pigmented lesions, but they employed the wavelength range between 320 and 1100 nm [64]. Their activities on modelling the diffuse-reflection spectra of various skin samples are very interesting, although initial results were only presented for the red wavelength region [65]. A histological assessment was carried out on a selection of sections of excised lesions. Some features were evaluated visually and scored on an arbitrary scale: for example, the degree of hyperkeratosis (which leads to a thickening of the stratum corneum), the amount of melanin, the size of melanocytes and others. A theoretical multilayer skin model was set up by using optical parameters from the literature. Reflectance spectra were calculated for normal skin and naevi by means of Monte Carlo simulations of the radiation transport within the scattering medium. Based on the histological assessment, the optical properties of melanoma were modified for improved agreement between simulation results and experimental spectra.

Our group has intensively studied near-infrared skin spectroscopy and its reproducibility. Some example effects, obtained with and without repositioning of the measuring probes, were presented for spectral data recorded from several individual people, as well as from a multi-person experiment [60]. The spectral variances observed under repositioning schemes were mainly due to water-concentration changes and temperature variations within the aqueous compartment probed. A more detailed principal-component analysis, reflecting the complexity of high-quality skin spectra, was recently published by us [66, 67]. It was based on single-person data obtained from the inner-lip mucosa. Several experiments, especially for testing a noninvasive blood-glucose assay, were carried out on single-person spectra recorded during a two-day trial and during a measurement campaign lasting two weeks. Only the blood-glucose level was changed deliberately, on other parameters, such as blood composition and physiological tissue changes, we had no directing influence, apart from sample thermostating. In addition, data sets were available that consisted of lip spectra recorded continuously with a time resolution of two spectra per second. Due to the smaller number of co-added interferograms compared with the other static measurements lasting each for one minute (9 versus 1200), the signal-to-noise ratio was decreased for the fast measurements by an order of magnitude.

Fig. 13.12 provides the first ten factor spectra from the principal component analysis of the logarithmised inner-lip spectra that were recorded during two-week

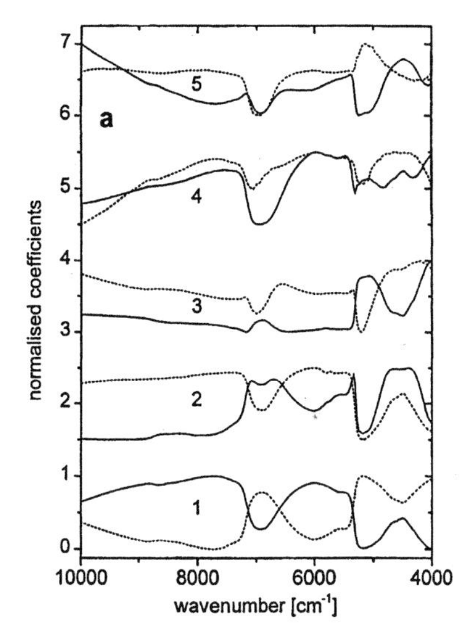

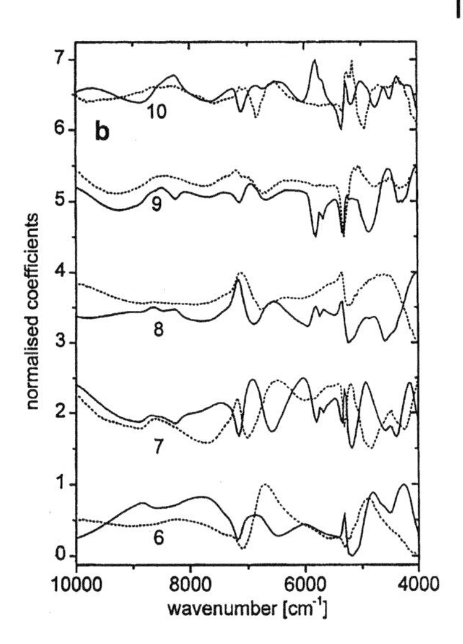

Fig. 13.12 Factor spectra from a principal-component analysis of the logarithmised inner lip spectra of a single person recorded during a 2-week period (solid lines) and those of a different test person obtained during a one-minute measurement with two spectra recorded per second (dashed lines); the first ten factor spectra (in the order of the decreasing respective singular values) are shown, each of which was min-max normalised and offset for better comparison [67]

and two-minute testing periods. The vectors were minimum-maximum normalised for better presentation. It must be noted that the first data set mentioned was recorded with repositioning of the diffuse-reflection probe, so that slight changes in tissue composition due to local heterogeneity within the mucosa tissue, blood composition changes etc. should also have an impact on the individual lip spectra recorded. It is interesting to see that all-factor spectra of both data sets, although quite different in time scale, are dominated by effects that can be assigned to water. There are different hydrogen-bonding effects when water is bound to different components or exists at different temperatures; it is known that a depth-dependent temperature gradient exists within the skin [68]. When plotting the logarithm of the singular values from the principal component analysis versus their indices, we see a definite slope change after nine principal components and again after a second package of the same size, before the singular values can be attributed to random noise within the spectra recorded at 0.5 s. The low-noise spectra obtained from the one-minute measurements provided about 30 independent factors, before random noise was definitely modelled by the following factors [66].

13.3.3 Noninvasive Metabolite Monitoring

Blood analysis, as described in the previous subchapter on near-infrared applications in clinical chemistry, is an important cornerstone in medical diagnostics. Usually, blood sampling and testing is a rather infrequent event; however, there are several cases where regular, or even continuous, monitoring of physiologically important parameters such as glucose is necessary. Such monitoring is essential in intensive care units or operating theatres. Another situation is that faced by diabetic patients, because self-monitoring of blood glucose is part of their daily routine to regulate their glucose level through insulin injections. Indeed, insulin therapy is life-saving for patients with type I or late-stage type II diabetes. However, the discomfort and inconvenience of the invasive techniques have led many diabetics to neglect blood-glucose testing; this puts them at risk of serious health problems, finally culminating in kidney failure, stroke or blindness, owing to the failure to normalise blood-glucose concentrations. Generally, a high blood-glucose level will lead to an increase in protein glycosylation and later to biochemically unfavourable advanced glycation end products.

To escape from the current practice of blood sampling by finger pricking, noninvasive measurement technology is needed. Recent advances in optical glucose sensing include optical absorption spectroscopy, polarimetry, Raman spectroscopy and glucose sensing by fluorescence spectroscopy. Details on recent spectroscopic approaches can be found in several reviews [67–70]. As shown for in vitro assays, there is potential in using near-IR spectroscopic methods that exploit specific optical glucose parameters, whether wavelength-dependent absorptivities or refractive indexes. Due to the complexity of the integrally probed tissue with the presence of many interfering compounds, only multivariate spectrum measurement and evaluation can be successful. Further complications have their origin in the heterogeneous distribution of glucose in intravascular, interstitial and intracellular space, which are connected by active and passive transport processes, whereas the measurement of capillary blood glucose is still considered to provide the "gold standard" for an intensified insulin therapy. Many factors have to be considered for an optimisation of an in vivo assay, such as tissue structure, blood volume, blood flow and physiological variations with respect to a glucose assay based on skin measurements, see [67, 71].

Worldwide, there are many activities towards the development of noninvasive methodologies by exploiting the near-infrared spectral range. Several different strategies exist by which this challenging task can be approached. For the short-wave near-infrared spectral range, it is possible to model the spectroscopic effects of several physiologically important factors and different analytes within certain measurement scenarios, by knowing their influence on the absorption and scattering properties of the tissue [72]. There are four different effects of increasing the glucose concentration: the tissue absorption coefficient is additively altered by an increase in intrinsic glucose absorption and a decrease in water absorption through its displacement by glucose (the latter is certainly not selective enough for glucose monitoring). The other observable optical effects are a change in the refractive index, and consequently also in the wavelength-dependent scattering coefficient.

The last effect was utilised for the development of portable instrumentation consisting of four LEDs of different wavelengths in the red and short-wave near-infrared spectral range and six differently located photodetectors. The tissue-glucose concentration was followed during several glucose clamp experiments with rapid changes in blood glucose [73]. The specificity is of greatest concern, since many other effects, not necessarily correlated to glucose, may influence the signals. Further experiments were recently carried out under the conditions of slow changes as observed during oral glucose-tolerance tests, undergone by five healthy subjects and 13 patients with type II diabetes [74]. Two sensor heads simultaneously recorded the individual traces that were obtained from evaluating the signals from scattering changes occurring in tissue. However, due to residual drifts and artefacts, correlation between the scattering properties of tissues and blood-glucose concentration was not acceptable in one quarter of the experiments.

The photo-acoustic near-IR measurement technique has also been employed, and promising results were published by MacKenzie and co-workers [75]. The photo-acoustic technique involves the conversion of optical energy into acoustic energy by nonradiative processes, in which the absorbed energy leads to localised sample heating with thermal volumetric expansion. The resulting pressure pulse is the basis of the photo-acoustic signal generation, which can be detected by a piezoelectric sensor. A paper on in vitro and in vivo experiments including tests on eight subjects was recently published by this group [76]. The need for multivariate spectral measurements became obvious from the in vitro experiments carried out in the spectral range between 800 and 1200 nm.

The majority of the research groups concerned with noninvasive glucose assays use absorption spectroscopy, combined with either transmission or diffuse-reflection measurements. Until now, the near-IR absorption approach has been shown to be the most effective method for providing opportunities for in vivo blood-glucose monitoring. For tissue sensing, the fingertip, earlobe, tongue, lip and forearm have been proposed. A publication on the evaluation of different sites suitable for transmission measurements recently appeared, which was concerned with the noise level in spectra of various tissues and their percentage of body fat [77]. Our group focused on other criteria, being particularly concerned with the blood volume inside the skin-tissue volume investigated by diffuse-reflection spectroscopy [51]; see also Fig. 13.9(b) for the haemoglobin absorption in diffuse-reflection spectra of different skin types.

The spectral range in previous investigations of noninvasive glucose assays varied with data from the short-wave near-infrared range down to a long-wave interval from 2100 to 2400 nm. An extremely important aspect is the glucose selectivity achieved with data from particular spectral intervals, which has been tested by us on a population of blood-plasma samples [5]. The other essential condition is to reach a spectral noise level that allows a quantification even at the lowest glucose concentration levels during hypoglycaemia observed in the tissue of diabetic patients. Blood-glucose levels could be decreased down to 30 mg dL^{-1}, which is the lower limit of the clinically relevant range. However, the average tissue glucose concentration can be much lower and also depends on the blood volume probed within the tissue under investigation; see also [67, 71].

There are further recent publications concerned with noninvasive assays based on near-infrared spectroscopy. Transmission measurements across the human tongue, by using first overtone spectra between 6500 and 5500 cm^{-1}, were presented by Burmeister et al. [78]. Experimental protocols were designed to minimise chance correlation with blood-glucose concentrations as pointed out earlier [79]. The testing period for each of the five diabetic subjects was 39 days, and the maximum number of measurements was six per day. The standard error of prediction for the best calibration model was 3.4 mM (61 mg dL^{-1}), but clinically acceptable results cannot be achieved at the current stage.

In other studies, the diffuse reflection measurement technique was employed. Arnold et al. [79] illustrated some pitfalls resulting from chance correlations within the field of statistical calibrations, as found for the conditions of oral glucose tolerance testing (OGTT), when continuous testing is used. Many papers published previously on noninvasive blood-glucose assays must be critically examined by applying stricter rules for model validation than purely leave-one-out cross-validation.

Our results were obtained by using cross-validation with one sample or larger packets of samples left out for independent testing [51]. In Fig. 13.13(a) the blood-glucose profiles of a diabetic person during testing periods within two subsequent days are shown. In the top diagram, the predictions were calculated by using a calibration model based on 24 specific spectral variables and leave-one-out cross-validation. Further calibration models were recently calculated from the data of a single day, and the data set of the second day was used for model validation. Results from this strategy with day-to-day testing are presented in the bottom diagram of Fig. 13.13(a). Effects of integral tissue glucose concentrations on the calibration results, when based on capillary blood-glucose profiles, are discussed in a paper to be submitted [80]. The drastic effects from overfitting are illustrated with the calibration results by using full spectral interval data (9000–5400 cm^{-1}) with 115 spectral variables to predict the sample index number. Whereas the leave-one-out strategy leads to a perfect fit, the model building with data from a single day and the validation of this model, by using the data of another day, show the limitations of the calibration models calculated.

Our test demonstrated that any fictitious concentration profile can often be fitted with good precision by using a large number of spectral variables under the leave-one-out cross-validation strategy, when small analyte signals must compete with noise. Gabriely et al. recently published transcutaneous glucose measurements by using near-infrared spectroscopy during hypoglycaemia [81]. The standard errors of prediction were around 5 mg dL^{-1}, but since their results were not properly validated, the paper leaves many questions unanswered. Such low prediction errors have never been achieved before, not even with far less complex in vitro samples, so that we had to comment on their results [82].

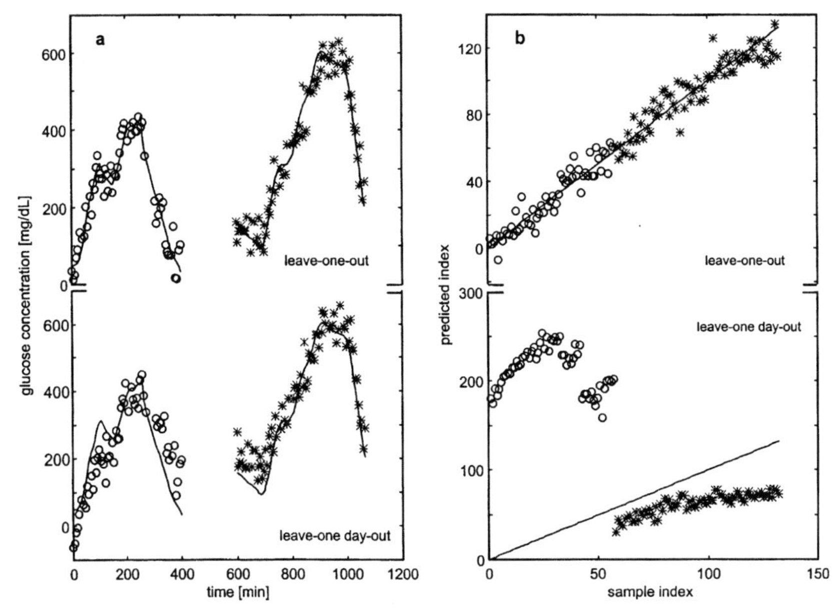

Fig. 13.13 Comparison of time-dependent capillary blood-glucose concentration values during a two-day experiment (solid curves) and prediction results (circles and asterisks) based on PLS calibrations by using diffuse-reflectance lip spectra of a type I diabetic subject: (a) cross-validated results from a PLS model with 24 wavenumbers specially selected (upper traces; the time gap between the two daily profiles is arbitrary), and predictions within one day from calibration models by using the same wavenumbers as above, but calculated from the data recorded during the other day with crosswise repeated procedure (lower traces); (b) comparison of deliberately constructed calibration data with concentration values replaced by sample numbers (solid curves) and their cross-validated prediction results (circles and asterisks) from a PLS model with full spectrum evaluation (115 spectral variables, same spectra and calibration model building strategies as given in (a))

Malin and co-workers reported on results from noninvasive glucose measurements by using OGTT for three nondiabetic test subjects [83]. A total of five daily experiments was carried out under identical conditions, so that the shapes of the glucose concentration profiles for the individuals were rather similar. For calibration, the data sets from two days, for validation one data set – timely surrounded by the calibration data or appended to those – was used. The authors noticed a strong clustering of spectral data for complete days, which they believed to be related to variations in skin hydration, skin surface smoothness and day-to-day variations in skin-surface temperature. Spectral similarity testing was therefore inevitable for data selection by hierarchical and K means clustering. Standard errors of prediction were between 44 mg dL^{-1} and 34 mg dL^{-1}. The identical OGTT experiments for calibration and validation were a weak point. To avoid chance correlations with other physiological parameters, for example circadian changes, it would have been wise to change the experimental conduct with respect to the resulting blood-glucose profiles in the validation set, compared to those of the calibration data. Such an experimental design could better prove that glucose absorptions had been the physical basis of their calibration models.

Investigations into the long-term accuracy and stability of calibration models obtained from neural networks and based on radial basis functions were carried out by Sämann et al. [84]. Whereas standard errors of prediction for blood-glucose concentrations for individual diabetic patients, obtained with leave-one-out cross-validation at the beginning of the study, ranged between 1 and 3 mM, the average results for tests were between 4 and 6 mM, which is unacceptably large. The tests lasted between 84 and 169 days.

Besides the long term stability of individual calibration models, one might be interested in developing a global, multivariate calibration model that predicts the analyte satisfactorily both for a single patient and across different subjects. A strategy based on the development of a so-called generic calibration model, which works successfully within the individual subject environment and which can be adapted to each other subject separately, was outlined by Thomas [85]. Such adaptable multivariate calibration models were investigated for the case of noninvasive blood-glucose measurement by using diffuse-reflection spectra of the underside of the forearm from 7200 to 4200 cm^{-1}. Promising long-term average results over a period of seven weeks were presented. For two subjects, the period of observation even spanned more than six months. Other aspects than statistical calibration adaptation are certainly worth considering and a few topics related to blood micro-circulation and dynamic glucose-transport processes within the different physiological compartments in the tissue were discussed in [67].

13.4
Short-Wave Near-Infrared Spectroscopy for Medical Monitoring

Optical spectroscopy and imaging uses radiation emitted into different tissues in order to determine their chemical composition and interior structure. The principles of optical spectroscopy for in vivo studies with respect to haemoglobin oxygenation were developed from laboratory work on in vitro samples. The spectra of haemoglobin in both its reduced and oxygenated states had been the basis of such assays (see also Fig. 13.7(a)). Since the early 1930s, considerable efforts have been invested in improving noninvasive optical techniques that provide information about the in vivo state of tissue oxygen supply. The interested reader will find literature on historically important papers on the subject of visible biospectroscopy in ref. [86].

For wavelengths between 600 and 1300 nm (16700–7700 cm^{-1}), the so-called "therapeutic window", opportunities exist for measurements of intact body tissue of centimetre thickness due to its favourable absorption and scattering characteristics. The monitoring of tissue physiology with respect to blood and tissue oxygenation, respiratory status or ischaemic damage is in particular possible by short-wave near-IR spectroscopy. As mentioned above, haemoglobin in blood is a key compound in such studies. The myoglobin protein, which is important for oxygen metabolism in muscle tissue, carries a haem group similar to haemoglobin. A chromophore of great interest is cytochrome c oxidase in the cellular mitochondrial membranes, because it is the terminal compound in the respiratory chain. Also of great interest is, for example, the acquisition of knowledge on haemodynamics, as well as on the intracellular cytochrome redox state in the brain of human neonates. Besides neurology studies with emphasis on paediatrics and neonatology, activities for cerebral blood-flow analysis or in brain monitoring during cardio-thorac-vascular surgery must be mentioned. Functional cerebral imaging is nowadays a research field tackled by many groups. Near-IR tomography for spatial tissue characterisation is another hot issue with a promising potential. The areas listed will be described in greater detail below.

13.4.1
Noninvasive Pulsatile Near-IR Spectroscopy

At the beginning of short-wave near-IR biospectroscopy came whole tissue probing, but further developments of noninvasive assay technology were encouraged by subsecond spectroscopic measurements, which allowed the probing of intravascular space through dynamic monitoring of tissue absorption. So-called photoplethysmography can be employed to measure the cardiovascular pulse wave spectrometrically, which is correlated to periodic changes in blood volume. During heart systole, the blood is maximally diffused through the vascular system, whereas at diastole, blood pressure is minimal. If the photometric measurement of a chromophore present in blood and tissue is used, a time-dependent periodic signal, such as shown in Fig. 13.14, can be observed. The analyte of interest could be, for example, an administered water-soluble dye. In the case in which the chromophore compound is only found in the vascular system, like in the red cells' haemoglobin, a part of the static tissue signal would be due to photon flux attenuation by scattering. For quantitative measurements within the arterial compartment, the signals at the highest and lowest pulsation points are evaluated preferably.

At present, the major clinical application of near IR spectroscopy with exploitation of the cardiac blood-volume modulation is in pulse oximetry, which allows the determination of the arterial haemoglobin oxygen saturation [87]. Particularly for patients under anaesthesia, monitoring of this parameter has been incorporated into the generally accepted standard of care. For noninvasive blood-gas analysis, pulse oximetry is also fundamental in the support of critical-care medicine applied for adult and neonatal monitoring. A review on theory and applications, including practical limitations of such a technique, was given by Mendelson [88]. The problems mentioned were concerned with low peripheral vascular perfusion, motion artefacts, systematic errors induced by different haemoglobin variants and derivatives (for example, HbCO), electromagnetic compatibility and stray light. Interestingly, a report on applications and limitations in clinical anaesthetic practice has been published, which was based on the analysis of 2000 incident reports [89].

One of the major problems with pulse oximetry is still measurement error introduced by patient movement, but special signal processing with nonlinear meth-

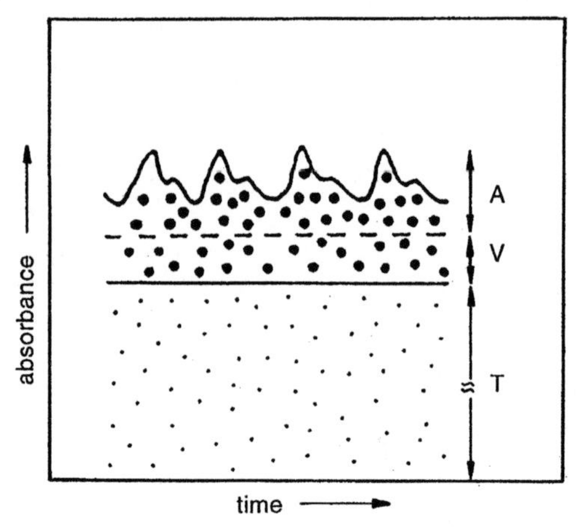

Fig. 13.14 Schematic pulsatile spectrometric signal from a tissue measurement over a period of four heartbeats with contributions from arterial blood (A), venous blood (V) and tissue (T): attenuation of the incident radiation may be caused, for example, by a water-soluble endogenous dye or an exogenous tracer within the vascular, interstitial and cellular tissue compartments; see also text

odology has been proposed for eliminating such motion artefacts [90]. Furthermore, the technology appears to be moving towards miniaturisation, which allows pulse oximetry to be used outside the critical-care area. Another problem investigated was lateral tissue inhomogeneity [91]. Near-IR LEDs with wavelengths centred on 820, 1000, 1220 and 1300 nm were employed in a set-up that largely avoided ambient-light interference and motion-induced probe-coupling artefacts. The fundamental concern of the study was that the sampled tissue volume, the fingertip with and without applied pressure, was not the same for all wavelengths tested. The use of 1300 nm and 800 nm wavelengths in a photoplethysmographic device had been described earlier by Schmitt et al. for noninvasive haematocrit measurement [92]. The haematocrit is defined as the volume percentage of red cells in whole blood that contain the haemoglobin protein. An application of near-IR laser diodes, with narrow spectral bandwidth and suitable for high frequency modulation, was reported for pulse oximetry, substituting the LEDs normally employed [93].

One application of pulse oximetry for obstetrics currently under development is maternal trans-abdominal monitoring of foetal arterial-blood oxygenation [94]. Additionally, the authors presented an informative oximetry review. The selection of optimal wavelengths was discussed in detail, which were suggested from the intervals of 670–720 nm and 825–925 nm. Their experiments proved that it was possible to discriminate between the foetal pulse rate and the mother's heart-beat frequency. By Fourier transformation of the measured signals, it was elucidated that the typical foetal pulse rates lay between 1.5 and 2.5 Hz, whereas the maternal rates were found within the narrow interval of 1.0 and 1.25 Hz. In another report, basic investigations of photon migration through the foetal head in utero were described by using continuous wave spectroscopy at 760 and 850 nm for the measurement of cerebral blood oxygenation [95]. Clinically studies from the maternal abdomen and laboratory tissue phantoms followed [96].

Low-cost instrumentation for multiwavelength measurements is now available for oximetry in the visible and near-IR spectral range. The performance of a diode-array-based VIS/NIR spectrometer with skin illumination through an integrating sphere and detection by means of a combination of lens and fibre-optics was tested by us for photoplethysmography of skin tissue. The spectrum sampling frequency was 4 Hz. In Fig. 13.15(a) the mean spectrum from a one-minute measurement of the finger tip is shown, whereas the time-dependent signals for the two absorbance maxima below the wavelengths of 600 nm were followed in (b). The Fourier transform of such wavelength-dependent time series leads to corresponding vectors with frequency domain coordinates (the Fourier amplitudes, displayed in the inset of Fig. 13.15(a), were obtained by the power function for the pulsating absorbance signal at 580 nm). The latter diagram clearly shows the cardiac pulse frequency. Aliasing from overtone components can be reduced by spec-

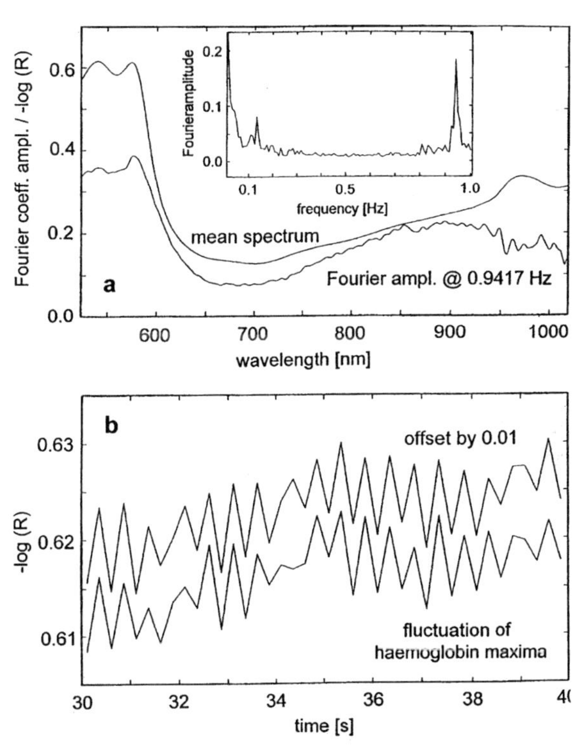

Fig. 13.15 Results from pulsatile spectroscopy by using a full-spectrum recording with a sampling frequency of 4 Hz: (a) mean spectrum from a one-minute measurement of diffuse-reflection finger-tip spectra and pulsatile component (arbitrary intensities) with inset showing a section of the Fourier transform of one of the signals below, recorded over one minute, and (b) fluctuating signals of the oxyhaemoglobin absorption maxima below 600 nm

tral measurements at a higher time resolution, but the separation of such peaks is often good enough to isolate the fundamental frequency component from the overtones. Lower frequencies can frequently be assigned to rhythmic arteriolar contractions called vasomotion [97]. If all wavelength-dependent Fourier coefficients selected at the cardiac frequency are plotted versus wavelength, the pulsatile spectrum is obtained (see also Fig. 13.15(a)). It bears analytical information on the arterial blood composition.

In Fig. 13.16(a), the mean spectra measured at the finger tip and the outer lip are presented. Apart from the oxyhaemoglobin band doublet, the prominent water absorption band at 960 nm is evident. A principal component analysis of 240 tissue spectra recorded subsequently over a one-minute period was carried out [98]. The individual squared singular values s_i^2 represent the contribution of a single factor component to the total variance, and V_R is defined by:

$$V_R = \sum_{i=1}^{R} s_i^2 \left\{ \sum_{j=1}^{N} s_i^2 \right\}^{-1}$$

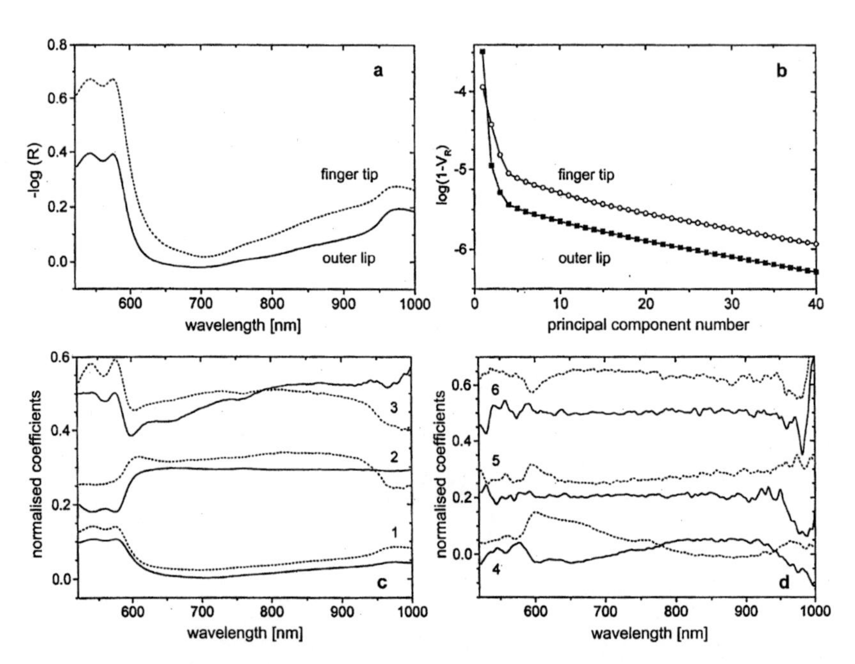

Fig. 13.16 Results from a principal component analysis of diffuse-reflection spectra of human skin of the index finger tip and the outer lip recorded over one minute with a sampling frequency of 4 Hz: (a) mean spectra, (b) normalised variance explained by a step-wise increased number of factors (for definition, see text), (c) and (d) first six factor spectra in the order of decreasing respective singular values (offset and min-max normalised for better comparison; solid curves: outer lip data, and dashed curves: finger tip data)

which is the normalised variance explained by R factor components, whereas N is the total number obtained from the singular value decomposition of the spectral matrix. Thus, the expression $(1-V_R)$ is the residual normalised variance after implementation of R factors (see Fig. 13.16(b)). Four significant factors were found without spectral mean centering, and the residual factors more or less represent noise components. In Fig. 16(c) and (d) the first six factor spectra, normalised to give the same maximum-minimum difference, are presented for finger-tip and lip-tissue spectra, respectively. By neglecting higher factors, improved signal-to-noise ratios can be achieved with spectrum reconstruction.

The same procedure was studied for analysing near-IR spectra of metabolites in the arterial blood of skin tissue. Due to limitations in signal-to-noise ratio, such a measurement principle – although proposed in patents – has not yet been applied for measurements in practice. However, first results from time-resolved measurements on human oral mucosa by using a custom-made diffuse reflection accessory have been presented [66]. Tissue spectra were analysed for their mean, and

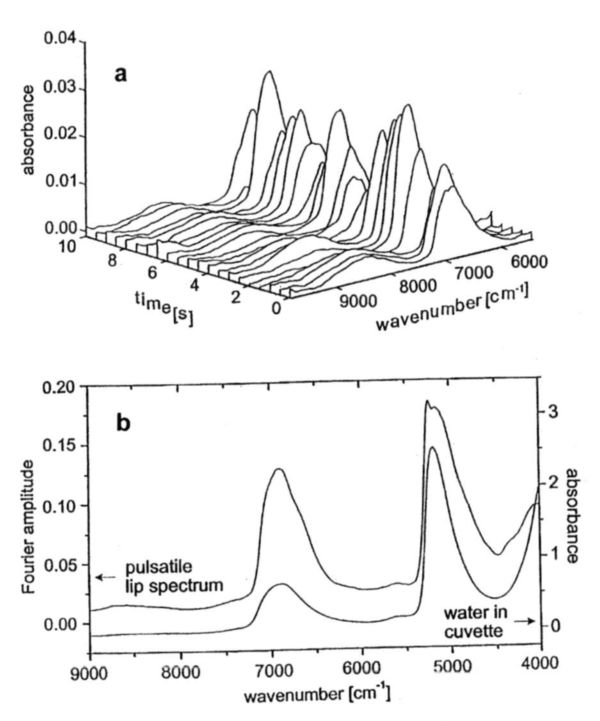

Fig. 13.17 (a) Subsecond diffuse-reflection spectra of inner lip mucosa of a person shown as differences versus the first measured lip spectrum after Savitzky-Golay smoothing; (b) pulsatile lip spectrum with arbitrary intensities obtained from the spectral Fourier amplitudes at heartbeat frequency after Fourier transformation of the individual spectral variable time series from the cardiac-pulse-modulated near-infrared lip spectra; for comparison an absorbance spectrum of water is given, as measured in a cell of 0.5 mm pathlength (offset for clarity; adapted from [51])

the pulsatile components were extracted by Fourier analysis of the dynamically recorded spectra obtained with subsecond time resolution (see also Fig. 13.17(a)). The arterial compartment constitutes only a small fraction of the probed tissue volume. Since only small blood-volume changes exist for the arterial vessels, the pulsatile spectral signal – to be evaluated for specific component absorptions – represents only a very minor fraction of the integrally measured tissue.

In Fig. 13.17(b), the pulsatile spectrum, originating from the heart-beat-modulated aqueous compartment of lip tissue, is compared to a water absorbance spectrum that had been recorded with a transmission cell of 0.5 mm pathlength. The absorbance pulsations for the water band maximum at 6900 cm^{-1} are due to cardiac blood-pressure changes and the amplitudes are about 20 mAU (AU = absorbance units), which is equivalent to an aqueous layer of 15 μm thickness (about a factor of 50 smaller than the observed signal for the water manifested in a lip spectrum that was obtained by a one-minute measurement). Relative to the spectrum that was recorded with a constant optical cell pathlength, the difference in the intensity ratio of bands from the pulsatile spectrum can be explained by the significantly different penetration depths for the near-infrared radiation that could be realised for those band maximum wavelengths. Further investigation is needed to improve the signal-to-noise ratio in such pulsatile spectra. These could provide the basis for a noninvasive analysis of substrates in blood, which have already been accessed by in vitro IR-spectrometric assays, as discussed above.

13.4.2
Monitoring of Blood-Tissue Oxygenation and Cytochrome Redox Status

Thanks to recent developments, the regional and temporal variations in skin-tissue oxygenation can be continuously assessed by using a near-IR spectroscopic imaging system [99]. Further developments of liquid-crystal tuneable filter-based instrumentation for spectroscopic imaging in the visible and near-IR spectral range were presented by Mansfield et al. [100]. Tissue viability testing, particularly important after plastic surgery, was also tackled by reflection measurements in the range between 650 and 900 nm [101–103]. For presentation, a form of data processing – fuzzy C-means clustering – was developed for clustering those spectra with similar intensity patterns [104]. Near-IR spectroscopic reflectance images from the forearm of a subject were investigated during ischaemia-reperfusion events. One, two and six wavelengths were used for clustering the spectroscopic "imaging cube". The six wavelengths were necessary to follow concentration changes in deoxyhaemoglobin, oxyhaemoglobin and cytochrome c oxidase (for the latter compound, see also below). Spatial variations in the tissue response to ischaemic events, for example, by venous occlusion could thus be monitored successfully [105]. One application of this reflectance-imaging technique for burn injury assessment was given by workers from the same research group [106]. A further publication on the noninvasive monitoring of the perfusion and the oxygenation within a skin flap after surgical elevation of a rat's skin was presented recently [107]. Principal component analysis of the spatially dependent spectra revealed well-perfused skin areas and compromised sections of the flap immediately following surgery, whereas visual inspection allowed a critical assessment only after six hours.

Noninvasive transmission experiments for the determination of haemoglobin oxygenation in hand tissue were carried out by Sato et al. [108] by using an electronically tuned Ti:Sapphire laser in the spectral range between 700 and 1000 nm. A different task is the determination of muscle oxygenation that can be measured by near-IR reflection spectroscopy, and many tissue investigations in the past were based on continuous wave (CW) spectroscopy. For example, quantitative measurements on muscle oxygenation were done by Niwayama et al. [109] who used two LEDs of 760 and 840 nm peak wavelength and an array of eight photodiodes that were located for detection at different distances from the IR-radiation sources. Their paper dealt with the influences of the subcutaneous fat layer and skin on the measurement results. Monte Carlo simulations were evaluated for a four-layered tissue model. Parameters of interest were changes in total haemoglobin, as well as in oxy- and deoxyhaemoglobin concentrations. In contrast to such CW measurements, time-resolved near-infrared spectroscopy was tested for ischaemic muscle deoxygenation by Hamaoka et al., who employed 780 and 830 nm laser diodes with a pulse half-width of 50 ps [110].

An example of the application of a frequency-domain oximeter that employed a modulation frequency of 110 MHz for the laser diodes (wavelengths at 758 and 830 nm) can also be given [111]. The experimental set-up consisted of an arrangement of 16 emitting fibres and two detectors. The same instrumentation was successfully employed for peripheral vascular disease diagnosis in 95 human subjects [112]. With the manifestation of arteriosclerosis, this disease can cause significant morbidity owing to increased problems in coronary blood circulation. A comprehensive literature review on similar applications of such near-IR instrumentation was given by these authors.

A noninvasive in vivo assay for haematocrit exploited spectral data of the visible/near-IR range between 580 and 1000 nm. The partial least-squares multivariate calibration algorithm was utilised with diffuse-reflection spectra taken from patients' forearm skin. Whereas single-person calibrations led to a 1% haematocrit prediction error, multisubject models showed significant patient-specific offsets in the calibration curves. After offset removal, the cross-validated error estimate was still double the results of single-patient calibrations [113].

Another compound of interest, similar to haemoglobin in blood, is myoglobin in muscle tissue. The absorption bands of the haem moiety of the protein provide useful markers for skeletal-muscle oxygenation. A further important chromophore for quantitative tissue spectroscopy is contained in cytochrome c oxidase, which is the terminal compound in the respiratory chain and found in the cellular mitochondrial membrane. However, only the oxidised form absorbs weakly at 830 nm, showing an unusually broad band [114]. The cytochrome enzyme catalyses more than 90% of the oxygen utilisation, so that monitoring of its redox state and of myoglobin oxygenation by visible/near-IR spectroscopy can provide valuable physiological information. Visible spectra of important cytochromes and myoglobin were recently presented in a study that used reflection spectroscopy of haemoglobin-free perfused heart tissue. In contrast to near-IR, the technique rendered only a millimetre probing depth within the epicardial surface layer [86]. In a similar context, the source/sink relationship of oxygen in different tissues and the transfer of in vitro data of the cytochrome c oxidase to the in vivo milieu was critically discussed [115].

Kupriyanov et al. [116] investigated buffer-perfused isolated pig hearts in the spectral range of 500 to 1000 nm to determine intracellular partial oxygen pressure, which depends on the oxygen diffusion from the vascular to the interstitial space before it finally reaches the intracellular mitochondria. Comparison was made between beating and arrested hearts. In a different study, myocardial oxygenation was examined in open chest dogs with brief artery occlusions, resulting in differently lasting ischaemic conditions [117]. Total oxygenated and deoxygenated haemoglobin, as well as myoglobin concentrations were monitored.

An important field is the noninvasive monitoring of oxygen metabolism in the living human brain. Here, the endogenous haemoglobin, in both its reduced and oxygenated states, is the key component, although many investigations have also been concerned with the bioenergetics of cytochrome c oxidase [118]. Cerebral monitoring, for example for neonates, has been achieved by monitoring a number of additional parameters, such as blood flow, oxygen delivery and blood volume, by using special equipment based on a small number of pulsed laser diodes (typical selections are for 775, 825, 850 and 905 nm, average power is about 2 mW) [119, 120]. A noninvasive measurement of the regional cerebral blood flow in the intact pig skull, with an intravenously infused dye (indocyanine green) as a tracer and near-IR photometric instrumentation as just described, was presented by Kuebler et al. [121]. Owing to the diagnostic and prognostic relevance of cerebral blood flow, there is an urgent demand for a rapid and noninvasive monitoring technique for patients suffering from neurotrauma or cerebral ischaemia.

For newborn infants it is possible to detect photons with head transillumination; however, adult measurements are usually carried out in reflection mode (reflection oximetry) with different source-detector separations. Mathematical simulations of the photon transport in tissue by so-called Monte Carlo techniques have shown that the anatomy of the layers surrounding the brain, especially the clear cerebrospinal fluid, affect the distribution of photons [122]; this explains in part the large variability of data published by different authors. Further investigations on the scattering and absorbing effects of the latter fluid on the photon "random walk" in the tissue have been reported [123]. In this context, a study on the noninvasive near-IR measurement of the width of the subarachnoid space, which contains the cerebrospinal fluid around the brain, must be mentioned [124]. Changes in the geometry of this compartment may play an important role in cranio-spinal physiology and pathology.

Frequency-domain and CW spectroscopy for monitoring of optical properties of newborn piglet brain based on the two wavelengths at 758 and 830 nm were compared by Fantini et al. [125]. By using an acquisition time of 160 ms, they were able to follow the optical signals associated with the arterial pulse and with respiration. Besides oxygen saturation, quantitative cerebral haemoglobin concentration could be determined by the frequency domain measurement variant.

A review on cerebral near-IR spectroscopy in neonatal intensive care was published recently, stating that a lot of credible and important data have been accumulated so far [126]. On the other hand, it has been difficult to take advantage of the continuous and noninvasive nature of the methodology, as the interpretation of the data, especially on the redox-state measurements of cytochrome c oxidase, is still to be assessed critically. Experimental and spectral evaluation algorithm problems connected to clinical trials have been reported and discussed by Macnab et al. [127]. From the same area of neonatology, recent advances in foetal monitoring by using near-IR spectroscopy and specially designed optical probes were described by D'Antona et al. [128]. The feasibility of measuring foetal cerebral blood oxygenation and blood volume through transabdominal near-IR spectrometry was presented by Ramanujam and Chance [129] (see also the previous section on pulsatile spectroscopy).

Recent developments in spatially resolved near-IR spectroscopy, by using a multidistance approach for the determination of local cerebral cortex haemoglobin oxygen saturation, were published by Quaresima et al. [130]. The tissue oxygenation index obtained, representing a regional mixture from venous (~75%) and arterial/capillary compartments (~25%), was compared for validation with the cerebral venous oxygenation saturation measured by a conventional CW near-IR photometer. In the measuring protocol, a partial occlusion of the jugular vein is needed, by which a rise in oxyhaemoglobin and total haemoglobin concentration in the tissue is induced.

Other developments for a portable small-tissue oximeter based on a two-wavelength LED set-up (760 and 840 nm) have been described. Special coefficients for the concentration estimation algorithm were deduced empirically by varying blood volume and scattering intensity in a tissue-like phantom [131].

Nowadays, oximeters that are capable of functional imaging generate great interest. One type of instrumentation was described by Maki and co-workers [132, 133]. The group developed a so-called "optical topography" system based on 24 measurement channels that used intensity-modulated near-IR spectroscopy at wavelengths of 780 and 830 nm. A temporal resolution of 0.5 s could be achieved for a measurement area of 90 mm x 90 mm. The system was used to visualise spatial and temporal changes of blood oxygenation in the human brain caused by cortical activity, for example, testing for different language function [134]. Pioneering work in the field of trans-cranial dynamic near-IR imaging has been reviewed by Koizumi et al. [135]. For diagnostics, the regional cerebral blood-volume dynamics were studied recently by scientists from the same group during seizures in patients with medically intractable epilepsy [136].

A different optical imaging system was developed by other researchers [137]. The basic unit is a multichannel frequency-domain tissue oximeter, as described above, that uses a modulation frequency of 110 MHz with dual wavelength measurements at 758 and 830 nm. The set-up was used for on-line imaging of an area of human brain $4 \times 9\ cm^2$ with a time resolution of 160 ms. Functional imaging of cerebral haemodynamics during rest and motor stimulation was presented in detail [138]. Results on hypoxic-ischaemic brain injury in neonatal piglets by using similar instrumentation were reported by Stankovic et al. [139]. The acquisition time for complete images of a 4×4 cm^2 area was 192 ms.

So far, devices have been described that are at least two wavelengths for two- or more-component systems to be studied. As shown by van Huffel and co-workers, the multivariate determination of haemoglobin, oxyhaemoglobin and cytochrome c oxidase can be improved by implementing many wavelengths with wavelength selection, in addition to the use of total least squares, by taking the different reference data precision into account for regression analysis [140]. Full-spectrum visible/near-IR diffuse-reflection spectroscopy, measured with interoptode distances of 20 mm within the interval of 614 to 900 nm, was used to determine changes in cerebral haemoglobin concentration and oxygen saturation in new born babies [141]. As the pathlength of the photons inside the tissue is wavelength dependent, an approach was chosen to apply a multiplicative correction term obtained from the photon diffusion equation. With such an absorbance definition adapted to the Lambert-Beer law, the component spectra measured from solution work were in accordance with wavelength-dependent molecular absorptivity data used for fitting the experimental spectrum. Their results indicated that dynamic changes occur in cerebral circulation and oxygenation as part of the physiology changes taking place after birth.

13.4.3
Near-Infrared Tomography

A broad engagement in the field of medical spectroscopy is expected from people developing optical tomography techniques based on time- or frequency-domain technology. Such instrumentation can especially be used for the noninvasive detection of breast cancer [142, 143]. The simplest technique relies on CW spectroscopy, that is, employing an infrared source of constant intensity and measuring the intensity attenuation at some distance from the source. An introduction to the rather complex photonic techniques and mathematical reconstruction algorithms can be found in ref. [144]. Time-of-flight measurements with streak-cameras or phase- and intensity-modulation techniques have been evaluated for media in which the photon transport is influenced by the absorption coefficient, which depends on the chemical composition – in particular on the concentration of various chromophores, the scattering coefficient and the scattering anisotropy. The forward scattering characteristics in soft tissue, expressed by the scattering anisotropy coefficient, were already mentioned above. Photons entering the tissue experience multiple scattering, so that on excitation three components are observed, that is, ballistic "snakelike" and diffuse.

A time-resolved optical imaging system has been developed by Japanese scientists [145]. The instrument has three pulsed infrared laser sources with wavelengths at 761, 791 and 830 nm (pulse duration 100 ps, peak power >500 mW and a pulse frequency of 5 MHz, resulting in an average power of about 0.25 mW). The detecting channels were connected to 64 optical fibre bundles. Their tomographic imaging system was recently employed to construct quantitative images of haemoglobin concentration changes associated with the neuronal activation of the human brain [146]. To list other groups' activities, a charge coupled device-based instrument was recently presented that is capable of making parallel multisite and multiwavelength measurements with a maximum sampling rate of 150 Hz [147]. Target illumination was achieved with two laser diodes (785 and 810 nm). This underlined the importance of information on tissue function derived from noninvasive haemoglobin oxygenation studies.

As mentioned above, most in vivo imaging instrumentation aims at the provision of diagnostic screening technology for the detection of breast cancer. Both the scattering and absorption properties of tumours and the surrounding tissue can be exploited for sensing tissue abnormalities. A systematic study of the optical properties of normal and diseased breast tissue was undertaken by Troy et al. using three wavelengths [148]. Cancerous tumours are known to show increased angiogenesis and therefore an increased haemoglobin concentration. In addition, the partial oxygen pressure was found to be lower than in normal breast tissue. In this context, the more extended spectral analysis of various biological tissues and biofluids by Khairullina et al. [149] must also be mentioned. Optical imaging by spatial near-IR measurements at wavelengths of 750 and 830 nm as an adjunct to sonography was presented for discriminating between benign and malignant lesions in several mammography patients [150]. By this approach, ultrasound specificity will be improved and will hopefully lead to a reduction in the frequency of unnecessary biopsies. Methods for the determination of absorption and reduced scattering coefficients were recently reviewed, and a novel method that used noncontact Fourier-transform hyperspectral imaging was presented for broad-range measurements [151].

Despite the promising advances, it is a fact, however, that the spatial resolution of optical tomography, when compared to conventional X-ray mammography, will always be inferior due to tissue scattering. For example, a time-domain optical mammograph developed by Grosenik et al. [152] showed some limitations. Because of the expected influence of vascular dynamic changes, a higher time resolution than can be obtained with the current system would be needed for further improvements. Time-series optical data were collected with different recently developed instruments, to allow sampling rates of up to 150 Hz [153]. The field of optical tomography has now advanced to the state that instrumentation for clinical studies is now available. Further improvements are needed, but the potential of this photonic technology for screening purposes, although technically complex, is enormous.

13.5 Concluding Remarks

There are many interdisciplinary research activities worldwide searching for new and more efficient analytical methods and techniques for expanding the field of medical diagnostics. The novel approaches can now be based on gentle, reagent-free and automation-capable near-IR spectroscopy. Clinical chemistry will profit from the less expensive miniaturised spectrometers that hover just over the horizon. Assays for blood and urine have already been developed, and the analysis of other body fluids will be undertaken. There is still a plethora of conventional or classical assays to be replaced by faster and more accurate spectroscopic methodology. Direct diagnostics and classification of diseases may be derived from the spectroscopic fingerprints of biomedical samples. Similar expectations exist in near-IR pathology, given the sampled material of biopsies or cell smears. Noninvasive and nondestructive diagnostic tools, made possible by fibre-optic probes, should be available in the near future, in particular for skin-tissue analysis, including dynamic tests for blood-flow and micro-circulation problems.

Over the last decade, instruments for noninvasive monitoring of tissue oxygenation, in particular of the brain, have become increasingly common in medicine. Absolute quantification is still difficult due to signal contamination from the overlying tissues of the scalp and skull. However, the random walk of photons will possibly also reduce the number of biopsies, for example for the detection of malignant breast lesions, due to sound and noninvasive diagnostics. Image reconstruction can be improved, despite the image blurring caused by photon scattering in the tissues. Optical spectroscopy and imaging are nearing clinical application, although near-IR spectroscopy must be still considered as a research technique. For the future, the marriage of near-IR spectroscopy and biomedical expertise will lead to more progress in our health care arena, which will certainly contribute to improving our quality of life.

13.6 References

[1] J. W. Hall, A. Pollard, *Clin. Chem.* **1992**, *38*, 1623.

[2] M. Jackson, H. H. Mantsch in: *Infrared Spectroscopy of Biomolecules* (Eds. D. Chapman, H. H. Mantsch), Wiley-Liss, New York, **1996**, pp. 311–340.

[3] F. F. Jöbsis, *Science* **1977**, *198*, 1264.

[4] F. F. Jöbsis-van der Vliet, *J. Biomed. Optics* **1999**, *4*, 392.

[5] H. M. Heise, R. Marbach, A. Bittner, Th. Koschinsky, *J. Near Infrared Spectrosc.* **1998**, *6*, 361.

[6] A. Bittner, R. Marbach, H. M. Heise, *J. Mol. Struct.* **1995**, *349*, 341.

[7] H. M. Heise, *Mikrochim. Acta* **1997**, *14 [Suppl.]*, 67.

[8] M. J. McShane, B. D. Cameron, G. L. Coté, C. H. Spiegelman, *Appl. Spectrosc.* **1999**, *53*, 1575.

[9] M. J. McShane, B. D. Cameron, G. L. Coté, C. H. Spiegelman, *Proc. SPIE* **1999**, *3599*, 101.

[10] H. M. Heise, A. Bittner, *Fresenius' J. Anal. Chem.* **1998**, *362*, 141.

[11] Q. Ding, G. W. Small, M. A. Arnold, *Anal. Chim. Acta* **1999**, *384*, 333.

[12] L. A. Sodickson, M. J. Block, *Clin. Chem.* **1994**, *40*, 1838.

[13] A. M. Helwig, M. A. Arnold, G. W. Small, *Appl. Optics* **2000**, *39*, 4715.

[14] M. R. Riley, M. A. Arnold, D. W. Murhammer, *Appl. Spectrosc.* **2000**, *54*, 255.

[15] N. A. Cingo, G. W. Small, M. A. Arnold, *Vibr. Spectrosc.* **2000**, *23*, 103.

[16] K. H. Hazen, M. A. Arnold, G. W. Small, *Anal. Chim. Acta* **1998**, *371*, 255.

[17] J. W. Hall, A. Pollard, *Clin. Biochem.* **1993**, *26*, 483.

[18] J. W. Hall, A. Pollard in *Leaping Ahead with Near Infrared Spectroscopy* (Eds. G. D. Batten, P. C. Flinn, L. A. Welsh, A. B. Blakeney), Royal Australian Chemical Institution, Melbourne, **1995**, pp. 421–430.

[19] K. H. Norris, J. T. Kuenstner in *Leaping Ahead with Near Infrared Spectroscopy* (Eds. G. D. Batten, P. C. Flinn, L. A. Welsh, A. B. Blakeney), Royal Australian Chemical Institution, **1995**, p. 431–436.

[20] M. R. Gatin, J. R. Long, P. W. Schmitt, P. J. Galley, J. F. Price in *Near Infrared Spectroscopy: The Future Waves* (Eds. A. M. C. Davies, P. Williams), NIR Publications, Chichester, **1996**, pp. 347–352.

[21] D. Lafrance, L. C. Lands, L. Hornby, D. H. Burns, *Appl. Spectrosc.* **2000**, *54*, 300.

[22] G. Domján, K. J. Kaffka, J. M. Jákó, I. T. Vályi-Nagy, *J. Near Infrared Spectrosc.* **1995**, *2*, 67.

[23] R. J. Dempsey, D. G. Davis, R. G. Buice, Jr., R. A. Lodder, *Appl. Spectrosc.* **1996**, *50*, 18A.

[24] K. Murayama, K. Yamada, R. Tsenkova, Y. Wang, Y. Ozaki, *J. Near Infrared Spectrosc.* **1998**, *6*, 375.

[25] K. Murayama, K. Yamada, R. Tsenkova, Y. Wang, Y. Ozaki, *Fresenius' J. Anal. Chem.* **1998**, *362*, 155.

[26] J. T. Kuenstner, K. H. Norris, W. F. McCarthy, *Appl. Spectrosc.* **1994**, *48*, 484.

[27] J. T. Kuenstner, K. H. Norris, *J. Near Infrared Spectrosc.* **1994**, *2*, 59.

[28] J. T. Kuenstner, K. H. Norris, *J. Near Infrared Spectrosc.* **1995**, *3*, 11.

[29] G. Yoon, S. Kim, Y. J. Kim, J. W. Kim, W. K. Kim in *Infrared Spectroscopy: New Tool in Medicine* (Eds. H. H. Mantsch, M. Jackson), *Proc. SPIE 3257*, **1998**, 126.

[30] A. Roggan, M. Friebel, K. Dörschel, A. Hahn, G. Müller, *J. Biomed. Optics* **1999**, *4*, 36.

[31] J. L. Gonczy, L. L. Gyarmati in *Making Light Work: Advances in Near Infrared Spectroscopy* (Eds. I. Murray, I. A. Cowe), VCH, Weinheim, **1992**, pp. 603–609.

[32] A. W. van Toorenenbergen, *Clin. Chem.* **1994**, *40*, 1788.

[33] R. A. Shaw, S. Kotowich, H. H. Mantsch, M. Leroux, *Clin. Biochem.* **1996**, *29*, 11.

[34] J. W. Hall, A. Pollard. *J. Near Infrared Spectrosc.* **1993**, *1*, 127.

[35] R. A. Shaw, S. Kotowich, H. H. Eysel, M. Jackson, G. T. D. Thomson, H. H. Mantsch, *Rheumatol. Int.* **1995**, *15*, 159.

[36] H. H. Eysel, M. Jackson, A. Nikulin, R. L. Somorjai, G. T. D. Thomson, H. H. Mantsch, *Biospectrosc.* **1997**, *3*, 161.

[37] K. Z. Liu, M. K. Ahmed, T. C. Dembinsky, H. H. Mantsch, *Int. J. Gynecol. Obstet.* **1997**, *57*, 161.

[38] R. A. Shaw, H. H. Eysel, K. Z. Liu, H. H. Mantsch, *Anal. Biochem.* **1998**, *259*, 181.

[39] R. Salzer, G. Steiner, H. H. Mantsch, J. Mansfield, E. N. Lewis, *Fresenius' J. Anal. Chem.* **2000**, *366*, 712.

[40] *Infrared and Raman Spectroscopy of Biological Materials* (Eds. H.-U. Gremlich, B. Yan), Marcel Dekker, New York, **2001**.

[41] J. M. Carney, W. Landrum, L. Mayes, Y. Zou, R. A. Lodder, *Anal. Chem.* **1993**, *65*, 1305.

[42] L. A. Cassis, R. A. Lodder, *Anal. Chem.* **1993**, *65*, 1247.

[43] L. A. Cassis, J. Yates, W. C. Symons, R. A. Lodder, *J. Near Infrared Spectrosc.* **1998**, *6*, A21.

[44] W. R. Windham, W. H. Morrison, *J. Near Infrared Spectrosc.* **1998**, *6*, 229.

[45] M. G. Sowa, J. R. Mansfield, M. Jackson, J. C. Docherty, R. Deslauriers, H. H. Mantsch, *Mikrochim. Acta* **1997**, *14 [Suppl.]*, 451.

[46] J. Wallon, S. He Yan, J. Tong, M. Meurens, J. Haot, *Appl. Spectrosc.* **1994**, *48*, 190.

[47] S. H. Yan, J. Mayaudon, J. Wallon, *J. Near Infrared Spectrosc.* **1998**, *6*, A273.

[48] Z. Ge, C. W. Brown, H. J. Kisner, *Appl. Spectrosc.* **1995**, *49*, 432.

[49] G. Downey, J. McElhinney, T. Fearn, *Appl. Spectrosc.* **2000**, *54*, 894.

[50] Z. Ge, K. T. Schomaker, N. S. Nishioka, *Appl. Spectrosc.* **1998**, *52*, 833.

[51] H. M. Heise, A. Bittner, R. Marbach, *Clin. Chem. Lab. Med.* **2000**, *38*, 137.

[52] J. R. Mourant, J. P. Freyer, A. H. Hielscher, A. A. Eick, D. Shen, T. M. Johnson, *Appl. Optics* **1998**, *37*, 3586.

[53] J. R. Mourant, M. Canpolat, C. Brocker, O. Esponda-Ramos, T. M. Johnson, A. Matanock, K. Stetter, J. P. Freyer, *J. Biomed. Optics* **2000**, *5*, 131.
[54] J. G. Fujimoto, B. Bouma, G. J. Tearney, S. A. Boppart, C. Pitris, J. F. Southern, M. E. Brezinski, *Ann. (N.Y.) Acad. Sci.* **1998**, *838*, 95.
[55] K. Hoffmann, M. Happe, B. Fricke, A. Knüttel, D. Böcker, M. Stücker, P. Altmeyer, M. von Düring in *Dermatologie* (Eds. C. Garbe, G. Rassner), Springer, Berlin, **1998**, pp. 4–8.
[56] A. Knüttel, M. Boehlau-Godau, *J. Biomed. Optics* **2000**, *5*, 83.
[57] R. Marbach, H. M. Heise, *Appl. Optics* **1995**, *43*, 610.
[58] J. M. Schmitt, H. Yang, J. N. Qu, *Proc. SPIE* **1998**, *3257*, 134.
[59] G. Kumar, J. M. Schmitt, *Appl. Optics* **1997**, *36*, 2286.
[60] A. Bittner, S. Thomassen, H. M. Heise, *Mikrochim. Acta* **1997**, *14 [Suppl.]*, 429.
[61] K. Martin, *Appl. Spectrosc.* **1998**, *52*, 1001.
[62] M. G. Sowa, J. Wang, C. P. Schultz, M. K. Ahmed, H. H. Mantsch, *Vibr. Spectrosc.* **1995**, *10*, 49.
[63] T. Bui, S. Stewart, L. Rintoul, B. Thomas, P. Fredericks in *Fourier Transform Spectroscopy, 12th International Conference* (Eds. K. Itoh, M. Tasumi), Waseda University Press, Tokyo, **1999**, p. 271.
[64] V. P. Wallace, *Spectrophotometry for the Assessment of Pigmented Lesions*, Ph.D. Thesis, University of London, England, **1997**.
[65] V. P. Wallace, P. S. Mortimer, R. J. Ott, J. C. Bamber, *Proc. SPIE* **1999**, *3597*, 130.
[66] H. M. Heise, A. Bittner, R. Marbach, *J. Near Infrared Spectrosc.* **1998**, *6*, 349.
[67] H. M. Heise in *Encyclopedia of Analytical Chemistry: Instrumentation and Application, Vol. I* (Ed. R. A. Meyers), Wiley, Chichester, **2000**, pp. 56–83.
[68] H. M. Heise in *Biosensors in the Body: Continuous in vivo Monitoring* (Ed. D. M. Fraser), Wiley, Chichester, **1997**, pp. 79–116.
[69] O. S. Khalil, *Clin. Chem.* **1999**, *45*, 165.
[70] R. J. McNichols, G. L. Coté, *J. Biomed. Optics* **2000**, *5*, 5.
[71] J. N. Roe, B. R. Smoller, *Crit. Rev. Therap. Drug Carrier Syst.* **1998**, *15*, 199.
[72] J. Qu, B. C. Wilson, *J. Biomed. Optics* **1997**, *2*, 319.
[73] L. Heinemann, G. Schmelzeisen-Redeker, M. Berger, T. Heise, F. A. Gries, T. Koschinsky, H. Orskov, J. Sandahl Christiansen, D. Böcker, M. Essenpreis, *Diabetologia* **1998**, *41*, 848.
[74] L. Heinemann, U. Krämer, H.-M. Klötzer, M. Hein, D. Volz, M. Hermann, T. Heise, K. Rave, *Diabetes Techn. Therapeutics* **2000**, *2*, 211.
[75] H. A. MacKenzie, H. S. Ashton, Y. C. Shen, J. Lindberg, P. Rae, K. M. Quan, S. Spiers in *Biomedical Optical Spectroscopy and Diagnostics/Therapeutic Laser Applications* (Eds. E. M. Sevick-Muraca, J. A. Izatt, M. N. Ediger), OSA Trends in Optics and Photonics Series *22*, Optical Society of America, Washington, **1998**, pp. 156–159.
[76] H. A. MacKenzie, H. S. Ashton, S. Spiers, Y. Shen, S. S. Freeborn, J. Hannigan, J. Lindberg, P. Rae, *Clin. Chem.* **1999**, *45*, 1587.
[77] J. J. Burmeister, M. A. Arnold, *Clin. Chem.* **1999**, *45*, 1621.
[78] J. J. Burmeister, M. A. Arnold, G. W. Small, *Diabetes Technol. Therap.* **2000**, *2*, 5.
[79] M. A. Arnold, J. J. Burmeister, G. W. Small, *Anal. Chem.* **1998**, *70*, 1773.
[80] P. Lampen, H. M. Heise, *Appl. Spectrosc.* to be submitted.
[81] I. Gabriely, J. Kaplan, R. Wozniak, Y. Aharon, M. Mevorach, H. Shamoon, *Diabetes Care* **1999**, *22*, 2026.
[82] H. M. Heise, P. Lampen, *Diabetes Care* **2000**, *23*, 1208.
[83] S. F. Malin, T. L. Ruchti, T. B. Blank, S. N. Thennadil, S. L. Monfre, *Clin. Chem.* **1999**, *45*, 1651.
[84] A. Sämann, Ch. Fischbacher, K.-U. Jagemann, K. Danzer, J. Schüler, L. Papenkordt, U. A. Müller, *Exp. Clin. Endocrinol. Diabetes* **2000**, *108*, 406.
[85] E. V. Thomas, *Anal. Chem.* **2000**, *72*, 2821.
[86] J. Hoffmann, D. W. Lübbers, H. M. Heise, *Phys. Med. Biol.* **1998**, *43*, 3571.
[87] J. P. de Kock, L. Tarassenko, *Med. Biol. Eng. Comp.* **1993**, *31*, 291.

[88] Y. Mendelson, *Clin. Chem.* **1992**, *38*, 1601.
[89] W. B. Runciman, R. K. Webb, L. Barker, M. Currie, *Anaesth. Intens. Care* **1993**, *21*, 543.
[90] M. J. Hayes, P. R. Smith, *Appl. Optics* **1998**, *37*, 7437.
[91] L. A. Sodickson, *Clin. Chem.* **1999**, *45*, 1687.
[92] J. M. Schmitt, Z. Guan-Yiong, J. Miller, *Proc. SPIE* **1992**, *1641*, 150.
[93] S. M. Lopez Silva, R. Giannetti, M. L. Dotor, J. R. Sendra, J. P. Silveira, F. Briones, *Proc. SPIE* **1999**, *3570*, 294.
[94] A. Zourabian, A. Siegel, B. Chance, N. Ramanujan, M. Rode, D. A. Boas, *J. Biomed. Optics* **2000**, *5*, 391.
[95] G. Vishnoi, A. H. Hielscher, N. Ramanujan, B. Chance, *J. Biomed. Optics* **2000**, *5*, 163.
[96] N. Ramanujan, G. Vishnoi, A. H. Hielscher, M. Rode, I. Forouzan, B. Chance, *J. Biomed. Optics* **2000**, *5*, 173.
[97] M. Stücker, J. Steinbrügge, C. Ihrig, K. Hoffmann, D. Ihrig, A. Röchling, D. W. Lübbers, H. Jungmann, P. Altmeyer, *Acta Derm. Venereol. (Stockh.)* **1998**, *78*, 1.
[98] A. Bittner, Ph.D. Thesis, University of Essen, Germany, **1997**.
[99] M. G. Sowa, J. R. Mansfield, G. B. Scarth, H. H. Mantsch, *Appl. Spectrosc.* **1997**, *51*, 143.
[100] J. R. Mansfield, M. G. Sowa, H. H. Mantsch, *Proc. SPIE* **2000**, *3920*, 99.
[101] J. R. Payette, M. G. Sowa, S. L. Germscheid, M. F. Stranc, B. Abdulrauf, H. H. Mantsch, *Am. Clin. Lab.* **1999**, *18*, 4.
[102] M. F. Stranc, M. G. Sowa, B. Abdulrauf, H. H. Mantsch, *Brit. J. Plastic. Surg.* **1998**, *51*, 210.
[103] J. R. Mansfield, M. G. Sowa, J. Payette, B. Abdulrauf, M. F. Stranc, H. H. Mantsch, *IEEE Trans. Med. Imaging* **1998**, *6*, 1011.
[104] B. Abdulrauf, M. F. Stranc, M. G. Sowa, S. L. Germscheid, H. H. Mantsch, *Plastic Reconstr. Surg.* **2000**, *8*, 68.
[105] J. R. Mansfield, M. G. Sowa, H. H. Mantsch, *Proc. SPIE* **1999**, *3597*, 270.
[106] L. Leonardi, M. G. Sowa, J. R. Payette, M. Hewko, B. Schattka, A. Matas, H. H. Mantsch, *Proc. SPIE* **2000**, *3918*, 83.
[107] M. G. Sowa, J. R. Payette, M. D. Hewko, H. H. Mantsch, *J. Biomed. Optics* **1999**, *4*, 474.
[108] H. Sato, S. Wada, M. Ling, H. Tashiro, *Appl. Spectrosc.* **2000**, *54*, 1163.
[109] M. Niwayama, L. Lin, J. Shao, T. Shiga, N. Kudo, K. Yamamoto, *Proc. SPIE* **1999**, *3597*, 291.
[110] T. Hamaoka, T. Katsumura, N. Murase, S. Nishio, T. Osada, T. Sako, H. Higuchi, Y. Kurosawa, T. Shimomitsu, M. Miwa, B. Chance, *J. Biomed. Optics* **2000**, *5*, 102.
[111] L. A. Paunescu, C. Casavola, M. A. Franceschini, S. Fantini, L. Winter, J. Kim, D. Wood, E. Gratton, *Proc. SPIE* **1999**, *3597*, 317.
[112] D. J. Wallace, B. Michener, D. Choudhury, M. Levi, P. Fennelly, D. M. Hueber, B. Barbieri, *Proc. SPIE* **1999**, *3597*, 300.
[113] S. Zhang, B. R. Soller, S. Kaur, K. Perras, Th. J. van der Salm, *Appl. Spectrosc.* **2000**, *54*, 294.
[114] C. E. Cooper, M. Cope, V. Quaresima, M. Ferrari, E. Nemoto, R. Springett, S. Matcher, P. Amess, J. Penrice, L. Tyszczuk, J. Wyatt, D. T. Delpy in *Optical Imaging of Brain Function and Metabolism II* (Eds. A. Villringer, U. Dirnagl), Plenum, New York, **1997**, pp. 63–73.
[115] F. F. Jöbsis-vander Vliet, P. D. Jöbsis, *J. Biomed. Optics* **1999**, *4*, 397.
[116] V. V. Kupriyanov, R. A. Shaw, B. Xiang, H. H. Mantsch, R. Deslauriers, *J. Mol. Cell. Cardiol.* **1997**, *29*, 2431.
[117] W. J. Parsons, J. C. Rembert, R. P. Baumann, J. C. Greenfield, Jr., C. A. Piantadosi, *J. Biomed. Optics* **1998**, *3*, 191.
[118] C. E. Cooper, S. J. Matcher, J. S. Wyatt, M. Cope, G. C. Brown, E. M. Nemoto, D. T. Delpy, *Biochem. Soc. Trans.* **1994**, *22*, 974.
[119] *Transcranial Cerebral Oximetry* (Eds. G. Litscher, G. Schwarz), Pabst Science, Lengerich, **1997**.

[120] C. E. Elwell, *A Practical Users Guide to Near Infrared Spectroscopy.* Hamamatsu Photonics, Joko-cho, **1995**.

[121] W. M. Kuebler, A. Sckell, O. Habler, M. Kleen, G. E. H. Kuhnle, M. Welte, K. Messmer, A. E. Goetz, *J. Cereb. Blood Flow Metab.* **1998**, *18*, 445.

[122] E. Okada, M. Firbank, M. Schweiger, S. R. Arridge, M. Cope, D. T. Delpy, *Appl. Optics* **1997**, *36*, 21.

[123] H. Dehghani, D. T. Delpy, *Appl. Optics* **2000**, *39*, 4721.

[124] J. Plucinski, A. F. Frydrychowski, J. Kaczmarek, W. Juzwa, *J. Biomed. Optics* **2000**, *5*, 291.

[125] S. Fantini, D. Hueber, M. A. Franceschini, E. Gratton, W. Rosenfeld, P. G. Stubblefield, D. Maulik, M. R. Stankovic, *Phys. Med. Biol.* **1999**, *44*, 1543.

[126] L. Skov, N. C. Brun, G. Greisen, *J. Biomed. Optics* **1997**, *2*, 7.

[127] A. J. Macnab, R. E. Gagnon, F. A. Gagnon, *J. Biomed. Optics* **1998**, *3*, 386.

[128] D. D'Antona, C. J. Aldrich, P. O'Brien, S. Lawrence, D. T. Delpy, J. S. Wyatt, *J. Biomed. Optics* **1997**, *2*, 15.

[129] N. Ramanujam, B. Chance, *Proc. SPIE* **1999**, *3597*, 661.

[130] V. Quaresima, S. Sacco, R. Totaro, M. Ferrari, *J. Biomed. Optics* **2000**, *5*, 201.

[131] T. Shiga, K. Yamamoto, K. Tanabe, Y. Nakase, B. Chance, *J. Biomed. Optics* **1997**, *2*, 154.

[132] A. Maki, Y. Yamashita, E. Watanabe, T. Yamamoto, K. Kogure, F. Kawaguchi, H. Koizumi, *Proc. SPIE* **1999**, *3597*, 202.

[133] Y. Yamashita, A. Maki, H. Koizumi, *J. Biomed. Optics* **1999**, *4*, 414.

[134] T. Yamamoto, Y. Yamashita, H. Yoshizawa, A. Maki, M. Iwata, E. Watanabe, H. Koizumi, *Proc. SPIE* **1999**, *3597*, 230.

[135] H. Koizumi, Y. Yamashita, A. Maki, T. Yamamoto, Y. Ito, H. Itagaki, R. Kennan, *J. Biomed. Optics* **1999**, *4*, 403.

[136] E. Watanabe, A. Maki, F. Kawaguchi, Y. Yamashita, H. Koizumi, Y. Mayanagi, *J. Biomed. Optics* **2000**, *5*, 287.

[137] M. A. Franceschini, V. Toronov, M. E. Filiaci, E. Gratton, S. Fantini, *Optics Express* **2000**, *6*, 49.

[138] V. Toronov, M. A. Franceschini, M. E. Filiaci, S. Fantini, M. Wolf, A. Michalos, E. Gratton, *Med. Phys.* **2000**, *27*, 801.

[139] M. R. Stankovic, D. Maulik, W. Rosenfeld, P. G. Stubblefield, A. D. Kofinas, S. Drexler, R. Nair, M. A. Franceschini, D. Hueber, E. Gratton, S. Fantini, *J. Perinat. Med.* **1999**, *27*, 279.

[140] S. van Huffel, P. Casaer, P. van Mele, G. Willems, *Proc. SPIE* **1995**, *2389*, 743.

[141] K. Isobe, T. Kusaka, Y. Fujikawa, M. Kondo, K. Kawada, S. Yasuda, S. Itoh, K. Hirao, S. Onishi, *J. Biomed. Optics* **2000**, *5*, 283.

[142] J. C. Hebden, M. Tziraki, D. T. Delpy, *Appl. Optics* **1997**, *36*, 3802.

[143] S. Fantini, M. A. Franceschini, G. Gaida, E. Gratton, H. Jess, W. M. Mantulin, K. T. Moesta, P. M. Schlag, M. Kaschke, *Med. Phys.* **1996**, *23*, 149.

[144] *Medical Optical Tomography: Functional Imaging and Monitoring, Vol. IS11* (Eds. G. Müller, B. Chance, R. Alfano, S. Arridge, J. Beuthan, E. Gratton, M. Kaschke, B. Masters, S. Svanberg, P. van der Zee), SPIE Opt. Eng. Press, Bellingham, **1993**.

[145] H. Eda, I. Oda, Y. Ito, Y. Wada, Y. Oikawa, Y. Tsunazawa, Y. Tsuchiya, Y. Yamashita, M. Oda, A. Sassaroli, Y. Yamada, M. Tamura, *Rev. Sci. Instr.* **1999**, *70*, 3595.

[146] Y. Hoshi, I. Oda, Y. Wada, Y. Ito, Y. Yamashita, M. Oda, K. Ohta, Y. Yamada, M. Tamura, *Cogn. Brain Res.* **2000**, *9*, 339.

[147] C. H. Schmitz, H. L. Graber, H. Luo, I. Arif, J. Hira, Y. Pei, A. Bluestone, S. Zhong, R. Andronica, I. Soller, N. Ramirez, S.-L. S. Barbour, R. L. Barbour, *Appl. Optics* **2000**, *39*, 6466.

[148] T. L. Troy, D. L. Page, E. M. Sevick-Muraca, *J. Biomed. Optics* **1996**, *1*, 342.

[149] A. Ya. Khairullina, T. V. Oleinik, L. M. Bui, N. I. Artishevskaya, N. P. Prigoun, Ya. I. Sevkovsky, T. V. Mokhort, I. A. Lomonosova, M. A. Savchenko, *J. Opt. Technol. 6*, **1997**, *4*, 198.

[150] Q. Zhu, E. Conant, B. Chance, *J. Biomed. Optics* **2000**, *5*, 229.

[151] T. H. Pham, F. Bevilacqua, T. Spott, J. S. Dam, B. J. Tromberg, S. Andersson-Engels, *Appl. Optics* **2000**, *39*, 6487.
[152] D. Grosenik, H. Wabnitz, H. H. Rinneberg, K. T. Moesta, P. M. Schlag, *Appl. Optics* **1999**, *38*, 2927.
[153] V. Quaresima, S. J. Matcher, M. Ferrari, *Photochem. Photobiol.* **1998**, *67*, 4.

Appendix

Tab. 1 Group Frequencies in the Near-Infrared Region

Overtone, combination	*Wavelength/nm*		*Wavenumber/cm⁻¹*		*Remarks*
(1) Groups containing only C and H atoms					
1) $-CH_3$ methyl					
combination	2275	2285	4400	4380	CH stret.+CH bend.
	1355	1365	7380	7330	2×CH stret.+CH bend.
	1010	1020	9900	9800	2×CH stret.+3×CH bend.
first overtone	1710	1730	5850	5780	first overtone of asym.stret.
	1770	1785	5650	5600	first overtone of sym.stret.
second overtone	1150	1165	8700	8580	second overtone of asym.stret.
	1190	1200	8400	8330	second overtone of sym.stret.
third overtone	870	885	11490	11300	third overtone of asym.stret.
	900	910	11110	10990	third overtone of sym.stret.
2) $-CH_2$ methylene					
combination	2320	2330	4310	4290	CH stret.+CH bend.
	2305	2315	4340	4320	
	1410	1420	7090	7040	2×CH stret.+CH bend.
	1390	1400	7190	7140	
	1050	1060	9520	9430	
first overtone	1735	1750	5760	5710	first overtone of anti-sym.stret.
	1780	1795	5620	5570	first overtone of sym.stret.
second overtone	1170	1180	8550	8470	second overtone of anti-sym.stret.
	1200	1210	8330	8260	second overtone of sym.stret.
third overtone	885	895	11300	11170	third overtone of anti-sym.stret.
	910	920	10990	10870	third overtone of sym.stret.

Tab. 1 (continued)

Overtone, combination	***Wavelength/nm***		***Wavenumber/cm^{-1}***		***Remarks***
3) –CH					
first overtone	1755	1775	5700	5630	
second overtone	1185	1195	8440	8370	
third overtone	900	910	11110	10990	
4) C=C alkenes (vinyl, vinylidine, vinylene)					
combination	2340	2350	4270	4260	CH_2 stret.+=CH_2 bend.
	2185	2195	4580	4560	CH_2 stret.+C=C stret.
	2135	2145	4680	4660	=CH stret.+C=C stret.
first overtone	1675	1695	5970	5900	
	1645	1660	6080	6020	vinyl group
second overtone	1130	1145	8850	8730	
	1110	1120	9010	8930	vinyl group
third overtone	860	870	11630	11490	
	840	850	11900	11760	vinyl group
5) –C=CH alkynes (ethynyle)					
first overtone	1535	1545	6510	6470	
second overtone	1035	1045	9660	9570	
third overtone	780	790	12820	12660	
6) CH (aromatic)					
combination	1440	1450	6940	6900	2×CH stret.+CH bend.
	1410	1420	7090	7040	2×CH stret.+CH bend.
	1070	1085	9350	9220	2×CH stret.+2×C–C stret.
first overtone	1680	1690	5950	5920	
second overtone	1130	1140	8850	8770	
third overtone	850	860	11760	11630	
(2) O atom containing groups					
7) H_2O					
combination	1930	1940	5180	5150	OH stret.+OH bend.
	1375	1385	7270	7220	OH anti-sym. stret.+OH sym. stret.
first overtone	1450	1460	6900	6850	
second overtone	975	985	10260	10150	
third overtone	740	750	13510	13330	
8) free –OH alcohol					
combination	2060	2090	4850	2060	OH stret.+OH bend.
first overtone	1395	1425	7170	7020	
second overtone	2370	2390	4220	4180	second overtone of OH bend.
	940	955	10640	10470	second overtone of OH stret.
third overtone	730	745	13700	13420	

Tab. 1 (continued)

Overtone, combination	*Wavelength/nm*		*Wavenumber/cm^{-1}*		*Remarks*
9) bound – OH alcohol					
first overtone	1435	1480	6970	6760	intermolecular hydrogen bond
	1500	1595	6670	6270	intramolecular hydrogen bond
second overtone	980	990	10200	10100	intermolecular hydrogen bond
	1035	1045	9660	9570	intramolecular hydrogen bond
10) COOH carboxylic acids, COOR esters					
second overtone	1890	1920	5290	5210	2×C=O stret. (carboxylic acids)
	1930	1950	5180	5130	2×C=O stret. (esters)
11) C=O ketones					
second overtone		1950		5130	
12) CHO aldehydes					
combination	2190	2210	4570	4520	CH stret.+C=O stret.
13) epoxides					
first overtone	1640	1650	6100	6060	first overtone of CH stret.
(3) N atom containing groups					
14) – NH_2 primary amines					
combination	1970	2010	5080	4980	NH stret.+NH bend.
first overtone	1520	1540	6580	6490	first overtone of NH_2 sym.stret.
	1500	1520	6670	6580	first overtone of NH_2 anti-sym.stret.
	1450	1480	6900	6760	$ArNH_2$
second overtone	1020	1040	9800	9620	second overtone of NH_2 sym.stret.
	1000	1020	10000	9800	second overtone of NH_2 anti-sym.stret.
	980	1020	10200	9800	$ArNH_2$
third overtone	800	820	12500	12200	third overtone of NH_2 sym.stret.
	780	800	12820	12500	$ArNH_2$
	770	790	12990	12660	third overtone of NH_2 anti-sym.stret.
15) – NH secondary amines					
first overtone	1490	1545	6710	6470	
second overtone	1010	1040	9900	9620	

Tab. 1 (continued)

Overtone, combination	*Wavelength/nm*		*Wavenumber/cm^{-1}*		*Remarks*
16) -$CONH_2$ primary amides					
combination	2140	2170	4670	4610	2×amide+amide
	2100	2130	4760	4690	NH stret.+amide
	2040	2060	4900	4850	NH stret.+amide
	1950	1970	5130	5080	NH stret.+amide
first overtone	1600	1620	6250	6170	intermolecular hydrogen bond
	1510	1530	6620	6540	intermolecular hydrogen bond
	1490	1510	6710	6620	first overtone of NH_2 sym.stret.
	1440	1460	6940	6850	first overtone of NH_2 anti-sym.stret.
second overtone	2020	2040	4950	4900	second overtone of amide
	1070	1090	9350	9170	intramolecular hydrogen bond
	1015	1035	9850	9660	intramolecular hydrogen bond
	1000	1020	10000	9800	second overtone of NH_2 sym.stret.
	970	990	10310	10100	second overtone of NH_2 anti-sym.stret.
17) –CONH– secondary amides					
combination	2150	2170	4650	4610	2×amide+amide
	2100	2120	4760	4720	NH stret.+amide
	1990	2010	5030	4980	NH stret.+amide
first overtone	1530	1670	6540	5990	hydrogen bond
	1460	1510	6850	6620	free
second overtone	1910	1930	5240	5180	second overtone of amide
	1035	1120	9660	8930	hydrogen bond
	1000	1050	10000	9520	free
(4) S or P atom containing group					
18) – SH (thiols)					
first overtone	1735	1745	5760	5730	
19) P–OH (phosphorus acids, phosphonic acids, phosphinic acids)					
first overtone	1900	1910	5260	5240	first overtone of OH stret.
20) PH (phosphines, phosphorus acid esters)					
first overtone	1890	1900	5290	5260	

References

[1] B.G. Osborne, T. Fearn, P. H. Hindle: Practical NIR Spectroscopy with Application in Food and Beverage Analysis (Longman Scientific and Technical, Harlow, **1993**).

[2] P. Williams, K. Norris: Near-Infrared Technology in the Agricultural and Food Industries (Academic Association of Cereal Chemists, St. Paul, **1987**).

Subject Index